## DATE DUE

| | | | |
|---|---|---|---|
| DEC 1 3 2000 | | | |
| AUG 0 6 2007 | | | |
| | | | |
| | | | |
| | | | |
| | | | |
| | | | |
| | | | |
| | | | |
| | | | |
| | | | |
| | | | |
| | | | |
| | | | |
| | | | |
| | | | |
| | | | |

# Fermat's Last Theorem
# for Amateurs

**Springer**
*New York*
*Berlin*
*Heidelberg*
*Barcelona*
*Hong Kong*
*London*
*Milan*
*Paris*
*Singapore*
*Tokyo*

FERMAT AND FRIEND

Paulo Ribenboim

# Fermat's Last Theorem
# for Amateurs

 Springer

Paulo Ribenboim
Department of Mathematics
   and Statistics
Queen's University
Kingston, Ontario, K7L 3N6
Canada

With 2 Illustrations

Mathematics Subject Classification (1991): 11Axx

Library of Congress Cataloging-in-Publication Data
Ribenboim, Paulo.
      Fermat's last theorem for amateurs / Paulo Ribenboim.
         p.      cm.
      Includes bibliographical references and index.
      ISBN 0-387-98508-5 (hc. : alk. paper)
      1. Fermat's last theorem.   I. Title.
   QA244.R53   1999
   512′.74—dc21                                      98-41246

Printed on acid-free paper.

Production managed by MaryAnn Cottone; manufacturing supervised by Thomas King.
Camera-ready copy prepared from the author's LaTeX files.
Printed and bound by R.R. Donnelley and Sons, Harrisonburg, VA.
Printed in the United States of America.

9 8 7 6 5 4 3 2 1

ISBN 0-387-98508-5 Springer-Verlag New York Berlin Heidelberg   SPIN 10674267

# Preface

It is now well known that Fermat's last theorem has been proved. For more than three and a half centuries, mathematicians — from the great names to the clever amateurs — tried to prove Fermat's famous statement. The approach was new and involved very sophisticated theories. Finally the long-sought proof was achieved. The arithmetic theory of elliptic curves, modular forms, Galois representations, and their deformations, developed by many mathematicians, were the tools required to complete the difficult proof.

Linked with this great mathematical feat are the names of TANI-YAMA, SHIMURA, FREY, SERRE, RIBET, WILES, TAYLOR. Their contributions, as well as hints of the proof, are discussed in the Epilogue. This book has not been written with the purpose of presenting the proof of Fermat's theorem. On the contrary, it is written for amateurs, teachers, and mathematicians curious about the unfolding of the subject. I employ exclusively elementary methods (except in the Epilogue). They have only led to partial solutions but their interest goes beyond Fermat's problem. One cannot stop admiring the results obtained with these limited techniques.

Nevertheless, I warn that as far as I can see — which in fact is not much — the methods presented here will not lead to a proof of Fermat's last theorem for all exponents.

The presentation is self-contained and details are not spared, so the reading should be smooth.

Most of the considerations involve ordinary rational numbers and only occasionally some algebraic (non-rational) numbers. For this reason I excluded Kummer's important contributions, which are treated in detail in my book, *Classical Theory of Algebraic Numbers* and described in my *13 Lectures on Fermat's Last Theorem* (new printing, containing an Epilogue about recent results).

There are already — and there will be more — books, monographs, and papers explaining the ideas and steps in the proof of Fermat's theorem. The readers with an extended solid background will profit more from reading such writings. Others may prefer to stay with me.

In summary, if you are an amateur or a young beginner, you may love what you will read here, as I made a serious effort to provide thorough and clear explanations.

On the other hand, if you are a professional mathematician, you may then wonder why I have undertaken this task now that the problem has been solved. The tower of Babel did not reach the sky, but it was one of the marvels of ancient times. Here too, there are some admirable examples of ingenuity, even more remarkable considering that the arguments are strictly elementary. It would be an unforgivable error to let these gems sink into oblivion. As Jacobi said, all for "l'honneur de l'esprit humain."

August, 1997                                     Paulo Ribenboim

# Reader

You may feel tempted to write your own (simpler) proof of Fermat's last theorem.

I have strong views about such a project. It should be written in the Constitution of States and Nations, in the Chapter of Human Rights:

It is an inalienable right of each individual to produce his or her own proof of Fermat's last theorem.

However, such a solemn statement about Fermat's last theorem (henceforth referred to as THE theorem) should be tempered by the following articles:

> Art. 1. *No attempted proof of THE theorem should ever duplicate a previous one.*

> Art. 2. *It is a criminal offense to submit false proofs of THE theorem to professors who arduously earn their living by teaching how not to conceive false proofs of THE theorem.*

Infringement of the latter, leads directly to Hell. Return to Paradise only after the said criminal has understood and is able to reproduce Wiles' proof. (Harsh punishment.)

# Contents

# Acknowledgment

Karl Dilcher has supervised the typing of this book, read carefully the text, and made many valuable suggestions. I am very grateful for his essential help.

I am also indebted to various colleagues who indicated necessary corrections in early versions of this book. My thanks go especially to the late Kustaa Inkeri as well as to Takashi Agoh, Vinko Botteri, Hendrik Lenstra, Tauno Metsänkylä, and Guy Terjanian.

For the epilogue I received advice from Gerhard Frey, Fernando Gouvêa, and Ernst Kani, to whom I express my warmest thanks.

# The Problem

In the margin of his copy of Bachet's edition of the works of Dio-phantus, [1] Fermat wrote:

"It is impossible to separate a cube into two cubes, or a biquadrate into two biquadrates, or in general any power higher than the second into powers of like degree; I have discovered a truly marvelous proof, which this margin is too small to contain."

In modern language, this means:

If $n$ is any natural number greater than 2, the equation

$$X^n + Y^n = Z^n$$

has no solutions in integers, all different from 0 (i.e., it has only the trivial solution, where one of the integers is equal to 0).

The above statement has been called *Fermat's last theorem*, or *conjecture*, or *problem*.

We begin with the following remarks.

In order to prove Fermat's theorem for all exponents greater than 2, it suffices to prove it for the exponent 4 and every odd prime

---

[1] This copy is now lost, but the remarks appeared in the 1679 edition of the works of Fermat, edited in Toulouse by his son Samuel de Fermat.

exponent $p$. Indeed, if $n$ is composite, $n > 2$, it has a factor $m$ which is 4 or an odd prime $p$. If the theorem fails for $n = ml$, where $m = 4$ or $p$, $l > 1$, if $x, y, z$ are non-zero integers such that $x^n + y^n = z^n$ then $(x^l)^m + (y^l)^m = (z^l)^m$ and the theorem would fail for $m$.

Occasionally, we shall also indicate some results and proofs for even exponents or prime-power exponents.

The following general remarks are quite obvious and henceforth will be taken for granted.

If $n$ is odd then $X^n + Y^n = Z^n$ has a non-trivial solution if and only if $X^n + Y^n + Z^n = 0$ has a non-trivial solution.

If $x, y, z$ are non-zero integers such that $x^n + y^n = z^n$, if $d = \gcd(x, y, z)$ and $x_1 = x/d, y_1 = y/d, z_1 = z/d$ then $x_1^n + y_1^n = z_1^n$, where the non-zero integers $x_1, y_1, z_1$ are pairwise relatively prime. So, if we assume that Fermat's equation has a non-trivial solution, then it has one with pairwise relatively prime integers.

Moreover, if $x, y, z$ are non-zero pairwise relatively prime integers such that $x^n + y^n = z^n$ then $x + y, z - x, z - y$ are also pairwise relatively prime. Indeed, if a prime $p$ divides $x + y$ and $z - x$ then $x \equiv z \pmod{p}$, hence $x^n \equiv z^n \equiv x^n + y^n \pmod{p}$ therefore $y^n \equiv 0 \pmod{p}$, so $p$ divides $y$, and since $p$ divides $x + y$ then $p$ divides $x$, which is contrary to the hypothesis. This shows that $\gcd(x + y, z - x) = 1$. In a similar way, we may show that $\gcd(x + y, z - y) = 1$ and $\gcd(z - x, z - y) = 1$.

Following tradition, we say that the *first case of Fermat's theorem* is true for the odd prime exponent $p$ when: if $x, y, z$ are (non-zero) integers, not multiples of $p$, then $x^p + y^p \neq z^p$.

The *second case* is true for the odd prime exponent $p$ when: if $x, y, z$ are non-zero pairwise relatively prime integers, and $p$ divides $xyz$ then $x^p + y^p \neq z^p$. As said above, in this case $p$ divides one and only one of the integers $x, y, z$.

More generally, for an arbitrary integer $n = 2^u m$, $u \geq 0$, $m$ odd, we say that the *first case* of Fermat's theorem is true for the exponent $n$ when: if $x, y, z$ are (non-zero) integers and $\gcd(m, xyz) = 1$ then $x^n + y^n \neq z^n$.

Similarly, the *second case* is true for the exponent $n$ when: if $x, y, z$ are (non-zero) pairwise relatively prime integers and $\gcd(m, xyz) \neq 1$ then $x^n + y^n \neq z^n$.

# I
# Special Cases

This chapter is devoted to the proof of special cases of Fermat's theorem: exponents 4, 3, 5, and 7. However, we begin by considering the exceptional case of exponent 2.

## I.1. The Pythagorean Equation

We study briefly the Pythagorean equation

$$(1.1) \qquad\qquad X^2 + Y^2 = Z^2.$$

A triple $(x, y, z)$ of positive integers such that $x^2 + y^2 = z^2$ is called a *Pythagorean triple*, for example, $(3, 4, 5)$ since $3^2 + 4^2 = 5^2$.

If $x, y, z$ are nonzero integers such that $x^2 + y^2 = z^2$ then $|x|$, $|y|$, $|z|$ also satisfy the same equation. Note that $x, y$ cannot be both odd, otherwise $z^2 \equiv 1 + 1 \pmod 4$, which is impossible. Moreover, if $d = \gcd(x, y, z)$ then $x/d$, $y/d$, $z/d$ also satisfy the equation. Thus, it suffices to determine the *primitive solutions* $(x, y, z)$ of (1.1), namely those such that $x > 0$, $y > 0$, $z > 0$, $x$ is even, and $\gcd(x, y, z) = 1$, hence $y$ and $z$ are odd.

It is stated in Dickson's (1920) *History of the Theory of Numbers,* Vol. II, pp. 165–166, that Pythagoras and Plato gave methods to find solutions of equation (1.1). In Lemma 1 to Proposition 29 of Book X of *The Elements*, Euclid gave a geometric method to find solutions of (1.1).

Diophantus indicated how to find all solutions, as in the following result.

Leonardo di Pisa (Fibonacci) also gave in 1225 an interesting method to find solutions.

**(1A)**   *If $a, b$ are integers such that $a > b > 0$, $\gcd(a, b) = 1$, $a, b$ of different parity, then the triple $(x, y, z)$, given by*

$$\begin{cases} x = 2ab, \\ y = a^2 - b^2, \\ z = a^2 + b^2, \end{cases}$$

*is a primitive solution of (1.1). This establishes a one-to-one correspondence between the set of pairs $(a, b)$ satisfying the above conditions, and the set of primitive solutions of (1.1).*

PROOF. If $a, b$ are integers satisfying the conditions of the statement, let $x, y, z$ be defined as indicated. Then

$$x^2 + y^2 = 4a^2b^2 + (a^2 - b^2)^2 = (a^2 + b^2)^2 = z^2.$$

Clearly $x > 0$, $y > 0$, $z > 0$, $x$ is even, and $\gcd(x, y, z) = 1$ because if $d$ divides $x$, $y$, and $z$ then $d$ divides $2a^2$ and $2b^2$, so $d = 1$ or $d = 2$ (since $\gcd(a, b) = 1$); but $d \neq 2$ because $y$ is odd ($a, b$ do not have the same parity).

Different pairs $(a, b)$ give different triples $(x, y, z)$.

Conversely, let $(x, y, z)$ be a primitive solution of (1.1), so $x^2 + y^2 = z^2$. From $\gcd(x, y, z) = 1$ we have $\gcd(x, z) = 1$. Since $x$ is even then $z$ is odd hence $\gcd(z - x, z + x) = 1$. Since $y^2 = (z - x)(z + x)$, it follows from their decomposition into prime numbers that $z - x$, $z + x$ are squares of integers, say $z + x = t^2$, $z - x = u^2$, and $t, u$ must be positive odd integers, with $t > u > 0$. Let $a, b$ be integers such that $2a = t + u$, $2b = t - u$, hence $t = a + b$, $u = a - b$ with $a > b > 0$. So

$$\begin{cases} x = ((a + b)^2 - (a - b)^2)/2 = 2ab, \\ y^2 = u^2 t^2 = (a - b)^2(a + b)^2 = (a^2 - b^2)^2 \quad \text{so} \quad y = a^2 - b^2, \\ z = ((a + b)^2 + (a - b)^2)/2 = a^2 + b^2. \end{cases}$$

We note that $\gcd(a, b) = 1$ because $\gcd(z - x, z + x) = 1$ and finally $a + b = t$ is odd so $a, b$ are not both odd.  □

For example, the smallest primitive solutions for (1.1), ordered according to increasing values of $z$, are the following:

$$(4,3,5), \quad (12,5,13), \quad (8,15,17), \quad (24,7,25),$$
$$(20,21,29), \quad (12,35,37), \quad (40,9,41), \quad (28,45,53),$$
$$(60,11,61), \quad (56,33,65), \quad (16,63,65), \quad (48,55,73).$$

In view of (1A), to find the primitive solutions of (1.1) amounts to determining which odd positive integers are sums of two squares, and in each case, to write all such representations. Fermat proved: $n > 0$ is a sum of two squares of integers if and only if every prime factor $p$ of $n$, such that $p \equiv 3 \pmod 4$, appears to an even power in the decomposition of $n$ into prime factors (see the proof below). For every integer $n$ which is the sum of two squares of integers, let $r(n)$ be the number of ordered pairs $(a,b)$ such that $a^2 + b^2 = n$, $a,b$ integers *not* necessarily positive. For example, $r(1) = 4$, $r(5) = 8$. It was proved by Jacobi, and independently by Gauss, that

$$r(n) = 4\left(d_1(n) - d_3(n)\right),$$

where $d_1(n)$ (respectively, $d_3(n)$) is the number of divisors of $n$ which are congruent to 1 modulo 4 (respectively, congruent to 3 modulo 4) (see Hardy and Wright (1938, p. 241)).

With this information, it is possible to determine explicitly the primitive Pythagorean triples $(x, y, z)$. Now we paraphrase Fermat's proof which is of historical importance. We begin with a very easy identity:

$$(1.2) \qquad (a^2 + b^2)(c^2 + d^2) = (ac + bd)^2 + (ad - bc)^2$$
$$= (ac - bd)^2 + (ad + bc)^2.$$

Now we show

**(1B)**     *A prime number $p$ is a sum of two squares if and only if $p = 2$ or $p \equiv 1 \pmod 4$.*

PROOF. If $p \neq 2$ and $p = a^2 + b^2$, then $a, b$ cannot both be even — otherwise 4 divides $p$. If $a, b$ are both odd, then $p \equiv 1 + 1 = 2 \pmod 4$, since every odd square is congruent to 1 modulo 4. Thus $p = 2$. If, say, $a$ is odd and $b$ is even, then $p \equiv 1 + 0 = 1 \pmod 4$.

Conversely, $2 = 1^2 + 1^2$, so let $p \equiv 1 \pmod 4$. From the theory of quadratic residues, $-1$ is a square modulo $p$, so there exists $x$, $1 \leq x \leq p - 1$, such that $x^2 + 1 \equiv 0 \pmod p$, thus $x^2 + 1 =$

$mp$, with $1 \leq m \leq p - 1$. Hence the set $\{m \mid 1 \leq m \leq p - 1$, such that $mp = x^2 + y^2$ for some integers $x, y\}$ is not empty. Let $m_0$ be the smallest integer in this set, so $1 \leq m_0 \leq p - 1$. We show that $m_0 = 1$, hence $p$ is a sum of two squares. Assume, on the contrary, that $1 < m_0$. We write

$$\begin{cases} x = cm_0 + x_1, \\ y = dm_0 + y_1, \end{cases}$$

with $-m_0/2 < x_1, y_1 \leq m_0/2$, and integers $c, d$. We observe that $x_1$ or $y_1$ is not 0. Otherwise $m_0^2$ divides $x^2 + y^2 = m_0 p$, hence $m_0$ divides $p$, thus $m_0 = p$, which is absurd. We have $0 < x_1^2 + y_1^2 \leq m_0^2/4 + m_0^2/4 = m_0^2/2 < m_0^2$ and $x_1^2 + y_1^2 \equiv x^2 + y^2 \equiv 0 \pmod{m_0}$. Hence, $x_1^2 + y_1^2 = m_0 m'$ with $1 \leq m' < m_0$. But $m_0 p = x^2 + y^2$, $m_0 m' = x_1^2 + y_1^2$, hence $m_0^2 m' p = (x^2 + y^2)(x_1^2 + y_1^2) = (xx_1 + yy_1)^2 + (xy_1 - yx_1)^2$. We also have

$$\begin{aligned} xx_1 + yy_1 &= x(x - cm_0) + y(y - dm_0) \\ &= (x^2 + y^2) - m_0(c + yd) \\ &= m_0 t, \\ xy_1 - yx_1 &= x(y - dm_0) - y(x - cm_0) \\ &= -m_0(xd - yc) \\ &= m_0 u \end{aligned}$$

for some integers $t, u$. Hence $m' p = t^2 + u^2$, with $1 \leq m' < m_0$. This is a contradiction and concludes the proof. $\square$

**(1C)**    *A natural number $n$ is the sum of two squares of integers if and only if every prime factor $p$ of $n$, such that $p \equiv 3 \pmod 4$, appears to an even power in the decomposition of $n$ into prime factors.*

PROOF. Let $n = p_1^{k_1} \cdots p_j^{k_j}$ and assume that $k_j$ is even if $p_j \equiv 3 \pmod 4$. Then $n = n_0^2 n_1$ where $n_0 \geq 1$, $n_1 \geq 1$, and $n_1$ is the product of distinct primes which are either equal to 2, or congruent to 1 modulo 4. By (1B), each factor of $n_1$ is a sum of two squares; by the identity indicated in (1.2), $n_1$ and therefore also $n$, is a sum of two squares. Conversely, let $n = x^2 + y^2$; the statement is trivial if $x = 0$ or $y = 0$. Let $x, y$ be nonzero, let $d = \gcd(x, y)$, so $d^2$ divides $n$. Let $n = d^2 n'$, $x = dx'$, $y = dy'$, hence $\gcd(x', y') = 1$ and $n' = x'^2 + y'^2$. If $p$ divides $n'$, then $p$ does not divide $x'$ — otherwise

$p$ would also divide $y'$. Let $k$ be such that $kx' \equiv y' \pmod{p}$. Then $x'^2 + y'^2 \equiv x'^2(1 + k^2) \equiv 0 \pmod{p}$. Thus $p$ divides $1 + k^2$, that is, $-1$ is a square modulo $p$, so $p = 2$ or $p \equiv 1 \pmod 4$, by the theory of quadratic residues. It follows that if $p_j \equiv 3 \pmod 4$ then $p_j$ does not divide $n'$, hence $p_j$ divides $d$, so the exponent $k_j$ must be even.    $\square$

It is customary to say that a right triangle is a *Pythagorean triangle* when its sides are measured by integers $a, b, c$. If $c$ is the hypothenuse, then $c^2 = a^2 + b^2$. See also Mariani (1962).

On this matter, we recommend Shanks' book (1962) which contains an interesting chapter on Pythagoreanism and its applications, as well as the book by Sierpiński (1962).

In 1908, Bottari gave another parametrization for the solutions of (1.1). The following simpler proof is due to Cattaneo (1908):

**(1D)**    *If $a, b$ are odd natural numbers such that $\gcd(a, b) = 1$, if $s \geq 1$ then the triple $(x, y, z)$ given by*

$$\begin{cases} x = 2^{2s-1}a^2 + 2^s ab, \\ y = b^2 + 2^s ab, \\ z = 2^{2s-1}a^2 + b^2 + 2^s ab, \end{cases}$$

*is a primitive solution of (1.1). This establishes a one-to-one correspondence between the set of triples $(a, b, s)$ satisfying the above conditions and the set of primitive solutions of (1.1).*

PROOF. It is clear that if $x, y, z$ are defined as indicated then the triple $(x, y, z)$ is a primitive solution of (1.1).

Different triples $(a, b, s)$ give rise to different primitive solutions $(x, y, z)$, because

$$\begin{cases} b^2 = z - x, \\ 2^{2s-1}a^2 = z - y, \\ 2^s ab = x + y - z. \end{cases}$$

Finally, if $(x, y, z)$ is a primitive solution, $0 < x < z, 0 < y < z$, and $z < x + y$, because $z^2 = x^2 + y^2 < (x + y)^2$. We write

$$\begin{cases} x = z - u, \\ y = z - v, \\ z = x + y - w, \end{cases}$$

with $u, v, w > 0$. Then

$$\begin{cases} x = v + w, \\ y = u + w, \\ z = u + v + w. \end{cases}$$

From $x^2 + y^2 = z^2$ it follows that $w^2 = 2uv$, hence $w$ is even. Since $\gcd(u, v, w) = 1$ and $x$ is even and $y$ is odd then $v$ is even and $u$ is odd. Let $w = 2^s w'$, $v = 2^t v'$, where $v'$, $w'$ are odd, $s \geq 1$, $t \geq 1$. Then $2^{2s} w'^2 = 2u \cdot 2^t v'$ so $t = 2s - 1$ and $w'^2 = uv'$ with $\gcd(u, v') = 1$. Hence necessarily $u, v'$ are squares: $u = b^2$, $v' = a^2$, and therefore $x = 2^{2s-1} a^2 + 2^s ab$, $y = b^2 + 2^s ab$, $z = 2^{2s-1} a^2 + b^2 + 2^s ab$.  $\square$

It is also interesting to describe the solutions of

(1.3) $$X^2 + Y^2 = 1.$$

The solutions in integers are just $(\pm 1, 0)$, $(0, \pm 1)$.

We shall consider the solutions in rational numbers as well as, for each prime $p$, the solutions in the field with $p$ elements.

Let $\mathbb{Q}$ denote the field of rational numbers. For each prime $p$, let $\mathbb{F}_p$ be the set $\{\bar{0}, \bar{1}, \dots, \overline{p-1}\}$ of residue classes of $\mathbb{Z}$ modulo $p$. So, if $a, b \in \mathbb{Z}$, we have $\bar{a} = \bar{b}$ if and only if $a, b$ have the same remainder when divided by $p$. The operations of addition and multiplication in $\mathbb{F}_p$ are defined as follows: $\bar{x} + \bar{y} = \overline{x + y}$, $\bar{x} \bar{y} = \overline{xy}$. With these operations, which satisfy the usual properties, $\mathbb{F}_p$ becomes a field: if $\bar{a} \in \mathbb{F}_p$ and $\bar{a} \neq \bar{0}$, we have $\gcd(a, p) = 1$, so there exist $r, s \in \mathbb{Z}$ such that $ar + ps = 1$; then $\bar{a} \bar{r} = \bar{1}$. So $\bar{r}$ is the inverse of $\bar{a}$ in $\mathbb{F}_p$. For simplicity, we may use the notation $x$ instead of $\bar{x}$ for the elements of $\mathbb{F}_p$. We shall indicate a result that is valid for $\mathbb{Q}$ as well as for each field $\mathbb{F}_p$ for $p > 2$. Thus, let $F = \mathbb{Q}$ or $\mathbb{F}_p$ (for $p > 2$). (More generally, $F$ may be taken to be any field of characteristic different from 2, that is, $1 + 1 \neq 0$ in the field $F$.)

Let $\infty$ be a symbol, $\infty \notin F$, and let $T = \{\infty\} \cup \{t \in F \mid 1 + t^2 \neq 0\}$. Let $S = S_F = \{(x, y) \in F \times F \mid x^2 + y^2 = 1\}$. So the elements of $S$ are the solutions of (1.3) in the field $F$.

Let $\varphi : T \to F \times F$ be the following mapping:

(1.4) $$\begin{cases} \varphi(\infty) = (0, -1), \\ \text{if } t \in F \text{ then } \varphi(t) = \left( \dfrac{2t}{1 + t^2}, \dfrac{1 - t^2}{1 + t^2} \right). \end{cases}$$

We note that since $1+t^2 \neq 0$, then $1+t^2$ is invertible, so the mapping $\varphi$ is well defined.

**(1E)**    *With the above notations, $\varphi$ is a one-to-one mapping from $T$ onto $S$.*

PROOF. Since

$$\left(\frac{2t}{1+t^2}\right)^2 + \left(\frac{1-t^2}{1+t^2}\right)^2 = 1,$$

then $\varphi(t) \in S$ for each $t \in F$ such that $1+t^2 \neq 0$. Also, $(0,-1) \in S$, so $\varphi(T) \subseteq S$.

If $t \in F, 1+t^2 \neq 0$ then $(1-t^2)/(1+t^2) \neq -1$, because $1+1 \neq 0$. Also, if $t_1, t_2 \in F$, $1+t_1^2 \neq 0$, $1+t_2^2 \neq 0$ then $\varphi(t_1) \neq \varphi(t_2)$, as is easily seen. Thus the mapping $\varphi$ is one-to-one (because $1+1 \neq 0$).

Now we show that $\varphi(T) = S$. Clearly, $(0,-1) = \varphi(\infty)$. Let $(x,y) \in S$, $(x,y) \neq (0,-1)$. If $x = 0$ then $y = 1$ and $(0,1) = \varphi(0)$. If $x \neq 0$ let $t = (1-y)/x$, so

$$1+t^2 = \frac{2(1-y)}{x^2}, \qquad 1-t^2 = \frac{2y(y-1)}{x^2},$$

and

$$\frac{2t}{1+t^2} = x, \qquad \frac{1-t^2}{1+t^2} = y,$$

hence $(x,y) = \varphi(t)$, concluding the proof. $\square$

If $F = \mathbb{Q}$ then $1+t^2 \neq 0$ for all $t \in \mathbb{Q}$, so $T = \mathbb{Q} \cup \{\infty\}$. If $F = \mathbb{F}_p$ with $p > 2$ then $1+t^2 = 0$ if and only if $-1$ is a square modulo $p$. According to the result of Fermat already quoted, $-1$ is a square modulo $p > 2$ if and only if $p \equiv 1 \pmod 4$.

Let $N_p$ denote the number of elements of $S_{\mathbb{F}_p}$. We have

**(1F)**    $N_2 = 2$ *and if $p > 2$ then*

$$N_p = \begin{cases} p-1 & when \quad p \equiv 1 \pmod 4, \\ p+1 & when \quad p \equiv -1 \pmod 4. \end{cases}$$

PROOF. $S_{\mathbb{F}_2} = \{(0,1),(1,0)\}$, so $N_2 = 2$. Let $p > 2$. If $p \equiv 1 \pmod 4$, then there exist two elements $t_1, t_2$ such that $t_1^2+1 = t_2^2+1 = 0$. So $\#(T) = (p-2)+1 = p-1$, hence by (1E), $\#(S) = p-1$.

Similarly, if $p \equiv -1 \pmod 4$, then $\#(T) = p+1$ and $\#(S) = p+1$. $\square$

# Bibliography

Euclid, *The Elements*, Book X (editor T.L. Heath, 3 volumes), Cambridge University Press, Cambridge, 1908; reprinted by Dover, New York, 1956.

1621 Bachet, C.G., *Diophanti Alexandrini Arithmeticorum libri sex et de numeris multangulis liber unus*, S.H. Drovart, Paris, 1621; reprinted by S. de Fermat, with notes by P. de Fermat, 1670.

1676 Frénicle de Bessy, *Traité des Triangles Rectangles en Nombres*, Paris, 1676; reprinted in Mém. Acad. Roy. Sci. Paris, **5** (1729), 83–166.

1863 Gauss, C.F., *Zur Theorie der Complexen Zahlen* (I), Neue Theorie der Zerlegung der Cuben, *Collected Works*, Vol. II, pp. 387–391, Königliche Ges. Wiss., Göttingen, 1876.

1908 Bottari, A., *Soluzione intere dell'equazione pitagorica e applicazione alle dimostrazione di alcune teoremi della teoria dei numeri*, Period. Mat., (3), **23** (1908), 218–220.

1908 Cattaneo, P., *Osservazioni sopra due articoli del Signor Amerigo Bottari*, Period. Mat., (3), **23** (1908), 218–220.

1915 Carmichael, R.D., *Diophantine Analysis*, Wiley, New York, 1915.

1920 Dickson, L.E., *History of the Theory of Numbers*, Vol. II, Carnegie Institution, Washington, DC, 1920; reprinted by Chelsea, New York, 1971.

1938 Hardy, G.H. and Wright, E.M., *An Introduction to the Theory of Numbers*, Clarendon Press, Oxford, 1938.

1962 Mariani, J., *The group of Pythagorean numbers*, Amer. Math. Monthly, **69** (1962), 125–128; reprinted in *Selected Papers in Algebra*, Math. Assoc. of America, 1977, pp. 25–28.

1962 Shanks, D., *Solved and Unsolved Problems in Number Theory*, Vol. I, Spartan, Washington, DC, 1962; reprinted by Chelsea, New York, 1978.

1962 Sierpiński, W., *Pythagorean Triangles*, Yeshiva University, New York, 1962.

1972 Ribenboim, P., *L'Arithmétique des Corps*, Hermann, Paris, 1972.

## I.2. The Biquadratic Equation

Now we take up the case $n = 4$. Fermat considered the problem of whether the area of a Pythagorean triangle may be the square of an integer (observation to Question 20 of Diophantus, Book VI of *Arithmetica*).

He was led to study the equation

(2.1) $$X^4 - Y^4 = Z^2$$

and he showed (date unknown):

**(2A)**    *Equation (2.1) has no solutions in integers all different from 0.*

PROOF. If the statement is false, let $(x, y, z)$ be a triple of positive integers with smallest possible $x$, such that $x^4 - y^4 = z^2$. Then $\gcd(x, y) = 1$, because if a prime $p$ divides both $x, y$ then $p^4$ divides $z^2$, so $p^2$ divides $z$; letting $x = px'$, $y = py'$, $z = p^2 z'$ then $x'^4 - y'^4 = z'^2$, with $0 < x' < x$, which is contrary to the hypothesis.

We have $z^2 = x^4 - y^4 = (x^2 + y^2)(x^2 - y^2)$ and $\gcd(x^2 + y^2, x^2 - y^2)$ is equal to 1 or 2, as is easily seen, because $\gcd(x, y) = 1$. We distinguish two cases.

*Case* 1: $\gcd(x^2 + y^2, x^2 - y^2) = 1$.

Since the product of $x^2 + y^2$, $x^2 - y^2$ is a square then $x^2 + y^2$, $x^2 - y^2$ are squares; more precisely, there exist positive integers $s$, $t$, $\gcd(s, t) = 1$ such that

$$\begin{cases} x^2 + y^2 = s^2, \\ x^2 - y^2 = t^2. \end{cases}$$

It follows that $s, t$ must be odd (since $2x^2 = s^2 + t^2$ then $s, t$ have the same parity and they cannot both be even).

So there exist positive integers $u, v$ such that

$$\begin{cases} u = (s + t)/2, \\ v = (s - t)/2, \end{cases}$$

and necessarily $\gcd(u, v) = 1$, because $s, t$ are odd.

We have $uv = (s^2-t^2)/4 = y^2/2$ hence $y^2 = 2uv$. Since $\gcd(u,v) = 1$ then there exist positive integers $l, m$ such that

$$\left\{ \begin{array}{l} u = 2l^2, \\ v = m^2, \end{array} \right. \quad \text{or} \quad \left\{ \begin{array}{l} u = l^2, \\ v = 2m^2. \end{array} \right.$$

We just consider the first alternative, the other one being analogous. So $u$ is even, $\gcd(u, v, x) = 1$, and

$$u^2 + v^2 = \frac{(s+t)^2 + (s-t)^2}{4} = \frac{s^2 + t^2}{2} = x^2.$$

It follows from (1A) that there exist positive integers $a, b, 0 < b < a$, $\gcd(a, b) = 1$, such that

$$\left\{ \begin{array}{l} 2l^2 = u = 2ab, \\ m^2 = v = a^2 - b^2, \\ x = a^2 + b^2, \end{array} \right.$$

hence $l^2 = ab$. Thus there exist positive integers $c, d, \gcd(c, d) = 1$, such that

$$\left\{ \begin{array}{l} a = c^2, \\ b = d^2, \end{array} \right.$$

and so $m^2 = c^4 - d^4$. We note that $0 < c < a < x$ and the triple of positive integers $(c, d, m)$ would be a solution of the equation, which is contrary to the choice of $x$ as smallest possible.

Case 2: $\gcd(x^2 + y^2, x^2 - y^2) = 2$.

Now $x, y$ are odd and $z$ is even. By (1A) there exist positive integers $a, b, 0 < b < a, \gcd(a, b) = 1$, such that

$$\left\{ \begin{array}{l} x^2 = a^2 + b^2, \\ y^2 = a^2 - b^2, \\ z = 2ab. \end{array} \right.$$

Hence $x^2 y^2 = a^4 - b^4$ with $0 < a < x$ and this is contrary to the choice of $x$ as smallest possible.  □

The above argument is called the *method of infinite descent* and was invented by Fermat. It may also be phrased as follows: if $(x_0, y_0, z_0)$ were a solution in positive integers of (2.1) then we would obtain a new solution in positive integers $(x_1, y_1, z_1)$ with $z_1 < z_0$.

Repeating this procedure, we would produce an infinite decreasing sequence of positive integers

$$z_0 > z_1 > z_2 > \cdots$$

which is not possible.

As a corollary, we obtain the original statement of Fermat, proposed as a problem or mentioned in letters to Mersenne [for Sainte-Croix] (September 1636), to Mersenne (May ?, 1640), to Saint-Martin (May 31, 1643), to Mersenne (August 1643), to Pascal (25 September 1654), to Digby [for Wallis] (April 7, 1658), to Carcavi (August 1659):

**(2B)**     *The area of a Pythagorean triangle is not the square of an integer.*

PROOF. Let $a, b, c$ be the sides of the Pythagorean triangle, where $c$ is the hypotenuse. So $c^2 = a^2 + b^2$.

Assume that the area is the square of an integer $s$: $ab/2 = s^2$. Then

$$\begin{cases} (a+b)^2 = c^2 + 4s^2, \\ (a-b)^2 = c^2 - 4s^2. \end{cases}$$

Hence $(a^2 - b^2)^2 = c^4 - (2s)^4$, so the equation $X^4 - Y^4 = Z^2$ would have nontrivial solution in integers, contradicting (2A). $\square$

We also state explicitly (this is proposed as a problem or mentioned in letters to Mersenne [for Sainte-Croix] (September, 1636), to Mersenne (1638), to Mersenne (May ?, 1640)):

**(2C)**     *The equation*

(2.2) $$X^4 + Y^4 = Z^4$$

*has no solution in integers, all different from* 0.

PROOF. If $x, y, z$ are nonzero integers such that $x^4 + y^4 = z^4$ then $z^4 - y^4 = (x^2)^2$, which contradicts (2A). $\square$

The above results were also reproduced by Euler (1770) and Legendre (1808, 1830).

A companion result to (2A) is the following (see the explicit proof by Euler, 1770):

**(2D)**    *The equation*

$$(2.3) \qquad X^4 + Y^4 = Z^2$$

*has no solution in integers all different from 0.*

PROOF. If the statement is false, let $(x, y, z)$ be a triple of positive integers, with smallest possible $z$, such that $x^4 + y^4 = z^2$. As in (2A), we may assume $\gcd(x, y) = 1$. We also note that $x, y$ cannot be both odd, otherwise $z^2 = x^4 + y^4 \equiv 2 \pmod 4$ and this is impossible. So we may, for example, assume $x$ to be even.

From $(x^2)^2 + (y^2)^2 = z^2$ it follows that $(x^2, y^2, z)$ is a primitive solution of (1.1). By (1A), there exist integers $a, b$, such that $a > b > 0$, $\gcd(a, b) = 1$, $a, b$ are not both odd and

$$\begin{cases} x^2 &= 2ab, \\ y^2 &= a^2 - b^2, \\ z &= a^2 + b^2. \end{cases}$$

Moreover, $b$ must be even. For if $b$ is odd, then $a$ is even, $y^2 = a^2 - b^2 \equiv -1 \pmod 4$, which is impossible.

Now we consider the relation $b^2 + y^2 = a^2$, where $y, b, a$ are positive integers, $b$ is even, and $\gcd(b, y, a) = 1$. By (1A), there exist integers $c, d$ such that $c > d > 0$, $\gcd(c, d) = 1$, $c, d$ of different parity and

$$\begin{cases} b &= 2cd, \\ y &= c^2 - d^2, \\ a &= c^2 + d^2. \end{cases}$$

Therefore $x^2 = 2ab = 4cd(c^2 + d^2)$. But $c, d, c^2 + d^2$ are pairwise relatively prime. By the decomposition of $x^2$ into primes, we conclude that $c, d, c^2 + d^2$ are squares of positive integers, say:

$$\begin{cases} c &= p^2, \\ d &= q^2, \\ c^2 + d^2 &= r^2. \end{cases}$$

Hence

$$(2.4) \qquad p^4 + q^4 = r^2,$$

that is, the triple $(p, q, r)$ is a solution of (2.3). But $z = a^2 + b^2 = (c^2 + d^2)^2 + 4c^2d^2 > r^4 > r$ (since $r > 1$). This contradicts the choice of $z$ as minimal possible, and concludes the proof.  □

TABLE 1. FLT for the exponent 4.

| Author | Year |
|---|---|
| Frénicle De Bessy | 1676 |
| Euler | 1738 (publ. 1747), 1771 |
| Kausler | 1795/6 (publ. 1802) |
| Barlow | 1811 |
| Legendre | 1823, 1830 |
| Schopis | 1825 |
| Terquem | 1846 |
| Bertrand | 1851 |
| Lebesgue | 1853, 1859, 1862 |
| Pepin | 1883 |
| Tafelmacher | 1893 |
| Bendz | 1901 |
| Gambioli | 1901 |
| Kronecker | 1901 |
| Bang | 1905 |
| Bottari | 1908 |
| Rychlik | 1910 |
| Nutzhorn | 1912 |
| Carmichael | 1913 |
| Vrănceanu | 1966 |

Other proofs of Fermat's theorem for the exponent 4 are given by the authors listed in Table 1. Now we indicate a statement which is equivalent to Fermat's theorem for the exponent 4 (see Vrănceanu, 1979):

**(2E)**   *The following statements are equivalent:*
  (1) *Fermat's last theorem is true for the exponent 4.*
  (2) *For every integer $m \neq 0$ the only solutions in nonzero integers of $2X^4 = mY(m^2 + Y^2)$ are $(m, m)$ and $(-m, m)$.*

PROOF. $(1) \rightarrow (2)$   Let $m \neq 0$ and assume that there exist nonzero integers $u, t$ such that $2u^4 = mt(m^2 + t^2)$. Let $x = 2u$, $y = t - m$, $z = t + m$. Then $z^4 - y^4 = (z - y)(z + y)(z^2 + y^2) = 2m \cdot 2t(2t^2 + 2m^2) = 8mt(t^2 + m^2) = 16u^4 = x^4$.

By hypothesis, $xyz = 0$. If $x = 0$ then $y = \pm z$ hence $m = 0$, contrary to the hypothesis. If $y = 0$ then $t = m$, $x = \pm z$, so $u = \pm m$. If $z = 0$ then $x = 0$, again contrary to the hypothesis.

(2) → (1)   If $x^4 + y^4 = z^4$ then $2x^4 = 2(z^4 - y^4) = 2(z - y)(z + y)(x^2 + y^2) = (z - y)(z + y)[(z - y)^2 + (z + y)^2]$. So taking $m = z - y$ then $t = z + y$, $u = x$ satisfy the relation $2u^4 = mt(m^2 + t^2)$. If $m$ or $u$ or $t$ is equal to 0 then $x = 0$. If $m, t, u \neq 0$, by hypothesis $t = m$ hence $y = 0$, an absurdity.  □

We conclude this section by illustrating how the method may be applied to find solutions of some similar diophantine equations.

**(2F)**    *The equation*

$$X^4 - 4Y^4 = \pm Z^2$$

*has no solution in nonzero integers.*

PROOF. It suffices to consider the equation $X^4 - 4Y^4 = Z^2$. Because if $x, y, z$ are nonzero integers such that $x^4 - 4y^4 = -z^2$ then $4x^4 - (2y)^4 = -(2z)^2$, so $(2y)^4 - 4x^4 = (2z)^2$ and $(2y, x, 2z)$ would be a solution of the first equation.

Now, if $x, y, z$ are positive integers such that $x^4 - 4y^4 = z^2$ and $\gcd(x, y, z) = 1$ (as we may assume without loss of generality), there exist integers $a, b$ with $a > b > 0$ and

$$\begin{cases} 2y^2 = 2ab, \\ z = a^2 - b^2, \\ x^2 = a^2 + b^2. \end{cases}$$

Since $\gcd(a, b) = 1$ then $a, b$ are squares, say $a = c^2$, $b = d^2$. Hence $x^2 = c^4 + d^4$, and this relation is impossible by (2D).  □

Legendre proved:

**(2G)**    *If $x, y, z$ are nonzero integers and*

$$x^4 + y^4 = 2z^2,$$

*then $x^2 = y^2$ and $z^2 = x^4$.*

PROOF. We have

$$4z^4 = (x^4 + y^4)^2 = (x^4 - y^4)^2 + 4x^4y^4,$$

so

$$z^4 - x^4 y^4 = \left(\frac{x^4 - y^4}{2}\right)^2.$$

(in particular, $x^4 - y^4$ is even). Since $x, y, z$ are not zero, by (2A) we have $x^4 = y^4$, so $x^2 = y^2$. Then $z^2 = x^4$.  □

**(2H)**    *If $x, y, z$ are nonzero integers and $2x^4 + 2y^4 = z^2$, then $x^2 = y^2$, $z^2 = 4x^4$.*

PROOF. Multiplying by 8 we have $(2x)^4 + (2y)^4 = 2(2z)^2$. By (2G), $(2x)^2 = (2y)^2$ and $(2z)^2 = (2x)^4$, so $x^2 = y^2$, $z^2 = 4x^4$.  □

The next result is due to Lucas (1877). We present here an easier proof due to Obláth (1952):

**(2I)**    *The equation*

$$4X^4 - 1 = 3Y^4$$

*has only the trivial solutions $(\pm 1, \pm 1)$ in integers.*

PROOF. If $x, y$ are integers such that $3y^4 = 4x^4 - 1 = (2x^2 + 1)(2x^2 - 1)$, since $2x^2 - 1 \not\equiv 0 \pmod 3$ then there exist integers $a, b$ such that $2x^2 + 1 = 3a^4$, $2x^2 - 1 = b^4$. By (2G), the last equation is only satisfied when $x = \pm 1$, $b = \pm 1$, hence $y = \pm 1$.  □

## Bibliography

Fermat, P., Ad problema XX commentarii in ultimam questionem Arithmeticorum Diophanti. Area trianguli rectanguli in numeris non potest esse quadratus. *Oeuvres*, Vol. I, p. 340 (in Latin), Vol. III, pp. 271–272 (in French). Publiées par les soins de MM. Paul Tannery et Charles Henry. Gauthier-Villars, Paris, 1891, 1896.

Fermat, P., Commentaire sur la question 24 du Livre VI de Diophante. *Oeuvres*, Vol. I, pp. 336–338 (in Latin), Vol. III, pp. 270–271 (in French). Publiées par les soins de MM. Paul Tannery et Charles Henry. Gauthier-Villars, Paris, 1891, 1896.

1636 Fermat, P., Lettre à Mersenne [pour Sainte-Croix] (Sept. 1636). *Oeuvres*, Vol. III, pp. 286–292. Publiées par les soins de MM. Paul Tannery et Charles Henry. Gauthier-Villars, Paris, 1896.

1638 Fermat, P., Lettre de Fermat à Mersenne (début Juin, 1638). *Correspondance du Père Marin Mersenne*, Vol. 7, pp. 272–283. Commencée par Mme. Paul Tannery, publiée et annotée par Cornelis de Waard. Ed. du C.N.R.S, Paris, 1962.

1640 Fermat, P., Lettre à Mersenne (Mai ?, 1640). *Oeuvres*, Vol. II, pp. 194–195. Publiées par les soins de MM. Paul Tannery et Charles Henry. Gauthier-Villars, Paris, 1894.

1643 Fermat, P., Lettre à Saint-Martin (31 Mai, 1643). *Oeuvres*, Vol. II, pp. 258–260. Publiées par les soins de MM. Paul Tannery et Charles Henry. Gauthier-Villars, Paris, 1894.

1643 Fermat, P., Lettre à Mersenne (Août, 1643). *Oeuvres*, Vol. II, pp. 260–262. Publiées par les soins de MM. Paul Tannery et Charles Henry. Gauthier-Villars, Paris, 1894.

1654 Fermat, P., Lettre à Pascal (25 Septembre, 1654). *Oeuvres*, Vol. II, pp. 310–314. Publiées par les soins de MM. Paul Tannery et Charles Henry. Gauthier-Villars, Paris, 1894.

1658 Fermat, P., Lettre à Digby [for Wallis] (7 Avril, 1658). *Oeuvres*, Vol. II, pp. 374–378. Publiées par les soins de MM. Paul Tannery et Charles Henry. Gauthier-Villars, Paris, 1894.

1659 Fermat, P., Lettre à Carcavi (Août, 1659). Relation des nouvelles découvertes en la science des nombres. *Oeuvres*, Vol. II, pp. 431–436. Publiées par les soins de MM. Paul Tannery et Charles Henry. Gauthier-Villars, Paris, 1894.

1676 Frénicle de Bessy, *Traité des Triangles Rectangles en Nombres*, Vol. I, Paris, 1676; reprinted in Mém. Acad. Roy. Sci. Paris, **5** (1729), 1666–1699.

1738 Euler, L., *Theorematum quorundam arithmeticorum demonstrationes*, Comm. Acad. Sci. Petrop., **10** (1738) 1747, 125–146; also in *Opera Omnia*, Ser. I, *Commentationes Arithmeticae*, Vol. I, pp. 38–58. Teubner, Leipzig, 1915.

1770/1 Euler, L., *Vollständige Anleitung zur Algebra*, 2 volumes. Royal Acad. Sci., St. Petersburg. French translation with notes of M. Bernoulli and additions of M. de LaGrange, Kais. Acad. Wiss., St. Petersburg, 1770; English translation by Rev. J. Hewlitt, Longman, Hurst, Rees, Orme, London,

1822. Also in *Opera Omnia*, Ser. I, Vol. I, pp. 437ff. Teubner, Leipzig, 1911.

1777 Lagrange, J.L., *Sur quelques problèmes de l'analyse de Diophante*. Nouveaux Mém. Acad. Sci. Belles Lettres, Berlin, 1777; reprinted in *Oeuvres*, Vol. IV (publiées par les soins de M. J.-A. Serret), pp. 377–398, Gauthier-Villars, Paris, 1869.

1802 Kausler, C.F., *Nova demonstratio theorematis nec summam, nec differentiam duorum biquadratorum biquadratum esse posse*, Novi Acta Acad. Petrop., **13** (1795/6), 1802, 237–244.

1808 Legendre, A.M., *Essai sur la Théorie des Nombres* ($2^e$ édition), p. 343, Courcier, Paris, 1808.

1811 Barlow, P., *An Elementary Investigation of Theory of Numbers*, pp. 144–145, J. Johnson, St. Paul's Church-Yard, London, 1811.

1825 Schopis, *Einige Sätze aus der unbestimmten Analytik*, Programm, Gummbinnen, 1825.

1830 Legendre, A.M., *Théorie des Nombres* ($3^e$ édition), Vol. II, p. 5, Firmin Didot Frères, Paris, 1830; reprinted by A. Blanchard, Paris, 1955.

1846 Terquem, O., *Théorèmes sur les puissances des nombres*, Nouv. Ann. Math., **5** (1846), 70–87.

1851 Bertrand, J., *Traité Élémentaire d'Algèbre*, pp. 217–230 and 395, Hachette, Paris, 1851.

1853 Lebesgue, V.A., *Résolution des équations biquadratiques $z^2 = x^4 \pm 2^m y^4$, $z^2 = 2^m x^4 - y^4$, $2^m z^2 = x^4 \pm y^4$*, J. Math. Pures Appl., **18** (1853), 73–86.

1859 Lebesgue, V.A., *Exercices d'Analyse Numérique*, pp. 83–84 and 89, Leiber et Faraguet, Paris, 1859.

1862 Lebesgue, V.A., *Introduction à la Théorie des Nombres*, pp. 71–73, Mallet-Bachelier, Paris, 1862.

1877 Lucas, E., *Sur la résolution du système des équations $2v^2 - u^2 = w^2$ et $2v^2 + u^2 = 3z^2$*, Nouv. Ann. Math., 2e série, **36** (1877), 409–416.

1883 Pepin, T., *Étude sur l'équation indéterminée $ax^4 + by^4 = cz^2$*, Atti Accad. Naz. Lincei, **36** (1883), 34-70.

1893 Tafelmacher, W.L.A., *Sobre la ecuación $x^4 + y^4 = z^4$*, Ann. Univ. Chile, **84** (1893), 307–320.

1901 Bendz, T.R., *Öfver diophantiska ekvationen $x^n + y^n = z^n$*, Almqvist & Wiksells Boktrycken, Uppsala, 1901.

1901 Gambioli, D., *Memoria bibliografica sull'ultimo teorema di Fermat*, Period. Mat., **16** (1901), 145–192.

1901 Kronecker, L., *Vorlesungen über Zahlentheorie*, Vol. I, pp. 35–38,
Teubner, Leipzig, 1901; reprinted by Springer-Verlag, Berlin, 1978.

1905 Bang, A., *Nyt Bevis for at Ligningen $x^4 - y^4 = z^4$, ikke kan have rationale Løsinger*, Nyt Tidsskrift Mat., **16B** (1905), 35–36.

1908 Bottari, A., *Soluzione intere dell'equazione pitagorica e applicazione alla dimostrazione di alcune teoremi della teoria dei numeri*, Period. Mat., **23** (1908), 104–110.

1910 Bachmann, P., *Niedere Zahlentheorie*, Vol. II, Teubner, Leipzig, 1910; reprinted by Chelsea, New York, 1968.

1910 Rychlik, K., *On Fermat's last theorem for $n = 4$ and $n = 3$* (in Bohemian), Časopis Pěst. Mat., **39** (1910), 65–86.

1912 Nutzhorn, F., *Den ubestemte Ligning $x^4 + y^4 = z^4$*, Nyt Tidsskrift Mat., **23B** (1912), 33–38.

1913 Carmichael, R.D., *On the impossibility of certain Diophantine equations and systems of equations*, Amer. Math. Monthly, **20** (1913), 213–221.

1915 Carmichael, R.D., *Diophantine Analysis*, Wiley, New York, 1915.

1952 Obláth, R., *Über einige unmögliche diophantische Gleichungen*, Matem. Tidsskrift Ser. A (1952), 53–62.

1966 Vrănceanu, G., *Asupra teorema lui Fermat pentru $n = 4$*, Gaz. Mat. Ser. A, **71** (1966), 334–335; reprinted in *Opera Matematica*, Vol. 4, pp. 202–205, Edit. Acad. Rep. Soc. Romana, Bucureşti, 1977.

1977 Edwards, H.M., *Fermat's Last Theorem, A Genetic Introduction to Algebraic Number Theory*, Springer-Verlag, New York, 1977.

1979 Vrănceanu, G., *Une interprétation géométrique du théorème de Fermat*, Rev. Roumaine Math. Pures Appl., **24** (1979), 1137–1140.

## I.3. Gaussian Numbers

We shall prove that $X^4 + Y^4 = Z^2$ has no solution in nonzero integers of the Gaussian field. This result is explicitly proved in Hilbert's *Zahlbericht* (1897, Theorem 169); see also Sommer (1907) and Hancock (1931).

The set of complex numbers $\alpha = a + bi$, where $i = \sqrt{-1}$ and $a, b \in \mathbb{Q}$, constitutes the Gaussian field $K = \mathbb{Q}(i)$. The numbers $\alpha = a + bi$, where $a, b \in \mathbb{Z}$, are called the *Gaussian integers*. They constitute a ring, denoted by $A = \mathbb{Z}[i]$.

If $\alpha, \beta \in K$, $\beta$ divides $\alpha$ if there exists a Gaussian integer $\gamma$ such that $\alpha = \beta\gamma$. We write $\beta \mid \alpha$ when $\beta$ divides $\alpha$. Two nonzero Gaussian integers $\alpha, \beta$ are *associated* when $\alpha$ divides $\beta$ and $\beta$ divides $\alpha$; we write $\alpha \sim \beta$. The Gaussian integers associated with 1 are called the (Gaussian) *units*. It is easily seen that they are $\pm 1$, $\pm i$.

A nonzero Gaussian integer $\alpha$ is a *prime* if it is not a unit and the only Gaussian integers dividing $\alpha$ are units or associated with $\alpha$. In the field of Gaussian numbers, every nonzero Gaussian integer $\alpha$ is the product of prime Gaussian integers: $\alpha = \gamma_1\gamma_2 \cdots \gamma_s$. This decomposition is unique in the following sense: if we also have $\alpha = \delta_1\delta_2 \cdots \delta_t$, where each $\delta_i$ is a prime Gaussian integer, then $s = t$, and changing the order if necessary, $\gamma_i$ and $\delta_i$ are associated (for every $i = 1, \ldots, s$).

Therefore we may define, in an obvious way, the greatest common divisor of nonzero Gaussian integers, which is unique up to units.

If $\alpha, \beta, \gamma$ are Gaussian numbers and $\gamma \neq 0$, we write $\alpha \equiv \beta$ (mod $\gamma$) when $\gamma$ divides $\alpha - \beta$. The congruence relation $\equiv$ satisfies the same properties as the congruence for ordinary integers. The Gaussian integer $\lambda = 1 - i$ is a prime and $2 = i\lambda^2$, so $\lambda^2 \mid 2$ but $\lambda^3 \nmid 2$. We have $1 + i = i(1 - i) = i\lambda$.

There are precisely four distinct congruence classes modulo 2, namely the classes of 0, 1, $i$, and $\lambda$. Indeed, these numbers are pairwise incongruent modulo 2. On the other hand, according to the parity of $a, b$, we deduce that $a + bi$ is congruent to 0, 1, $i$, or $\lambda$, modulo 2. In particular, if $\lambda \nmid a = a + bi$ then $\alpha \equiv 1$ (mod 2) or $\alpha \equiv i$ (mod 2). Then, $\alpha^2 \equiv \pm 1$ (mod 4) and $\alpha^4 \equiv 1$ (mod 8), that is, $\alpha^4 \equiv 1$ (mod $\lambda^6$) since $8 = -i\lambda^6$.

Now we show:

**(3A)**    *The equation*

$$X^4 + Y^4 = Z^2$$

*has no solution in Gaussian integers all different from zero.*

PROOF. Let $\xi, \eta, \theta \in \mathbb{Z}[i]$ be nonzero and such that $\xi^4 + \eta^4 = \theta^2$. We may assume without loss of generality that $\gcd(\xi, \eta) = 1$. Indeed, if $\delta = \gcd(\xi, \eta)$, then $\xi = \delta\xi'$, $\eta = \delta\eta'$, with $\xi', \eta' \in \mathbb{Z}[i]$, $\gcd(\xi', \eta') = 1$; so $\delta^4$ divides $\theta^2$, hence $\delta^2$ divides $\theta$, we may write $\theta = \delta^2\theta'$, with $\theta' \in \mathbb{Z}[i]$. Hence $\xi'^4 + \eta'^4 = \theta'^2$ where $\gcd(\xi', \eta') = 1$.

From $\gcd(\xi, \eta) = 1$ it follows that $\xi, \eta, \theta$ are pairwise relatively prime. We consider two cases.

*Case 1:* $\lambda$ does not divide $\xi\eta$.

By a preceding remark $\xi^4 \equiv 1 \pmod{\lambda^6}$, $\eta^4 \equiv 1 \pmod{\lambda^6}$ so $\theta^2 = \xi^4 + \eta^4 \equiv 2 \pmod{\lambda^6}$. Since $2 = i\lambda^2$ then $\lambda^2 \mid \theta^2$, hence $\lambda \mid \theta$. However, $\lambda^2 \nmid \theta$, because $\lambda^4 \nmid 2$. We write $\theta = \lambda\theta_1$, where $\lambda \nmid \theta_1$. Thus $\lambda^2\theta_1^2 \equiv 2 = i\lambda^2 \pmod{\lambda^6}$ hence $\theta_1^2 \equiv i \pmod{\lambda^4}$, and therefore $\theta_1^4 \equiv -1 \pmod{\lambda^6}$ since $\lambda^4 \sim 4$, $\lambda^6 \sim 8$. However, $\lambda \nmid \theta_1$ hence $\theta_1^4 \equiv 1 \pmod{\lambda^6}$, which would imply by subtraction that $2 \equiv 0 \pmod{\lambda^6}$, and this is absurd.

*Case 2:* $\lambda$ divides $\xi$.

Hence $\lambda \nmid \eta\theta$. We write $\xi = \lambda^m\xi'$, with $m \geq 1$, $\xi' \in \mathbb{Z}[i]$, and $\lambda \nmid \xi'$. The essential part of the proof consists in showing the following assertion:

Let $n \geq 1$ and let $\varepsilon$ be a unit of $\mathbb{Z}[i]$ (so $\varepsilon = \pm 1$ or $\pm i$). If there exist $\alpha, \beta, \gamma \in \mathbb{Z}[i]$, pairwise relatively prime, not multiples of $\lambda$, and $\varepsilon\lambda^{4n}\alpha^4 + \beta^4 = \gamma^2$ then:

(a) $n \geq 2$; and
(b) there exists a unit $\varepsilon_1$ and $\alpha_1, \beta_1, \gamma_1 \in \mathbb{Z}[i]$, pairwise relatively prime, not multiples of $\lambda$, such that

$$\varepsilon_1\lambda^{4(n-1)}\alpha_1^4 + \beta_1^4 = \gamma_1^2.$$

The hypothesis is satisfied with $n = m$, $\varepsilon = 1$, $\alpha = \xi'$, $\beta = \eta$, $\gamma = \theta$. By repeated application of the above assertion, we would find a unit $\varepsilon'$ and $\alpha', \beta', \gamma' \in \mathbb{Z}[i]$, pairwise relatively prime, not multiples of $\lambda$, such that

$$\varepsilon'\lambda^4\alpha'^4 + \beta'^4 = \gamma'^2.$$

This contradicts (a) above.

First we show that $n \geq 2$. Indeed $\varepsilon\lambda^{4n}\alpha^4 + \beta^4 - 1 = \gamma^2 - 1$ and since $\lambda \nmid \beta$, then $\beta^4 \equiv 1 \pmod{\lambda^6}$, so $\gamma^2 \equiv 1 \pmod{\lambda^4}$. But $\lambda \nmid \gamma$ hence $\gamma \equiv i \pmod{\lambda^2}$ or $\gamma \equiv 1 \pmod{\lambda^2}$. In the first case, $\gamma^2 \equiv -1 \pmod{\lambda^4}$ hence $\lambda^4$ would divide 2, a contradiction. So $\gamma - 1 = \lambda^2\mu$ where $\mu \in \mathbb{Z}[i]$ and hence $\gamma + 1 = \lambda^2\mu + 2 = \lambda^2(\mu + i)$. But either $\lambda \mid \mu$ or if $\lambda \nmid \mu$ then $\mu \equiv i \pmod{\lambda}$ because $1 \equiv i \pmod{\lambda}$; so $\mu \equiv -i \pmod{\lambda}$. We have shown that in any case $\lambda \mid \mu(\mu + i)$ so $\lambda^5$ divides $\gamma^2 - 1 = \lambda^4\mu(\mu + i)$, hence $\lambda^5$ divides $\varepsilon\lambda^{4n}\alpha^4 + (\beta^4 - 1)$; but $\lambda^6$ divides $\beta^4 - 1$, $\lambda \nmid \alpha$, hence $\lambda^6 \mid \lambda^{4n}$ so $n \geq 2$.

Now we prove (b). We have $\varepsilon\lambda^{4n}\alpha^4 = \gamma^2 - \beta^4 = (\gamma - \beta^2)(\gamma + \beta^2)$. We note that $\gcd(\gamma - \beta^2, \gamma + \beta^2) = \lambda^2$. Indeed $\lambda$ must divide one of the factors in the right-hand side, hence it divides both factors, because $(\lambda + \beta^2) - (\lambda - \beta^2) = 2\beta^2$ is a multiple of $\lambda^2$. Since $\lambda^4$ divides the right-hand side, this implies necessarily that $\lambda^2$ divides both factors:

$$\begin{cases} \gamma \mp \beta^2 = \lambda^2\nu, \\ \gamma \pm \beta^2 = \gamma^{4n-2}\nu', \end{cases}$$

where $\nu, \nu' \in \mathbb{Z}[i]$ and $\gcd(\nu, \nu') = 1$. Hence $\varepsilon\lambda^{4n}\alpha^4 = \lambda^{4n}\nu\nu'$ so by the uniqueness of factorization up to units, $\nu, \nu'$ must be fourth powers, up to units $\nu = \omega\kappa^4$, $\nu' = \omega'\kappa'^4$, where $\kappa, \kappa' \in \mathbb{Z}[i]$, $\gcd(\kappa, \kappa') = 1$, $\omega, \omega'$ are units. Thus

$$\begin{cases} \gamma - \beta^2 = \omega\lambda^2\kappa^4, \\ \gamma + \beta^2 = \omega'\lambda^{4n-2}\kappa'^4, \end{cases}$$

and subtracting,

$$2\beta^2 = \omega'\lambda^{4n-2}\kappa'^4 - \omega\lambda^2\kappa^4.$$

Hence

$$\beta^2 = -i\omega'\lambda^{4n-4}\kappa'^4 + i\omega\kappa^4,$$

so

$$\omega_1'\lambda^{4(n-1)}\kappa'^4 + \omega_1\kappa^4 = \beta^2,$$

with units $\omega_1' = -i\omega'$, $\omega_1 = i\omega$. We show that $\omega_1 = 1$, which suffices to establish statement (b). Since $n \geq 2$ then $\lambda^4 \mid \beta^2 - \omega_1\kappa^4$; but $\lambda \nmid \beta$, hence $\lambda \nmid \kappa$ hence $\kappa^4 \equiv 1 \pmod{\lambda^6}$ so $\lambda^4 \mid \beta^2 - \omega_1$. But $\lambda^6$ divides $\beta^4 - 1 = (\beta^2 - 1)(\beta^2 + 1)$ hence $\beta^2 \equiv 1$ or $-1 \pmod{\lambda^4}$. This shows that $\omega_1 = \pm 1$. If $\omega_1 = -1$, by multiplication with $-1$, we obtain the relation

$$-\omega_1'\lambda^{4(n-1)}\kappa'^4 + \kappa^4 = (i\beta)^2.$$

So, in all cases, we have shown (b), proving the statement.    □

## Bibliography

1897 Hilbert, D., *Die Theorie der algebraischen Zahlkörper*, Jahresber. Deutsch. Math.-Verein., **4** (1897), 175–546; reprinted in *Gesammelte Abhandlungen*, Vol. I, Chelsea, New York, 1965.

1907 Sommer, J., *Vorlesungen über Zahlentheorie*, Teubner, Leipzig, 1907.

1931 Hancock, H., *Foundations of the Theory of Algebraic Numbers*, Vol. I., Macmillan, New York, 1931.

## I.4. The Cubic Equation

Fermat proposed the problem to show that a cube cannot be equal to the sum of two nonzero cubes. See letters to Mersenne [for Sainte-Croix] (September, 1636), to Mersenne (May ?, 1640), to Digby [for Wallis] (April 7, 1658), to Carcavi (August, 1659), all mentioned in the Bibliography of Section I.2; see also a letter to Digby [for Brouncker] (August 15, 1657).

Euler discovered a proof of this statement. It used the method of infinite descent and appeared in his book on Algebra, published in St. Petersburg in 1770, translated into German in 1802, and into English in 1822. A critical study of Euler's proof uncovered an important missing step, concerning the divisibility properties of integers of the form $a^2 + 3b^2$. We note that in his paper of 1760, Euler had already proved rigorously that if an odd prime number $p$ divides $a^2 + 3b^2$ (where $a, b$ are nonzero relatively prime integers) then there exist integers $u, v$ such that $p = u^2 + 3v^2$. Yet, Euler did not establish in full the Lemma 4.7 which is required in the proof. Legendre reproduced Euler's proof in his book (1808, 1830) without completing the details.

In 1875, Pepin published a long paper on numbers of the form $a + b\sqrt{-c}$ pointing out arguments which had been insufficiently justified by Euler concerning numbers of the form $a^2 + cb^2$, especially for $c = 1, 2, 3, 4, 7$. Schumacher (1894) noted explicitly the missing link in the proof. In 1901 Landau offered a rigorous proof; this was again the object of Holden's paper (1906) and, once more in 1915, a detailed proof appeared in Carmichael's book.[1] In 1966, Bergmann published a paper with historical considerations and a thorough analysis of Euler's proof. Once more, in 1972, R. Legendre pointed out that Euler's proof was not perfect. In his book, Edwards (1977) discusses also this proof.

**(4A)**    *The equation*

(4.1) $$X^3 + Y^3 + Z^3 = 0$$

*has only the trivial solutions in integers.*

PROOF. Assume that $x$, $y$, and $z$ are nonzero, pairwise relatively prime integers such that $x^3 + y^3 + z^3 = 0$. Then they must be distinct (because 2 is not a cube) and exactly one of these integers is even, say $x, y$ are odd and $z$ is even. Among all the solutions with above properties, we choose one for which $|z|$ is the smallest possible.

We shall produce nonzero pairwise relatively prime integers $l, m, n$ which are such that $l^3 + m^3 + n^3 = 0$, $n$ is even, and $|z| > |n|$. This will be a contradiction. Since $x + y$, $x - y$ are even, there exist integers $a, b$ such that $2a = x + y$, $2b = x - y$; so $x = a + b$, $y = a - b$ and therefore $a, b \neq 0$, $\gcd(a, b) = 1$ and $a, b$ have different parity.

Then $-z^3 = x^3 + y^3 = (a + b)^3 + (a - b)^3 = 2a(a^2 + 3b^2)$. But $a^2 + 3b^2$ is odd and $z$ is even, hence 8 divides $z^3$, so 8 divides $2a$, so $b$ is odd. We have $\gcd(2a, a^2 + 3b^2)$ equal to 1 or 3. In fact, if $p^k (k \geq 1)$ is a prime power dividing $2a$ and $a^2 + 3b^2$ then $p \neq 2$ so $p^k$ divides $a$, hence $3b^2$; but $p$ does not divide $b$, so $k = 1$ and $p = 3$.

Now we consider two cases.

*Case* 1: $\gcd(2a, a^2 + 3b^2) = 1$.

Then 3 does not divide $a$. From $-z^3 = 2a(a^2 + 3b^2)$ it follows from the unique factorization of integers into primes that $2a$ and $a^2 + 3b^2$ are cubes:

$$\begin{cases} 2a = r^3, \\ a^2 + 3b^2 = s^3, \end{cases}$$

where $s$ is odd and not a multiple of 3. At this point we make use of a fact to be justified later: if $s$ is odd and $s^3 = a^2 + 3b^2$ with

---

[1] A proposed simplification of Euler's proof by Pizá (1955) is wrong, as pointed out by Yf (1956).

$\gcd(a, b) = 1$, then $s$ also must be of the form $s = u^2 + 3v^2$, with $u, v \in \mathbb{Z}$, and

$$\begin{cases} a = u(u^2 - 9v^2), \\ b = 3v(u^2 - v^2). \end{cases}$$

Then $v$ is odd, $u$ is even (because $b$ is odd), $u \neq 0$, 3 does not divide $u$ (since 3 does not divide $a$) and $\gcd(u, v) = 1$. Therefore, $2u$, $u + 3v$, $u - 3v$ are pairwise relatively prime and from $r^3 = 2a = 2u(u - 3v)(u + 3v)$ it follows that $2u$, $u - 3v$, $u + 3v$ are cubes:

$$\begin{cases} 2u = -n^3, \\ u - 3v = l^3, \\ u + 3v = m^3, \end{cases}$$

with $l, m, n$ distinct from 0 (since 3 does not divide $u$) and pairwise relatively prime. We conclude that

$$l^3 + m^3 + n^3 = 0,$$

where $n$ is even. Now we show that $|z| > |n|$. In fact,

$$|z|^3 = |2a(a^2 + 3b^2)| = |n^3(u^2 - 9v^2)(a^2 + 3b^2)| \geq 3|n^3| > |n^3|$$

because $u^2 - 9v^2 = l^3 m^3 \neq 0$ and $b \neq 0$, since it is odd. This contradicts the minimality of $|z|$.

*Case 2:*  $\gcd(2a, a^2 + 3b^2) = 3$.

We write $a = 3c$. Thus, $c$ is even and indeed 4 divides $c$, while 3 does not divide $b$ (since $a, b$ are relatively prime). So $-z^3 = 6c(9c^2 + 3b^2) = 18c(3c^2 + b^2)$ where $\gcd(18c, 3c^2 + b^2) = 1$. Indeed, $c$ is even and $b$ is odd, therefore $3c^2 + b^2$ is odd, 3 does not divide $3c^2 + b^2$ and $\gcd(b, c) = 1$. By the unique factorization of integers into primes, $18c$ and $3c^2 + b^2$ are cubes:

$$\begin{cases} 18c = r^3, \\ 3c^2 + b^2 = s^3, \end{cases}$$

where $s$ is odd and 3 divides $r$. By the same result already quoted, $s = u^2 + 3v^2$ with $u, v \in \mathbb{Z}$ and

$$\begin{cases} b = u(u^2 - 9v^2), \\ c = 3v(u^2 - v^2). \end{cases}$$

Thus $u$ is odd, $v$ is even (since $b$ is odd), $v \neq 0$, $\gcd(u, v) = 1$. Also, $2v$, $u + v$, $u - v$ are pairwise relatively prime. From $r^3 = 18c = $

$54v(u+v)(u-v)$ we deduce that $(r/3)^3 = 2v(u+v)(u-v)$ and $2v$, $u+v$, $u-v$ are cubes:

$$\begin{cases} 2v = -n^3, \\ u+v = l^3, \\ u-v = -m^3. \end{cases}$$

Thus $l^3 + m^3 + n^3 = 0$ with $l, m, n$ different from 0, and $n$ even. Now we show that $|z| > |n|$. In fact,

$$\begin{aligned} |z|^3 &= 18|c|(3c^2 + b^2) \\ &= 54|v(u^2 - v^2)|(3c^2 + b^2) \\ &= 27|n|^3|u^2 - v^2|(3c^2 + b^2) \\ &> |n|^3. \end{aligned}$$

Since $u^2 - v^2 = -l^3 m^3 \neq 0$, $|3c^2 + b^2| \geq 1$. Again, this contradicts the choice of $|z|$ as minimal.   $\square$

We shall now justify the step concerning the expression of $s$ as $s = u^2 + 3v^2$. For this purpose, we use arguments, already known to Fermat, in connection with the study of integers of the form $u^2 + v^2$.

Let $S$ be the set of integers of the form $a^2 + 3b^2$ $(a, b \in \mathbb{Z})$. $S$ is closed under multiplication, because

$$(4.2) \qquad (a^2 + 3b^2)(c^2 + 3d^2) = (ac \pm 3bd)^2 + 3(ad \mp bc)^2$$

(the equality holds with corresponding signs).

LEMMA 4.1. *Let $p$ be a prime different from 2 and 3. Then the following conditions are equivalent:*

(1) $p \equiv 1 \pmod 6$.
(2) $-3$ *is a square modulo $p$.*
(3) *The polynomial $X^2 + X + 1$ has a root in $\mathbb{F}_p$.*

PROOF. For the equivalence of (1) and (2) we compute the Legendre symbol, using Gauss' reciprocity law:

$$\left(\frac{-3}{p}\right) = \left(\frac{-1}{p}\right)\left(\frac{3}{p}\right) = (-1)^{(p-1)/2}(-1)^{(p-1)/2}\left(\frac{p}{3}\right) = \left(\frac{p}{3}\right).$$

So $(-3/p) = +1$ if and only if $p \equiv 1 \pmod 3$, that is, $p \equiv 1 \pmod 6$. For the equivalence of (2) and (3), we write

$$X^2 + X + 1 = \left(X + \tfrac{1}{2}\right)^2 + \tfrac{3}{4}.$$

If there exists $\alpha \in \mathbb{F}_p$ such that $\alpha^2 + \alpha + 1 = 0$ then $-3 = 4\left(\alpha - \frac{1}{2}\right)^2$ and conversely, if $-3 = \beta^2$ with $\beta \in \mathbb{F}_p$, we take $\alpha = -\frac{1}{2} + \beta/2$ so $\alpha^2 + \alpha + 1 = 0$.  $\square$

LEMMA 4.2. *If $k$ is a nonzero integer, if $p$ is a prime, and $p = c^2 + 3d^2 \in S$, $pk = a^2 + 3b^2 \in S$ then $p$ divides $ac \pm 3bd$ and $ad \mp bc$ (with corresponding signs ) and*

$$k = \left(\frac{ac \pm 3bd}{p}\right)^2 + 3\left(\frac{ad \mp bc}{p}\right) \in S.$$

PROOF. We have

$$k = \frac{(a^2 + 3b^2)(c^2 + 3d^2)}{(c^2 + 3d^2)^2} = \left(\frac{ac \pm 3bd}{c^2 + 3d^2}\right)^2 + 3\left(\frac{ad \mp bc}{c^2 + 3d^2}\right)^2$$

by (4.2). But $(ac + 3bd)(ac - 3bd) = a^2 c^2 - 9b^2 d^2 = a^2(c^2 + 3d^2) - 3(a^2 + 3b^2)d^2 = (a^2 - 3kd^2)(c^2 + 3d^2)$. Since $c^2 + 3d^2 = p$ is a prime then, say, $p$ divides $ac + 3bd$, that is, $(ac + 3bd)/p \in \mathbb{Z}$. Hence also $3\left((ad - bc)/p\right)^2 \in \mathbb{Z}$ and therefore $(ad - bc)/p \in \mathbb{Z}$, thus $k \in S$.  $\square$

LEMMA 4.3. *If $p$ is a prime then $p \in S$ if and only if $p = 3$ or $p \equiv 1$ (mod 3).*

PROOF. If $p = a^2 + 3b^2$, $p \neq 3$, then $b \neq 0$, so $p \equiv a^2$ (mod 3), and $3 \nmid a$; thus $p \equiv a^2 \equiv 1$ (mod 3).

Clearly $3 \in S$. Let $p \equiv 1$ (mod 3). Since $(-3/p) = 1$ then there exists $t$ such that $0 < t < p/2$ and $-3 \equiv t^2$ (mod $p$). Then $mp = t^2 + 3 < (p/2)^2 + 3 < p^2$ so $0 < m < p$. Now we observe that for every $t \geq 1$ there exists at most one prime $p \neq 2, 3$ such that $p \mid t^2 + 3$ but $p \nmid u^2 + 3$ for every $u$, $1 \leq u < t$.

Indeed, we assume that there exist distinct primes $p, p'$ as above, $p < p'$. By the preceding remark, we must have $0 < t < p/2$, and $t^2 + 3 = pm$ with $0 < m < p$. Since $p' \mid t^2 + 3$ then $p' \mid m$ so $p' \leq m < p$, which is a contradiction.

Now we are ready to prove the statement. Suppose there exists a prime $p$, $p \equiv 1$ (mod 3), such that $p \notin S$. We take the smallest such prime $p$. Let $t \geq 1$ be the smallest integer such that $p \mid t^2 + 3$, so $0 < t < p/2$, $t^2 + 3 = mp$ with $0 < m < p$. If $p'$ is any prime dividing $m$, $m = p'm'$, then $p' \leq m < p$, so $p' \in S$. From $p'(pm') = pm = t^2 + 3 \in S$ it follows from Lemma 4.2 that $pm' \in S$. If $m' = 1$ then $p \in S$, as we intended to show. If $p''$ is a prime dividing $m'$, $m' = p''m''$, then $p'' \leq m' < p$ so $p'' \in S$, hence $p''(pm'') = pm' \in S$ and by

Lemma 4.2, $pm'' \in S$, where $m'' < m'$. Repeating this argument, we eventually arrive at $p \in S$.  □

It is worth giving another proof, using Dirichlet's pigeon-hole principle, of the fact that if $p \equiv 1 \pmod 3$ then $p \in S$.

From $p \equiv 1 \pmod 3$ there exists $t$, $1 \le t \le p - 1$, such that $-3 \equiv t^2 \pmod p$. We consider the set of all pairs $(m, n)$ such that $0 \le m, n \le [\sqrt{p}]$; since there are more than $p$ such pairs, then there exist two distinct pairs $(m, n)$, $(m', n')$ such that $m + nt \equiv m' + n't \pmod p$. So $m \ne m'$ and $n \ne n'$, say $n > n'$, hence $t \equiv (m' - m)/(n - n') \pmod p$; since $0 < n - n' < \sqrt{p}$ and $0 < |m' - m| < \sqrt{p}$, then $t \equiv \pm a/b \pmod p$ with $0 < a < \sqrt{p}$, $0 < b < \sqrt{p}$. Thus $a^2 + 3b^2 \equiv 0 \pmod p$ and we may write $a^2 + 3b^2 = kp$, with $0 < k < 4$. It follows that $a^2 \equiv k \pmod 3$ hence $k \equiv 0$ or $1 \pmod 3$, that is, $k = 1$ or $3$. If $k = 1$ then $p \in S$. If $k = 3$ it follows that $3 \mid a$, so $a = 3a'$ and dividing by 3, $p = b^2 + 3a'^2 \in S$.

LEMMA 4.4. *Let $m = u^2 + 3v^2$, with $u, v \ne 0$, $\gcd(u, v) = 1$. If $p$ is an odd prime dividing $m$ then $p \in S$.*

PROOF. $3 \in S$, so we may assume that $p \ne 3$. Since $p$ divides $m$ then $p$ does not divide $v$, otherwise it would also divide $u$, contrary to the hypothesis. Let $v'$ be such that $vv' \equiv 1 \pmod p$. So $(uv')^2 \equiv -3 \pmod p$ and $(-3/p) = 1$, that is, $p \equiv 1 \pmod 3$. By Lemma 4.3, $p \in S$.  □

We complete the above lemmas as follows:

LEMMA 4.5. *If $p$ is a prime, $p \in S$, then its representation in the form $p = a^2 + 3b^2$ (with $a \ge 0$, $b \ge 0$) is unique.*

PROOF. We apply Lemma 4.2 with $k = 1$, thus $p = a^2 + 3b^2 = c^2 + 3d^2$ (where $a, c \ge 0$, $b > 0$, $d > 0$). Hence

$$1 = \left( \frac{ac \pm 3bd}{p} \right)^2 + 3 \left( \frac{ad \mp bc}{p} \right)^2,$$

so $p = ac \pm 3bd$, $ad = \pm bc$. Therefore $pd = acd \pm 3bd^2 = \pm bc^2 \pm 3bd^2 = \pm b(c^2 + 3d^2) = \pm bp$. Hence $d = \pm b$ thus $b = d$, hence $a = c$.  □

LEMMA 4.6. *Let $m = 3$ or $m = u^2 + 3v^2$, with $u, v \ne 0$ and $\gcd(u, v) = 1$. If $m$ is odd and $m = \prod_{i=1}^{n} p_i^{e_i}$ (where $p_1, \ldots, p_n$ are primes* and $e_i \geq 1$) then there exist integers $a_i, b_i$ $(i = 1, \ldots, n)$ such that $p_i = a_i^2 + 3b_i^2$ and

$$u + v\sqrt{-3} = \prod_{i=1}^{n}(a_i + b_i\sqrt{-3})^{e_i}.$$

PROOF. The proof is by induction on $m$. It is trivial when $m = 3$. Let $m > 3$, so $m = u^2 + 3v^2$, with $u, v \neq 0$, $\gcd(u, v) = 1$. Let $p$ be a prime dividing $m$, and $m = pk$. By Lemma 4.4, $p = a^2 + 3b^2$, and by Lemma 4.2, $k = c^2 + 3d^2$ where $c = (ua \pm 3vb)/p$, $d = (ub \mp va)/p$ (with corresponding signs). We also have $(a \pm b\sqrt{3})(c \mp d\sqrt{-3}) = (ac + 3bd) \pm (bc - ad)\sqrt{-3}$ where

$$ac + 3bd = \frac{1}{p}(ua^2 \pm 3vab + 3ub^2 \mp 3vab) = u,$$

$$\pm(bc - ad) = \pm\frac{1}{p}(uab \pm 3vb^2 - uab \pm va^2) = v,$$

that is,

$$(a \pm b\sqrt{-3})(c \mp d\sqrt{-3}) = u + v\sqrt{-3}.$$

If $k = 1$, it is trivial. If $k \neq 1$ then either $k = 3$ or $k \neq 3$. In this case, $c \neq 0$ (otherwise $c = 0$, so $d$ divides $u, v$, hence $d = 1$ and $k = 3$, contrary to the hypothesis); similarly $d \neq 0$ (otherwise $d = 0$, so $c$ divides $u, v$ hence $c = 1$ and $k = 1$, contrary to the hypothesis); moreover, $\gcd(c, d) = 1$, because $\gcd(u, v) = 1$. By induction, the result is true for $k$, hence $c \mp d\sqrt{-3}$ is expressible in the form indicated. Since $(a \pm b\sqrt{-3})(c \mp d\sqrt{-3}) = u + v\sqrt{-3}$ then the result also holds for $m$. $\square$

LEMMA 4.7. Let $E$ be the set of all triples $(u, v, s)$ such that $s$ is odd, $\gcd(u, v) = 1$ and $s^3 = u^2 + 3v^2$. Let $F$ be the set of all pairs $(t, w)$ where $\gcd(t, w) = 1$ and $t \not\equiv w \pmod{2}$. The mapping $\Phi : F \longrightarrow E$ given by $\Phi(t, w) = (u, v, s)$ with

$$\begin{cases} u = t(t^2 - 9w^2), \\ v = 3w(t^2 - w^2), \\ s = t^2 + 3w^2, \end{cases}$$

is onto $E$.

PROOF. It is clear that $u^2 + 3v^2 = s^3$. Since $t, w$ have different parity, then $s$ is odd. Next we show that $\gcd(u, v) = 1$. Indeed, first we note that $\gcd(t^2 - 9w^2, t^2 - w^2) = 1$ because if a prime $p$

divides $t^2 - 9w^2$ and $t^2 - w^2$, it divides $9t^2 - 9w^2$ so also $8t^2$, hence $p = 2$ (since $p$ cannot divide $t$ because $\gcd(t, w) = 1$). Since $t, w$ have different parity, this is impossible. Now we assume that $p$ is a prime, $e \geq 1$ and $p^e$ divides $u$ and $v$ then $p \mid t$ or $p \mid t^2 - 9w^2$ hence $p \mid t$ in both cases; so $p \nmid w(t^2 - w^2)$ hence $p = 3$. From $3^e \mid v$ since $3 \mid t$ then $e = 1$ thus $\gcd(u, v) = 1$ or $3$. If $3 \mid u$, $3 \mid v$ then $3 \mid t$, $3 \nmid w$ so $3 \mid s$ but $3^2 \nmid s$. However $s^3 = u^2 + 3v^2$ so $3^2 \mid s^3$ hence $3^2 \mid s$, which is a contradiction. This shows that $\Phi(t, w) = (u, v, s) \in E$.

Conversely, given $(u, v, s) \in E$, let $s^3 = \prod_{i=1}^{n} p_i^{e_i}$ be the decomposition of $s^3$ into a product of primes $(p_1, \ldots, p_n$ distinct, $e_i \geq 1)$; so $e_i = 3e_i'$ for every $i$. By Lemma 4.6 there exist integers $a_i$, $b_i$ $(i = 1, \ldots, n)$ such that $p_i = a_i^2 + 3b_i^2$ and

$$u + v\sqrt{-3} = \prod_{i=1}^{n}(a_i + b_i\sqrt{-3})^{e_i}.$$

Let $t, w \in \mathbb{Z}$ be defined by the relation

$$\prod_{i=1}^{n}(a_i + b_i\sqrt{-3})^{e_i'} = t + w\sqrt{-3},$$

so $u + v\sqrt{-3} = (t + w\sqrt{-3})^3$. Computing explicitly the cube in the right-hand side, it follows that $u = t(t^2 - 9w^2)$, $v = 3w(t^2 - w^2)$. Finally, by taking conjugates, $u - v\sqrt{-3} = (t - w\sqrt{-3})^3$, so by multiplying, $s^3 = u^2 + 3v^2 = (t^2 + 3w^2)^3$, hence $s = t^2 + 3w^2$. It follows that $t, u$ have different parity and also $\gcd(t, w) = 1$, $\Phi(t, w) = (u, v, s)$.  □

In this way we have established all the steps in Euler's proof of (4A).

Now we prove the following result due to Kronecker (1859); see also Vrănceanu (1956, 1960). It is a consequence of Fermat's theorem for the exponent 3.

**(4B)**

    (1) *For every integer $m \neq 0$ the only solutions in integers of the equation $4X^3 - 3mY^2 = m^3$ are $(m, m)$ and $(m, -m)$.*

    (2) *The only rational solutions of $4U^3 + 27T^2 = -1$ are $\left(-1, \frac{1}{3}\right)$ and $\left(-1, -\frac{1}{3}\right)$.*

    (3) *$X^3 - X \pm \frac{1}{3}$ are the only cubic polynomials with rational coefficients such that the sum of roots is equal to 0 and the discriminant equal to $-1$.*

(4) *If the discriminant of a cubic polynomial with rational coefficients is the sixth power of a nonzero rational number, then its roots are of the form $r + s\sqrt{3}\sin(\pi/9)$, $r + s\sqrt{3}\sin(2\pi/9)$, $r - s\sqrt{3}\sin(4\pi/9)$.*

PROOF. (1)    If $x, y$ are integers and $(x, y)$ satisfies $4x^3 - 3my^2 = m^3$ then letting $u = -2x$, $v = y + m$ then $u^3 + v^3 = -8x^3 + y^3 + 3y^2 m + 3ym^2 + y^3 = -2m^3 - 6my^2 + y^3 + 3y^2 m + 3ym^2 + y^3 = (y - m)^3$. Thus either $x = 0$ (which would imply $-3y^2 = m^2$, an absurdity) or $y = \pm m$; in this case we have necessarily $x = m$.

(2)    Let $u, t$ be rational numbers, such that $4u^3 + 27t^2 = -1$. We write $u = -x/m$, $t = y/3m$, so $-4x^3 + 3my^2 = -m^3$. By (1) we have $x = m$, $y = \pm m$, hence $u = -1$, $t = \pm\frac{1}{3}$.

(3)    If $X^3 + aX + b$ has rational coefficients and discriminant $\delta = -1$, since $\delta = 4a^3 + 27b^2$ then by (2), $a = -1$, $b = \pm\frac{1}{3}$.

(4)    If $f(X) = X^3 + a_1 X^2 + a_2 X + a_3$ has rational coefficients, if $g(X) = f(X - a_1/3)$ then $g(X)$, $f(X)$ have the same discriminant, and $g(X)$ is of the form $g(X) = X^3 + uX + t$ with rational coefficients.

If the discriminant is a sixth power of a nonzero rational number, say $-(4u^3 + 27t^2) = r^6$, then $4(u/2)^3 + 27(t/r^3)^2 = -1$. Hence $u = -r^2$, $t = \pm r^3/3$, so

$$g(X) = X^3 - r^2 X \pm \frac{r^3}{3} = r^3 \left[ \left(\frac{X}{r}\right)^3 - \left(\frac{X}{r}\right) \pm \frac{1}{3} \right].$$

The roots of the polynomials

$$X^3 - X \pm \tfrac{1}{3}$$

are $\pm(2\sqrt{3}/3)\sin(\pi/9)$, $\pm(2\sqrt{3}/3)\sin(2\pi/9)$, $\mp(2\sqrt{3}/3)\sin(4\pi/9)$, hence those of $f(X)$ are of the form indicated.  □

Conversely, in 1944 Schmid established the equivalence below and proved Fermat's theorem for the exponent 3 by showing directly the validity of (2); see also Vrănceanu (1956, 1960, 1979) where this fact is explicitly spelled out:

**(4C)**    *The following statements are equivalent:*

(1) *Fermat's last theorem is true for the exponent 3.*

(2) *For every integer $m \neq 0$ the only solutions in integers of the equation $4X^3 - 3mY^2 = m^3$ are $(m, m)$, $(m, -m)$.*

TABLE 2. FLT for the exponent 3.

| Author | Year |
|---|---|
| Kausler | 1795/6, publ. in 1802 |
| Legendre | 1823, 1830 |
| Calzolari | 1855 |
| Lamé | 1865 |
| Tait | 1872 |
| Günther | 1878 |
| Gambioli | 1901 |
| Krey | 1909 |
| Rychlik | 1910 |
| Stockhaus | 1910 |
| Carmichael | 1915 |
| van der Corput | 1915 |
| Thue | 1917 |
| Duarte | 1944 |

(3) *The only rational solutions of $4U^3 + 27T^2 = -1$ are $\left(-1, \frac{1}{3}\right)$, $\left(-1, -\frac{1}{3}\right)$.*

PROOF. We have seen in (4B) that (1) implies (2) and also that (2) implies (3). Now we assume that statement (3) is true and we shall derive that Fermat's theorem is true for the exponent 3.

Assume, on the contrary, that there exist nonzero, pairwise relatively prime integers $x, y, z$ such that $x^3 + y^3 = z^3$, so $y \neq z$. Let $u = x/(y - z)$, $t = (y + z)/(3(y - z))$. Then $4u^3 + 27t^2 = -1$, as is easily seen. So, by assumption, $u = -1$, $t = \pm 1/3$, hence $y - z = \pm(y + z)$; this leads to $y = 0$ or $z = 0$, contrary to the hypothesis. $\square$

In 1885, Perrin showed that if $X^3 + Y^3 + Z^3 = 0$ has a nontrivial solution in nonzero relatively prime integers then it would have an infinite number of such solutions, which are obtainable from the assumed solution by means of rational operations. Of course this statement is not of interest since there are no solutions of the type indicated.

Proofs of Fermat's theorem for the exponent 3 were also published by the authors listed in Table 2.

We conclude this section with the study of an equation similar to

(4.1). With the method of infinite descent, we show (see Legendre, 1808, 1830):

**(4D)**    *For every $m > 0$ the equation*

(4.3)    $$X^3 + Y^3 = 2^m Z^3$$

*has only the trivial solutions in integers; namely, the solutions are $(x, y, z)$ with $xyz = 0$ and if $m = 1$ also $(x, x, x)$ with any $x \neq 0$.*

PROOF. Assume that $x, y, z$ are nonzero integers, that $x, y, z$ are not equal if $m = 1$, and that $x^3 + y^3 = 2^m z^3$; we may assume without loss of generality that $\gcd(x, y, z) = \gcd(x, z) = \gcd(y, z) = 1$.

If $m = 3m'$ with $m' \geq 0$, $(x, y, 2^{m'} z)$ would be a nontrivial solution of the equation $X^3 + Y^3 = Z^3$, which contradicts (4A). Thus $3 \nmid m$.

Since $m \neq 0$ then $x, y$ have the same parity. If $x, y$ are not both odd, let $s \geq 1$ be the largest integer such that $2^s$ divides $x, y$. We write $x = 2^s x'$, $y = 2^s y'$ so $x'$ or $y'$ is odd and $x'^3 + y'^3 = 2^{m-3s} z^3$; thus $(x', y', z)$ is a nontrivial solution of an equation of the same type; then $x'$, $y'$ have the same parity, so both are odd. Thus, changing $m$ into $m - 3s$, we may assume without loss of generality that (4.4) has a nontrivial solution $(x, y, z)$ with $x, y$ odd, and also $\gcd(x, y, z) = \gcd(x, z) = \gcd(y, z) = 1$; therefore $\gcd(x, y) = 1$. With the same argument, we may also assume that $z$ is odd. So, we have

$$2^m z^3 = x^3 + y^3 = (x + y)(x^2 - xy + y^2),$$

with $\gcd(x + y, x^2 - xy + y^2) = 1$ or $3$. Indeed, if $p$ is a prime, $e \geq 1$ and $p^e$ divides both $x + y$ and $x^2 - xy + y^2$ then $x \equiv -y \pmod{p^e}$ so $x^2 - xy + y^2 \equiv 3x^2 \pmod{p^e}$; thus $p^e \mid 3x^2$; since $\gcd(x, y) = 1$ then $p \nmid x$, hence $p^3 \mid 3$, that is $p^e = 3$, proving the statement.

Moreover, since $x^2 - xy + y^2 = (x + y)^2 - 3xy$, then $3 \mid x + y$ is equivalent to $3 \mid x^2 - xy + yz$, which is in turn equivalent to $3 \mid z$ and again to $\gcd(x + y, x^2 - xy + y^2) = 3$.

We are led to two cases.

*Case 1: $3 \nmid z$.*

Since $\gcd(x + y, x^2 - xy + y^2) = 1$ and $x^2 - xy + y^2$ is odd there exist odd, relatively prime integers $a, b$ such that

$$\begin{cases} x + y = 2^m a^3, \\ x^2 - xy + y^2 = b^3, \end{cases}$$

with $z = ab$. Since $x + y$, $x - y$ are even, we may write

$$b^3 = \left(\frac{x+y}{2}\right)^2 + 3\left(\frac{x-y}{2}\right)^2.$$

It follows from Lemma 4.6 that there exist integers $t, w$ such that

$$\frac{x+y}{2} + \frac{x-y}{2}\sqrt{-3} = (t + w\sqrt{-3})^3$$

and

$$\begin{cases} (x+y)/2 = t(t^2 - 9w^2), \\ (x-y)/2 = 3x(t^2 - w^2), \end{cases}$$

$$b = t^2 + 3w^2.$$

If $t = 0$ or $t^2 - 9w^2 = 0$ then $x = -y$ so $z = 0$. If $w = 0$ or $t^2 - w^2 = 0$ then $x = y = 0$ or $x = y = z = \pm 1$ with $m = 1$. This was excluded by the hypothesis. It follows that $|t^2 - 9w^2| \neq 1$, otherwise $t + 3w = \pm 1$, $t - 3w = \pm 1$ which is easily seen to be impossible.

We have $3 \nmid t$, because otherwise $3 \mid b$ so $3 \mid z$, contrary to the hypothesis. Since $b$ is odd, so are $t + 3w$ and $t - 3w$. Hence $t, t + 3w, t - 3w$ are nonzero, pairwise relatively prime integers. From $2^{m-l}a^3 = t(t+3w)(t-3w)$ it follows that there exist nonzero integers $c, d, e$ such that

$$\begin{cases} t = 2^{m-1}c^3, \\ t + 3w = d^3, \\ t - 3w = e^3, \end{cases}$$

and $c, d, e$ are odd, pairwise relatively prime, with $a = cde$. Hence $d^3 + e^3 = 2^m c^3$ so $(d, e, c)$ is a solution of the given equation. But $s^{m-1}|c|^3 = |t| < |t| \times |t^2 - 9w^2| = 2^{m-1}|a|^3$ hence $|c| < |a|$. Also, $s^m|a|^3 = |x+y| \leq |x+y||x^2 - xy + y^2| = |x^3 + y^3| = 2^m|z|^3$ so $|c| < |z|$. Since $3 \nmid c$, repeating the argument with the solution $(d, e, c)$ this would yield a sequence of solutions $(d_1, e_1, c_1)$, $(d_2, e_2, c_2)$, $\ldots$ with $|z| > |c_1| > |c_2| > \cdots$, all the $c_i$ being nonzero integers, which is impossible.

*Case 2: $3 \mid z$.*
Now $\gcd(x+y, x^2 - xy + y^2) = 3$. We note that $3^2 \nmid x^2 - xy + y^2 = (x+y)^2 - 3xy$; otherwise $3 \mid xy$ so 3 divides both $x$ and $y$, contrary to the hypothesis.

Thus, there exist odd, relatively prime integers $a, b$ such that

$$\begin{cases} x + y = 2^m \times 3^2 a^3, \\ x^2 - xy + y^2 = 3b^3, \end{cases}$$

with $3 \nmid b$ and $z = 3ab$. Since $x+y$, $x-y$ are even and $3 \mid x+y$ we may write $3b^3 = ((x+y)/2)^2 + 3((x-y)/2)^2$ hence $b^3 = ((x-y)/2)^2 + 3((x+y)/6)^2$. It follows from Lemma 4.6 that there exist integers $t, w$ such that

$$\frac{x-y}{2} + \frac{x+y}{6}\sqrt{-3} = (t + w\sqrt{-3})^3$$

and

$$\begin{cases} (x-y)/2 = t(t^2 - 9w^2), \\ (x+y)/6 = 3w(t^2 - w^2), \\ b = t^2 + 3w^2. \end{cases}$$

If $t = 0$ or $t^2 - 9w^2 = 0$ then $x = y = 0$ or $x = y = z = \pm1$ with $m = 1$. If $w = 0$ or $t^2 - w^2 = 0$ then $x = -y$ so $z = 0$. This was excluded by the hypothesis. It follows that $|t^2 - w^2| \neq 1$ otherwise $t + w = \pm1$, $t - w = \pm1$, which is easily seen to be impossible.

Since $b$ is odd, so are $t+3w$, $t-3w$, hence also $t+w$, $t-w$. Therefore $w$, $t - w$, $t + w$ are nonzero pairwise relatively prime integers. From $2^{m-1}a^3 = (x+y)/18 = w(t-w)(t+w)$ it follows that there exist nonzero integers $c, d, e$ such that

$$\begin{cases} w = 2^{m-1}, \\ t + w = d^3, \\ t - w = e^3, \end{cases}$$

and $c, d, e$ are odd, pairwise relatively prime, with $a = cde$. Hence, $d^3 - e^3 = 2^m c^3$, so $(d, -e, c)$ is a solution of the given equation.

But $2^{m-1}|c|^3 = |w| < |w||t^2 - w^2| = |x+y|/18 = 2^{m-1}|a|^3$, thus $|c| < |a|$. Also

$$2^m 3^2 |a|^3 = |x+y|$$
$$\leq \frac{|x+y||x^2 - xy + y^2|}{3}$$
$$= \frac{|x^3 + y^3|}{3}$$
$$= \frac{2^m |z|^3}{3},$$

hence $|c| < |a| \leq |z|/3$.

Whether 3 divides $c$ or not, we repeat the argument in the first or second case, and this leads to a sequence of solutions $(d_1, e_1, c_1)$, $(d_2, e_2, c_2), \ldots$ with $|z| > |c_1| > |c_2| > \cdots$, all the $c_i$ being nonzero integers, which is impossible. $\square$

The theory of the equation $X^3 + Y^3 = AZ^3$ has been further developed by Legendre (1808, 1830), Pepin (1870, 1875, 1881), Lucas (1878, 1880), Sylvester (1856, 1879) and Hurwitz (1917) who proved the impossibility of the equation in integers, for many values of $A$ — but we shall not enter into this matter. Other interesting papers on ternary cubic diophantine equations are from Hurwitz (1917), Mordell (1956); see also Mordell (1969).

### Bibliography

1657 Fermat, P., Lettre à Digby (15 Août, 1657). *Oeuvres*, Vol. II, pp. 342–346. Publiées par les soins de MM. Paul Tannery et Charles Henry. Gauthier-Villars, Paris, 1894.

1760 Euler, L., *Supplementum quorundam theorematum arithmeticorum quae in non nullis demonstrationibus supponuntur.* Novi Comm. Acad. Sci. Petrop., **8** (1760/1), 1763, 105–128. Also in *Opera Omnia*, Ser. I, Vol. II, pp. 556–575. Teubner, Leipzig, 1915.

1770 Euler, L., *Vollständige Anleitung zur Algebra*, 2 volumes. Royal Acad. Sci., St. Petersburg, 1770. English translation by Rev. J. Hewlitt, Longman, Hurst, Rees, Orme, London, 1822. Also in *Opera Omnia*, Ser. I, Vol. I, pp. 486–490. Teubner, Leipzig, 1911.

1802 Kausler, C.F., *Nova demonstratio theorematis nec summam, nec differentiam duorum cuborum cubum esse posse*, Novi Acta Acad. Petrop., **13** (1795/6), 1802, 245–253.

1808 Legendre, A.M., *Essai sur la Théorie des Nombres* ($2^e$ édition), Courcier, Paris, 1808.

1823 Legendre, A.M., *Recherches sur quelques objets d'analyse indéterminée, et particulièrement sur le théorème de Fermat.* Mém. Acad. Roy. Sci. Institut France, **6** (1823), 1–60. Reprinted as the "Second Supplément" in 1825, to a printing of *Essai sur la Théorie des Nombres* ($2^e$ édition), Courcier, Paris; reprinted in Sphinx-Oedipe, **4** (1909), 97–128.

1830 Legendre, A.M., *Théorie des Nombres* ($3^e$ édition), Vol. II, pp. 357–360, Firmin Didot Frères, Paris, 1830; reprinted by A. Blanchard, Paris, 1955.

1855 Calzolari, L., *Tentativo per dimostrare il teorema di Fermat sull'equazione indeterminata* $x^n + y^n = z^n$, Ferrara, 1855.

1856 Sylvester, J.J., *Recherches sur les solutions en nombres entiers positifs ou négatifs de l'équation cubique homogène à trois variables*, Ann. Sci. Mat. Fis., B. Tortolini, **7** (1856), 398–400; reprinted in *Collected Math. Papers*, Vol. II, pp. 63–64, Cambridge University Press, Cambridge, 1908; reprinted by Chelsea, New York, 1976.

1859 Kronecker, L., *Über cubische Gleichungen mit rationalen Coefficienten*, J. Reine Angew. Math., **56** (1859), 188–189; also in *Werke*, Vol. I, pp. 121–122, Teubner, Leipzig, 1895.

1865 Lamé, G., *Étude des binômes cubiques* $x^3 \mp y^3$, C. R. Acad. Sci. Paris, **61** (1865), 921–924 and 961–965.

1870 Pepin, T., *Sur la décomposition d'un nombre entier en une somme de deux cubes rationnels*, J. Math. Pures Appl., (2), **15** (1870), 217–236.

1870 Tait, P.G., *Mathematical Notes*, Proc. Roy. Soc. Edinburgh, **7** (1872), 144.

1875 Pepin, T., *Sur certains nombres complexes compris dans la formule* $a + b\sqrt{-c}$, J. Math. Pures Appl. (3), **1** (1875), 317–372.

1878 Günther, S., *Über die unbestimmte Gleichung* $x^3 + y^3 = z^3$, Sitzungsber. Böhm Ges. Wiss., 1878, pp. 112–120.

1878 Lucas, E., *Sur l'équation indéterminée* $X^3 + Y^3 = AZ^3$, Nouv. Ann. Math., (2), **17** (1879), 425–426.

1879 Sylvester, J.J., *On certain ternary cubic form equations*, Amer. J. Math., **2** (1879), 280–285 and 357–393; also in *Collected Math. Papers*, Vol. III, 1909, Art. 39, pp. 312–391, Cambridge University Press, Cambridge, 1909; reprinted by Chelsea, New York, 1976.

1880 Lucas, E., *Théorèmes généraux sur l'impossibilité des équations cubiques indéterminées*, Bull. Soc. Math. France, **8** (1880), 173–182.

1880 Sylvester, J.J., *Sur les diviseurs des fonctions cyclotomiques*, C. R. Acad. Sci. Paris, **90** (1880), 287–289 and 345–347; also in *Collected Math. Papers*, Vol. III, 1909, Art. 44, pp. 428–432 and also p. 437; reprinted by Chelsea, New York,

1976.

1881 Pepin, T., *Mémoire sur l'équation indéterminée* $x^3 + y^3 = Az^3$, Atti Accad. Naz. Lincei, **34** (1881), 78–130.

1884 Perrin, R., *Sur l'équation indéterminée* $x^3 + y^3 = z^3$, Bull. Soc. Math. France, **13** (1885), 194–197.

1894 Schumacher, J., *Nachtrag zu Nr. 1077, XXIII, 269*, Z. Math. Naturwiss. Unterricht, **25** (1894), 350–351.

1898 Palmström, A., *Équation* $x^3 + y^3 = z^3$, L'Interm. Math., **5** (1898), 95.

1901 Gambioli, D., *Memoria bibliografica sull'ultimo teorema di Fermat*, Period. Mat., (2), **16** (1901), 145–192.

1901 Landau, E., *Sur une démonstration d'Euler d'un théorème de Fermat*, L'Interm. Mat., **8** (1901), 145–147.

1906 Holden, H., *On the complete solution in integers for certain values of p, of* $a(a^2 + pb^2) = c(c^2 + pd^2)$, Messenger Math., **36** (1906), 189–192.

1907 Sommer, J., *Vorlesungen über Zahlentheorie*, Teubner, Leipzig, 1907.

1909 Krey, H., *Neuer Beweis eines arithmetischen Satzes*, Math. Naturwiss. Blätter, **6** (1909), 179–180.

1910 Rychlik, K., *On Fermat's last theorem for n = 4 and n = 3* (in Bohemian), Časopis Pěst. Mat., **39** (1910), 65–86.

1910 Stockhaus, H., *Beitrag zum Beweis des Fermatschen Satzes*, Brandstetter, Leipzig, 1910, 90 pp.

1910 Welsch, *Réponse à une question de E. Dubouis*, L'Interm. Math., **17** (1910), 179–180.

1915 Carmichael, R.D., *Diophantine Analysis*, Wiley, New York, 1915.

1915 van der Corput, J.G., *Quelques formes quadratiques et quelques équations indéterminées*, Nieuw Archief Wisk., **11** (1915), 45–47.

1917 Hurwitz, A., *Über ternäre diophantische Gleichungen dritten Grades*, Vierteljahrschrift Natur. Gesells. Zürich, **62** (1917), 207–209; reprinted in *Math. Werke*, Vol. II, pp. 446–468, Birkhäuser, Basel, 1933.

1917 Thue, A., *Et bevis for at ligningen* $A^3 + B^3 = C^3$ *er umulig i hele tal fra nul forskjellige tal A, B og C*, Arch. Mat. Naturv., **34** (1917), no. 15, 5 pp.; reprinted in *Selected Mathematical Papers*, pp. 555–559, Universitetsforlaget, Oslo, 1977.

1929 Nagell, T., *L'Analyse Indéterminée de Degré Supérieur*, Mémoires des Sciences Math., Vol. 39, Gauthier-Villars, Paris, 1929.

1931 Hancock, H., *Foundations of the Theory of Algebraic Numbers*, Vol. I, Macmillan, New York, 1931.

1936 Nagell, T., *Bemerkungen über die Diophantische Gleichung* $x^3 + y^3 = Az^3$, Arkiv Mat. Astr. Fysik, **25B** (1936), no. 5, 6 pp.

1944 Duarte, F.J., *Sobre la ecuación* $x^3 + y^3 + z^3 = 0$, Bol. Acad. Ciencias Fis. Mat. Naturales, Caracas, **8** (1944), 971–979.

1944 Schmid, F., *Über die Gleichung* $x^3 + y^3 + z^3 = 0$, Sitzungsber. Akad. Wiss., Wien, IIa, **152** (1944), 7–14.

1955 Pizá, P.A., *On the case* $n = 3$ *of Fermat's last theorem*, Math. Mag., **28** (1955), 157–158.

1956 Mordell, L.J., *The diophantine equation* $x^3 + y^3 + z^3 + kxyz = 0$, Colloque sur la Théorie des Nombres, Bruxelles, 1955, Centre Belge Rech. Math., 1956, pp. 67–76.

1956 Vrănceanu, G., *Asupra unei teoreme echivalenti cu teorema lui Fermat* (On a theorem equivalent to Fermat's theorem), Gaz. Mat. Fiz. Bucuresti Ser. A, **8** (61) (1956), 23–24; reprinted in *Opera Matematică*, Vol. III, pp. 97–98, Edit. Acad. Rep. Soc. Romania, Bucureşti, 1973.

1956 Yf, P., *Comment on Pedro Pizá's "On the case* $n = 3$ *of Fermat's last theorem,"* Math. Mag., **29** (1956), 205–206.

1960 Vrănceanu, G., *Observatii asupra unei note precedente* (Remarks on a preceding note), Gaz. Mat. Fiz. (Bucureşti), Ser. A, **12** (1960), 1–2; reprinted in *Opera Matmatică*, Vol. III, pp. 463–464, Edit. Acad. Rep. Soc. Romania, Bucureşti, 1973.

1964 Sierpiński, W., *Elementary Theory of Numbers.*, Monografie Matematyczne, no. 42, 1942, Polska Akad. Nauk, Warszawa, 1964.

1966 Bergmann, G., *Über Eulers Beweis des großen Fermatschen Satzes für den Exponenten 3*, Math. Ann., **164** (1966), 159–175.

1966 Grosswald, E., *Topics from the Theory of Numbers*, Macmillan, New York, 1966.

1969 Mordell, L.J., *Diophantine Equations.* Academic Press, New

York, 1969.

1972 Legendre, R., *Sur la résolution par Euler de l'équation de Fermat pour l'exposant* 3, C. R. Acad. Sci. Paris, **275** (1972), 413–414.

1977 Edwards, H.M., *"Fermat's Last Theorem"*. *A Genetic Introduction to Algebraic Number Theory*, Springer-Verlag, New York, 1977.

1979 Vrănceanu, G., *Une interprétation géométrique du théorème de Fermat*, Rev. Roumaine Math. Pures Appl., **24** (1979), 1137–1140.

## I.5. The Eisenstein Field

We shall now give the proof of Gauss that Fermat's cubic equation has only trivial solutions in the Eisenstein field. The set of complex numbers $a + b\sqrt{-3}$, where $a, b$ are rational numbers, constitute a field, called the *Eisenstein field* and denoted by $K = \mathbb{Q}(\sqrt{-3})$. The numbers $\alpha = (a + b\sqrt{-3})/2$, where $a, b$ are ordinary integers of the same parity, are called the *integers* of $K$. They constitute a ring, denoted by $A$. If $\alpha, \beta \in K$, $\beta$ *divides* $\alpha$ if there exists an integer $\gamma \in A$ such that $\alpha = \beta\gamma$. We write $\beta \mid \alpha$ when $\beta$ divides $\alpha$. Two nonzero integers $\alpha, \beta$ are *associated* if $\alpha$ divides $\beta$ and $\beta$ divides $\alpha$; we write $\alpha \sim \beta$. The integers associated with 1 are called the *units* of $K$. It is easily shown that they are $\pm 1$, $\pm\zeta$, $\pm\zeta^2$, where $\zeta = (-1 + \sqrt{-3})/2$, $\zeta^2 = (-1 - \sqrt{-3})/2$. We note that $\zeta^3 = 1$, that is, $\zeta$ is a primitive cubic root of 1, and $1 + \zeta + \zeta^2 = 0$. A nonzero integer $\alpha \in A$ is a *prime* if it is not a unit and the only integers dividing $\alpha$ are units or associated with $\alpha$.

In the particular field $\mathbb{Q}(\sqrt{-3})$ under consideration, it is true that every nonzero integer $\alpha$ is the product of prime integers: $\alpha = \gamma_1\gamma_2 \cdots \gamma_s$. This decomposition is unique, in the following sense: if we also have $\alpha = \delta_1\delta_2 \cdots \delta_t$, where each $\delta_i$ is a prime of $K$, then $s = t$ and, changing the order if necessary, $\gamma_i$ and $\delta_i$ are associated (for every $i = 1, \ldots, s$). Therefore, we may define, in the obvious way, the greatest common divisor of nonzero integers of $A$ which is unique up to units of $A$.

The proofs of the following properties may be found in any standard text on algebraic numbers, for example, in Ribenboim (1999).

The *conjugate* of $\alpha = (a + b\sqrt{-3})/2$ is $\overline{\alpha} = (a - b\sqrt{-3})/2$. The *norm* of $\alpha$ is $N(\alpha) = \alpha\overline{\alpha} = (a^2 - 3b^2)/4$.

If $\alpha \in A$ then $A\alpha = \{\beta\alpha \mid \beta \in A\}$ is the ideal of multiples of $\alpha$. If $\alpha, \beta, \gamma \in A$, $\alpha \neq 0$, we write

$$\beta \equiv \gamma \pmod{\alpha},$$

when $\alpha$ divides $\beta - \gamma$; we say that $\beta$ and $\gamma$ are *congruent modulo* $\alpha$. This is an equivalence relation on the ring $A$ and the set of equivalence classes is denoted by $A/A\alpha$; the equivalence class of $\beta$ is denoted by $\overline{\beta}$ and called the *residue class* of $\beta$. We define the addition and multiplication of residue classes as follows: $\overline{\beta} + \overline{\gamma} = \overline{\beta + \gamma}$, $\overline{\beta} \cdot \overline{\gamma} = \overline{\beta\gamma}$. Then $A/A\alpha$ is a ring, called the *residue ring* of $A$ *modulo* $\alpha$. The residue ring $A/A\alpha$ is finite; its number of elements is equal to $|N(\alpha)|$.

Now we describe the decomposition of prime numbers $p$ as products of prime elements of the ring $A$.

(1) $p = 3$ is ramified, that is, $3 = (-\zeta^2)\lambda^2$, so $3 \sim \lambda^2$, where $\lambda = 1 - \zeta = (3 - \sqrt{-3})/2$, $\lambda$ is a prime element of $A$. There are three residue classes of $A$ modulo $\lambda$; the set $\{0, 1, -1\}$ is a system of representatives of the field $A/A\lambda$. The norm of $\lambda$ is $N(\lambda) = \lambda\overline{\lambda} = (1 - \zeta)(1 - \zeta^2) = 1 - \zeta - \zeta^2 + 1 = 3$, since $1 + \zeta + \zeta^2 = 0$.

(2) $p = 2$ is inert, that is, 2 is a prime of $A$. There are four residue classes of $A$ modulo 2, that is, $A/A2$ is the field with four elements; the norm of 2 is $N(2) = 4$.

(3) If $p \equiv 1 \pmod 3$ then $p \sim \lambda_1\lambda_2$, where $\lambda_1, \lambda_2$ are prime elements of $A$ which are not associated ($\lambda_1 \not\sim \lambda_2$); we say that $p$ *splits* (or is *decomposed*). Now $A/Ap$ has $p^2$ elements and it is the direct product of two copies of the field $\mathbb{F}_p$ with $p$ elements, and

$$N(\lambda_1) = N(\lambda_2) = p.$$

(4) If $p \equiv -1 \pmod 3$ then $p$ is a prime element, that is, $p$ is inert; $A/Ap$ is a field with $p^2$ elements, $N(p) = p^2$.

We shall not need (2), (3), (4) above in Gauss' proof.

We shall need the following precise congruence:

LEMMA 5.1. *If* $\alpha \in A$ *and* $\lambda$ *does not divide* $\alpha$ *then* $\alpha^3 \equiv \pm 1$ (mod $\lambda^4$).

PROOF. Since $\alpha \not\equiv 0$ (mod $\lambda$) then $\alpha \equiv \pm 1$ (mod $\lambda$). First, we assume $\alpha \equiv 1$ (mod $\lambda$), so $\alpha - 1 = \beta\lambda$ where $\beta \in A$. Then $\alpha - \zeta =$

$(\alpha - 1) + (1 - \zeta) = \beta\lambda + \lambda = \lambda(\beta + 1)$, $\alpha - \zeta^2 = (\alpha - \zeta) + (\zeta - \zeta^2) = \lambda(\beta + 1) + \zeta\lambda = \lambda(\beta - \zeta^2)$. Hence $\alpha^3 - 1 = (\alpha - 1)(\alpha - \zeta)(\alpha - \zeta^2) = \lambda^3\beta(\beta + 1)(\beta - \zeta^2)$. But $1 - \zeta^2 = (1 + \zeta)\lambda$, or $\zeta^2 \equiv 1 \pmod{\lambda}$. Hence $\beta$, $\beta + 1$, $\beta - \zeta^2$ are in three different classes modulo $\lambda$, and at least one is a multiple of $\lambda$. Therefore $\alpha^3 \equiv 1 \pmod{\lambda^4}$. If $\alpha \equiv -1 \pmod{\lambda}$ then $-\alpha^3 = (-\alpha)^3 \equiv 1 \pmod{\lambda^4}$, so $\alpha^3 \equiv -1 \pmod{\lambda^4}$. $\square$

The following result of Gauss implies (4A):

**(5A)**    *The equation*

(5.1) $$X^3 + Y^3 + Z^3 = 0$$

*has no solution in algebraic integers of $\mathbb{Q}(\sqrt{-3})$, all different from* 0.

PROOF. Assume that $\xi, \eta, \theta \in A$ are nonzero and satisfy $\xi^3 + \eta^3 + \theta^3 = 0$. If $\gcd(\xi, \eta, \theta) = \delta$ then $\xi/\delta, \eta/\delta, \theta/\delta$ satisfy the same equation and $\gcd(\xi/\delta, \eta/\delta, \theta/\delta) = 1$. So we may assume $\gcd(\xi, \eta, \theta) = 1$ and therefore, $\xi, \eta, \theta$ are pairwise relatively prime. So $\lambda$ cannot divide two of these elements $\xi, \eta, \theta$. We may assume, for example, that $\lambda \nmid \xi$, $\lambda \nmid \eta$.

*First Case*: We assume that $\lambda \nmid \theta$.
   Then
$$\begin{cases} \xi^3 \equiv \pm 1 \pmod{\lambda^3}, \\ \eta^3 \equiv \pm 1 \pmod{\lambda^3}, \\ \theta^3 \equiv \pm 1 \pmod{\lambda^3}, \end{cases}$$

so $0 = \xi^3 + \eta^3 + \theta^3 \equiv \pm 1 \pm 1 \pm 1 \pmod{\lambda^3}$. The eight combinations of signs give $\pm 1$ or $\pm 3$. These are congruent to 0 modulo $\lambda^3$, since $\pm 1$ are units, $\pm 3$ are associated with $\lambda^2$, hence not multiples of $\lambda^3$.

*Second Case*: We assume that $\lambda \mid \theta$.
   Let $\theta = \lambda^m\psi$, $\psi \in A$, $m \geq 1$, and $\lambda$ does not divide $\psi$. The essential part of the proof consists in establishing the following assertion:
   Let $n \geq 1$, and let $\varepsilon$ be a unit of $A$. If there exist $\alpha, \beta, \gamma \in A$, pairwise relatively prime, not multiples of $\lambda$, and $\alpha^3 + \beta^3 + \varepsilon\lambda^{3n}\gamma^3 = 0$, then:

   (a) $n \geq 2$; and
   (b) there exist a unit $\varepsilon_1$ and $\alpha_1, \beta_1, \gamma_1 \in A$, pairwise relatively prime, not multiples of $\lambda$, such that $\alpha_1^3 + \beta_1^3 + \varepsilon_1\lambda^{3(n-1)}\gamma_1^3 = 0$.

The hypothesis is satisfied with $n = m$, $\varepsilon = 1$, $\alpha = \xi$, $\beta = \eta$, $\gamma = \psi$. By repeated application of the above assertion, we would find a unit $\varepsilon'$, and $\alpha', \beta', \gamma' \in A$ not multiples of $\lambda$, such that $\alpha'^3 + \beta'^3 + \varepsilon'\lambda^3\gamma'^3 = 0$, and this contradicts (a) above.

First we show that $n \geq 2$. Indeed, $\lambda \nmid \alpha$ and $\lambda \nmid \beta$. So by Lemma 5.1, $\alpha^3 \equiv \pm 1 \pmod{\lambda^4}$, $\beta^3 \equiv \pm 1 \pmod{\lambda^4}$ and $\pm 1 \pm 1 = -\varepsilon\lambda^{3n}\gamma^3 \pmod{\lambda^4}$, $\lambda \nmid \gamma$. Since $\gamma \nmid \pm 2$ the left-hand side must be 0. From $\lambda \nmid \gamma$ we conclude that $3n \geq 4$, so $n \geq 2$.

Now we prove (b). We have

$$(5.2) \qquad -\varepsilon\lambda^{3n}\gamma^3 = \alpha^3 + \beta^3 = (\alpha + \beta)(\alpha + \zeta\beta)(\alpha + \zeta^2\beta).$$

Since $\lambda$ is a prime element dividing the right-hand side, then it must divide one of the factors. But $\alpha + \beta \equiv \alpha + \zeta\beta \equiv \alpha + \zeta^2\beta \pmod{\lambda}$ because $\lambda = 1 - \zeta$, $1 - \zeta^2 = -\zeta^2\lambda$ so $\lambda$ must divide all three factors; hence $(\alpha + \beta)/\lambda$, $(\alpha + \zeta\beta)/\lambda$, $(\alpha + \zeta^2\beta)/\lambda \in A$ and

$$-\varepsilon\lambda^{3(n-1)}\gamma^3 = \left(\frac{\alpha + \beta}{\lambda}\right)\left(\frac{\alpha + \zeta\beta}{\lambda}\right)\left(\frac{\alpha + \zeta^2\beta}{\lambda}\right).$$

Since $n \geq 2$, $\lambda$ divides the right-hand side, hence at least one factor. It cannot divide two of the factors, otherwise two among $\alpha+\beta$, $\alpha+\zeta\beta$, $\alpha+\zeta^2\beta$ are congruent modulo $\lambda^2$. We check that this is not possible: $(\alpha+\beta)-(\alpha+\zeta\beta) = \beta(1-\zeta) = \beta\lambda \equiv 0 \pmod{\lambda^2}$ implies $\lambda \mid \beta$, a contradiction; $(\alpha + \beta) - (\alpha + \zeta^2\beta) = \beta(1 - \zeta^2) = -\beta\zeta^2\lambda \equiv 0 \pmod{\lambda^2}$ implies $\lambda \mid \beta$ again; $(\alpha + \zeta\beta) - (\alpha + \zeta^2\beta) = \zeta\beta(1 - \zeta) = \zeta\beta\lambda \equiv 0 \pmod{\lambda^2}$ implies $\lambda \mid \beta$ again.

Let us assume that $\lambda$ divides $(\alpha+\beta)/\lambda$ (the other cases are treated by replacing $\beta$ by $\zeta\beta$ or $\zeta^2\beta$). Then $\lambda^{3(n-1)}$ divides $(\alpha+\beta)/\lambda$. Therefore

$$(5.3) \qquad \begin{cases} \alpha + \beta = \lambda^{3n-2}\kappa_1, \\ \alpha + \zeta\beta = \lambda\kappa_2, \\ \alpha + \zeta^2\beta = \lambda\kappa_3, \end{cases}$$

with $\kappa_1, \kappa_2, \kappa_3 \in A$, $\lambda$ not dividing $\kappa_1$, $\kappa_2$, $\kappa_3$. Multiplying, we have

$$(5.4) \qquad -\varepsilon\gamma^3 = \kappa_1\kappa_2\kappa_3.$$

We note that $\kappa_1, \kappa_2, \kappa_3$ are pairwise relatively prime. For example, if $\delta \in A$ divides $\kappa_1, \kappa_2$, then $\delta$ divides $(\alpha + \beta) - (\alpha + \zeta\beta) = \beta(1 - \zeta) = \beta\lambda$, and similarly when $\delta$ divides $\kappa_1, \kappa_3$ (or $\kappa_2, \kappa_3$). But $\lambda$ does not divide $\kappa_1, \kappa_2, \kappa_3$, so $\delta$ is not associated with $\lambda$; hence $\delta$ divides $\beta$ and therefore also $\alpha$, which is a contradiction.

By the unique factorization in the ring $A$, it follows from (5.4) that the elements $\kappa_1$, $\kappa_2$, $\kappa_3$ are associated with cubes, i.e., there exist units $\omega_i \in A$ and elements $\mu_i \in A$ such that $\kappa_i = \omega_i \mu_i^3$ $(i = 1, 2, 3)$. So

(5.5)
$$\begin{cases} \alpha + \beta = \lambda^{3n-2}\mu_1^3\omega_1, \\ \alpha + \zeta\beta = \lambda\mu_2^3\omega_2, \\ \alpha + \zeta^2\beta = \lambda\mu_3^3\omega_3. \end{cases}$$

We note again that $\mu_1$, $\mu_2$, $\mu_3$ are pairwise relatively prime and $\lambda$ does not divide $\mu_1$, $\mu_2$, $\mu_3$. Thus

$$0 = (\alpha + \beta) + \zeta(\alpha + \zeta\beta) + \zeta^2(\alpha + \zeta^2\beta)$$
$$= \lambda^{3n-2}\mu_1^3\omega_1 + \zeta\lambda\mu_2^3\omega_2 + \zeta^2\lambda\mu_3^3\omega_3,$$

so

$$\mu_2^3 + \tau\mu_3^3 + \tau'\lambda^{3(n-1)}\mu_1^3 = 0,$$

where $\tau$, $\tau'$ are units, $\mu_1$, $\mu_2$, $\mu_3 \in A$ are not zero, and $\gcd(\mu_2, \mu_3) = 1$. If $\tau = 1$, we have established (b). If $\tau = -1$, we replace $\mu_3$ by $-\mu_3$ and have again shown (b). To complete the proof we show that the unit $\tau$ cannot be equal to $\pm\zeta$ or $\pm\zeta^2$. In fact, $\mu_2^3 + \tau\mu_3^3 \equiv 0$ (mod $\lambda^2$). Since $\mu_2^3 \equiv \pm 1$ (mod $\lambda^4$), $\mu_3^3 \equiv \pm 1$ (mod $\lambda^4$) then $\mu_2^3 + \tau\mu_3^3 \equiv \pm 1 \pm \tau \equiv 0$ (mod $\lambda^2$). However, $\pm 1 \pm \zeta \not\equiv 0$ (mod $\lambda^2$) and $\pm 1 \pm \zeta^2 \not\equiv 0$ (mod $\lambda^2$), so $\tau \neq \pm\zeta, \pm\zeta^2$, and the proof of (b) is now complete.

As already explained, this suffices to prove the theorem. $\square$

We take this opportunity to indicate some results similar to (5A) that may be proved with the same methods. They may be attributed to Euler and Legendre.

**(5B)**     *Let $p$ be a prime, $p \equiv 2$ or $5$ (mod 9). If $\varepsilon$ is a unit of $K$, if there exists $x \in A$ such that $x^3 \equiv \varepsilon$ (mod $p$), then $\varepsilon = \pm 1$.*

PROOF. Assume that $\varepsilon = \pm\zeta$ or $\pm\zeta^2$ and that there exists $x \in A$ such that $x^3 \equiv \varepsilon$ (mod $p$). Since $p \equiv 2$ (mod 3) then $p$ is a prime element of $A$, $A/Ap$ is a field with $p^2$ elements, and $x^{p^2-1} \equiv 1$ (mod $p$). But $p^2 - 1 = (p+1)(p-1) = 3(r+1)(3r+1)$, where $p = 3r + 2$. Hence

$$\varepsilon^{r+1} \equiv \varepsilon^{(r+1)(3r+1)} \equiv x^{p^2-1} \equiv 1 \quad (\text{mod } p).$$

If $r \equiv 0$ or $1$ (mod $p$) then $p$ divides $1 \pm \zeta$ or $1 \pm \zeta^2$. Noting that $1 - \zeta^2 = (1+\zeta)(1-\zeta)$ and that $1 + \zeta = 1 - \zeta^2$, $1 + \zeta^2 = -\zeta$ are units, then $p$ divides $1 - \zeta$. This would imply that $p = 3$, which is contrary to the hypothesis. Hence $r \equiv 2 \pmod 3$ and $p \equiv 8 \pmod 9$, which is again a contradiction.   $\square$

Following Mordell, we prove the classical result:

**(5C)**   *Let $p$ be a prime, $p \equiv 2$ or $5 \pmod 9$, and let $\varepsilon$ be a unit of $\mathbb{Q}(\zeta)$. The equation*

$$(5.6) \qquad\qquad X^3 + Y^3 + \varepsilon a Z^3 = 0,$$

*with $a = p$ or $p^2$, $a \neq 2$, has only the trivial solution $(x, y, z)$ in $\mathbb{Q}(\zeta)$, namely $z = 0$, $x = -y$, or $-\zeta y$, or $-\zeta^2 y$. If $a = 2$ then there are also the solutions $x^3 = y^3 = z^3 = \pm 1$ when $\varepsilon = -1$.*

PROOF. To begin we note that if $(x, y, z)$ is a solution with $x = 0$ or $y = 0$ then necessarily $z = 0$. Indeed, if both $x, y$ are 0 then so is $z$. If, for example, $y \neq 0$ and $\varepsilon \neq 0$ we may assume $\gcd(y, z) = 1$. Then $p$ divides $y$, so $p^3 \mid az^3$, hence $p \mid z$, a contradiction.

Assume now that $x, y, z \in A$ are such that $x^3 + y^3 + \varepsilon az^3 = 0$, with $x, y, z \neq 0$. We may also assume that $\gcd(x, y, z) = 1$ from which it follows that $x, y, z$ are pairwise relatively prime.

Among all possible solutions, consider one for which the absolute value of the norm $|N(xyz)|$ is minimum. Note that $x^3 + y^3 \neq 0$ since $z \neq 0$; so $x \neq -y, -\zeta y, -\zeta^2 y$.

Consider the Lagrange resolvents

$$\begin{cases} \alpha = x + y, \\ \beta = \zeta x + \zeta^2 y, \\ \gamma = \zeta^2 x + \zeta y. \end{cases}$$

Then $\alpha, \beta, \gamma \in A$, $\alpha, \beta, \gamma = 0$ and

$$\begin{aligned} \alpha\beta\gamma &= \zeta^2 (x^2 + \zeta y^2 + (1 + \zeta)xy)(\zeta x + y) \\ &= \zeta^2 (\zeta x^3 + \zeta y^3 + (1 + \zeta + \zeta^2)x^2 y + (1 + \zeta + \zeta^2)xy^2) \\ &= x^3 + y^3 = -\varepsilon az^3. \end{aligned}$$

Let $\delta = \gcd(\alpha, \beta, \gamma) \in A$, so $\gcd(\alpha/\delta, \beta/\delta, \gamma/\delta) = 1$ and

$$\frac{\alpha}{\delta} \cdot \frac{\beta}{\delta} \cdot \frac{\gamma}{\delta} = -\varepsilon a \left( \frac{z}{\delta} \right)^3.$$

Hence $a$ divides one and only one of the factors in the left-hand side, say $\gamma/\delta$ (the other cases are similar). By the unique factorization theorem, which is valid in the ring $A$, we have

$$\begin{cases} \alpha/\delta = \varepsilon_1 \alpha_1^3, \\ \beta/\delta = \varepsilon_2 \beta_1^3, \\ \gamma/\delta = \varepsilon_3 a \gamma_1^3, \end{cases}$$

where $\varepsilon_1, \varepsilon_2, \varepsilon_3$ are units of $A$, and $\alpha_1, \beta_1, \gamma_1 \in A$, $\alpha_1, \beta_1, \gamma_1 \neq 0$, $\gcd(\alpha_1, \beta_1, \gamma_1) = 1$. Hence

$$\varepsilon_1 \alpha_1^3 + \varepsilon_2 \beta_1^3 + \varepsilon_3 a \gamma_1^3 = 0.$$

Let $\varepsilon' = \varepsilon_2/\varepsilon_1, \varepsilon'' = \varepsilon_3/\varepsilon_1$; hence

(5.7)
$$\alpha_1^3 + \varepsilon' \beta_1^3 + \varepsilon'' a \gamma_1^3 = 0.$$

We have $p \nmid \beta_1$ otherwise $p \mid \alpha_1$, so $p^3 \mid a\gamma_1{}^3$, hence $p \mid \gamma_1$ and this is impossible because $\gcd(\alpha_1, \beta_1, \gamma_1) = 1$. Taking the classes modulo $p$, (5.7) yields

$$\alpha_1^3 + \varepsilon' \beta_1^3 \equiv 0 \pmod{p},$$

hence $\varepsilon'$ is a cube modulo $p$. By (5B), $\varepsilon' = \pm 1$, so

$$\alpha_1^3 + (\pm \beta_1)^3 + \varepsilon'' a \gamma_1^3 = 0$$

and $(\alpha_1, \pm\beta_1, \gamma_1)$ is another nontrivial solution of (5.6), with $\gcd(\alpha_1, \pm\beta_1, \gamma_1) = 1$. By assumption,

$$|N(xyz)|^3 \leq |N(\alpha_1\beta_1\gamma_1)|^3 = \left| N\left( \frac{\alpha\beta\gamma}{\delta^3 a} \right) \right| = \left| N\left( \frac{z}{\delta} \right) \right|^3,$$

hence $|N(\delta xy)| \leq 1$. This implies that $x, y, \delta$ are units. Hence $x^3 = \pm 1$, $y^3 = \pm 1$, and $\pm 1 \pm 1 + \varepsilon a z^3 = 0$.

If $a \neq 2$ then $z = 0$, a contradiction.

If $a = 2$, we must have either $z = 0$, or $x^3 = y^3 \neq 0$, $x^3 + \varepsilon z^3 = 0$; then $\varepsilon$ is a cube modulo $p$, therefore by (5B), $\varepsilon = \pm 1$. But clearly $\varepsilon = +1$ would imply $z = 0$, so $\varepsilon = -1$ and $x^3 = y^3 = z^3 = \pm 1$. □

As an immediate corollary we have:

**(5D)**   *The equation*

(5.8)
$$X^3 + 4Y^3 = 1$$

*has no solution in nonzero integers.*

PROOF. Consider the equation

(5.9)                         $X^3 + T^3 + 4Y^3 = 0.$

If $(x, y)$ is a nontrivial solution of (5.8) then $(x, -1, y)$ is a nontrivial solution of (5.9), which is impossible, by (5C).  □

We also note other consequences:

**(5E)**    *The equations*

(5.10)                    $X^6 - 27Y^6 = 2Z^3,$
(5.11)                $X^6 - 16 \times 27Y^6 = Z^3,$
(5.12)                    $16X^6 - 27Y^6 = Z^3,$

*have no solutions in nonzero integers.*

PROOF. Assume that $x^6 - 27y^6 = 2z^3$, with $x, y, z$ nonzero integers. Then $(x^2)^3 + (-3y^2)^3 - 2z^3 = 0$. By (5C), $x^6 = y^6 = z^6 = \pm 1$ and therefore $x = y = z = 1$, which is impossible.

If $x^6 - 16 \times 27y^6 = z^3$, then multiplying with $2^3$ we have $(2x^2)^3 + (-2z)^3 - 2(2^2 \times 3y^2)^3 = 0$. By (5C), $(2x^2)^3 = (-2z)^3 = (2^2 \times 3y^2)^3 = \pm 1$ and therefore $2x^2 = \pm 1$, a contradiction.

Finally, if $16x^6 - 27y^6 = z^3$ then $(3y^2)^3 + z^3 - 2(2x^2)^3 = 0$, so by (5C), $(3y^2)^3 = z^3 = (2x^2)^3 = \pm 1$, thus $2x^2 = \pm 1$, a contradiction.  □

Furthermore, Legendre showed:

**(5F)**    *The equation*
$$X^3 + Y^3 = 3Z^3$$
*has no solution in integers different from zero.*

PROOF. The proof may be conducted following the same lines, with appropriate changes.  □

In 1856, Sylvester also announced:

**(5G)**    *The equation*
$$X^3 + Y^3 + Z^3 + 6XYZ = 0$$
*has no solution in integers different from zero.*

## Bibliography

1856 Sylvester, J.J., *Recherches sur les solutions en nombres en-tiers positifs ou négatifs de l'équation cubique homogène à trois variables*, Ann. Sci. Matem. Fis. (Tortolini), **7** (1856), 398–400. Reprinted in *Mathematical Papers*, Vol. II, pp. 14–15; Cambridge University Press, Cambridge, 1908.

1876 Gauss, C.F., *Zur Theorie der complexen Zahlen.* (I) Neue Theorie der Zerlegung der Cuben; *Werke*, Vol. II, pp. 387–391, Königl. Ges. Wiss., Göttingen, 1876.

1998 Ribenboim, P., *Classical Theory of Algebraic Numbers*, Springer-Verlag, New York, 1999.

## I.6. The Quintic Equation

The case $n = 5$ was first settled by Dirichlet. His paper was read at the Academy of Sciences of Paris in 1825, but his proof, published in 1828, did not consider all the possible cases. Legendre then published a complete and independent proof, while Dirichlet was able to settle the last remaining case. We reproduce Dirichlet's proof in modern language, using a few facts about the arithmetic of the quadratic field $K = \mathbb{Q}(\sqrt{5})$. The proofs may be found, for example, in Ribenboim (1999).

Let $A$ be the ring of integers of $\mathbb{Q}(\sqrt{5})$. The elements of $A$ are of the form $(a + b\sqrt{5})/2$, where $a, b$ are integers of the same parity. The invertible elements of $A$, i.e., the units of $K$, form a multiplicative group. $(a + b\sqrt{5})/2$ is a unit if and only if its norm

$$\left(\frac{a + b\sqrt{5}}{2}\right)\left(\frac{a - b\sqrt{5}}{2}\right) = \frac{a^2 - 5b^2}{4}$$

is equal to $\pm 1$ (i.e., $a^2 - 5b^2 = \pm 4$). It may be shown that the units of $A$ are precisely the elements $\pm((1 + \sqrt{5})/2)^e$, where $e$ is any integer.

An important fact required in the proof is that every element of $A$ may be written (up to a unit) in a unique way as a product of powers of prime elements. Or, equivalently, every ideal of $A$ is principal. Among the prime elements of $A$ there are the numbers $2, \sqrt{5}$.

We begin by establishing a property concerning certain principal ideals of $A$ which are fifth powers.

**(6A)**

(1) *Let $a, b$ be nonzero integers such that $\gcd(a, b) = 1$, $a \not\equiv b$ (mod 2), $5 \nmid a$, $5 \mid b$. If $a^2 - 5b^2$ is the fifth power of an element of $A$ then there exist nonzero integers $c, d$ such that*

(6.1)
$$\begin{cases} a = c(c^4 + 50c^2 d^2 + 125 d^4), \\ b = 5d(c^4 + 10 c^2 d^2 + 5 d^4), \end{cases}$$

*and $\gcd(c, d) = 1$, $c \not\equiv d$ (mod 2), $5 \nmid c$.*

(2) *Let $a, b$ be integers such that $\gcd(a, b) = 1$, $a, b$ both odd, $5 \nmid a$, $5 \mid b$. If $(a^2 - 5b^2)/4$ is the fifth power of an element of $A$, then there exist nonzero integers $c, d$ such that*

(6.2)
$$\begin{cases} a = c(c^4 + 50 c^2 d^2 + 125 d^4)/16, \\ b = 5d(c^4 + 10 c^2 d^2 + 5 d^4)/16, \end{cases}$$

*and $\gcd(c, d) = 1$, $c, d$ are both odd, $5 \nmid c$.*

PROOF. If the nonzero integers $c, d$ satisfy (6.1) (respectively, (6.2)) then $\gcd(c, d) = 1$, $5 \nmid c$, $c, d$ cannot be both odd, otherwise $a, b$ would be both even (respectively, $c, d$ cannot have different parity, otherwise 16 would divide $c$ and $d$, so $c, d$ must be odd).

Now we prove the existence of $c, d$, in both cases.

(1)  We first observe that if $a + b\sqrt{5} = ((h + k\sqrt{5})/2)^5$ with $h \equiv k$ (mod 2), then $h, k$ are even. Indeed, $2^5 b = 5k(h^4 + 10h^2 k^2 + 5k^4)$, so $2^5$ divides $h^4 + 10h^2 k^2 + 5k^4$. If $h, k$ are odd then $k \equiv \pm 1, \pm 3$ (mod 8), so $h^2 \equiv 1, 9$ (mod 16), $h^4 \equiv 1, 17$ (mod 32) and similarly for $k$, $k^2$, $k^4$. Hence $h^4 + 10h^2 k^2 + 5k^4$ is congruent modulo 32 to either

$$1 + 10k^2 + 5k^4 \equiv \begin{cases} 1 + 10 + 5 \equiv 16, \\ 1 + 90 + 85 \equiv 16, \end{cases}$$

or

$$17 + 26k^2 + 5k^4 \equiv \begin{cases} 17 + 26 + 5 \equiv 16, \\ 17 + 234 + 85 \equiv 16, \end{cases}$$

which is a contradiction.

Now we show that $\gcd(a + b\sqrt{5}, a - b\sqrt{5}) = 1$. In fact, if a prime element $\alpha \in A$ divides $a + b\sqrt{5}$ and $a - b\sqrt{5}$ then $\alpha$ divides $2a$ and $2b\sqrt{5}$. If $\alpha \mid \sqrt{5}$ then $\alpha = \sqrt{5}\omega$ ($\omega$ a unit of $A$); so $\sqrt{5}$ divides $2a, 5$

divides $4a^2$ (in $\mathbb{Z}$), and $5 \mid a$, contrary to the hypothesis. Thus $\alpha$ divides $2b$; since there exist integers $s, t$ such that $2as + 2bt = 2$ then $\alpha$ would divide 2. But 2 is a prime in $A$, then $\alpha = 2\omega$ ($\omega$ a unit of $A$). So 2 divides both $a + b\sqrt{5}$ and $a - b\sqrt{5}$, hence 4 divides $a^2 - 5b^2 = (a + b\sqrt{5})(a - b\sqrt{5})$. But $a, b$ have different parity, so $a^2 - 5b^2$ is odd, a contradiction.

Since $\gcd(a + b\sqrt{5}, a - b\sqrt{5}) = 1$ and $a^2 - 5b^2$ is the fifth power of an element of $A$, it follows from the unique factorization in $A$ that $a + b\sqrt{5}$ is the fifth power of an element, say $a + b\sqrt{5} = ((m + n\sqrt{5})/2)^5$, where $m \equiv n \pmod 2$. Hence

$$a + b\sqrt{5} = \left(\frac{m + n\sqrt{5}}{2}\right)^5 \left(\frac{t + u\sqrt{5}}{2}\right),$$

where $t \equiv u \pmod 2$, $(t + u\sqrt{5})/2$ is a unit of $A$, and so $t^2 - 5u^2 = \pm 4$. Let $((m + n\sqrt{5})/2)^5 = (m' + n'\sqrt{5})/2$ so that $16m' \equiv m^5 \pmod 5$, $16n' \equiv 0 \pmod 5$ and hence $5 \mid n'$. Also $4a = m't + 5n'u$, $4b = m'u + n't$; hence $5 \nmid m'$ (otherwise $5 \mid a$, contrary to the hypothesis) and therefore $5 \nmid m$. Since $5 \mid n'$, $5 \mid b$ then $5 \mid m'u$ so $5 \mid u$. If $u = 0$ then $t = \pm 2$ and $a + b\sqrt{5} = \pm((m + n\sqrt{5})/2)^5$. From the remark at the beginning, $m, n$ are even and we put $c = \pm m/2$, $d = \pm n/2$ and it is clear that $c, d$ satisfy the relation (6.1).

If $u \neq 0$ then $(t + u\sqrt{5})/2 = \pm 1$ and hence $(t + u\sqrt{5})/2 = \pm((1 + \sqrt{5})/2)^e$ with some exponent $e \neq 0$. Replacing, if necessary, $(1 + \sqrt{5})/2$ by its inverse, $-(1 - \sqrt{5})/2$, we may assume $e > 0$ and actually $e > 1$ (otherwise $u = \pm 1$ contrary to the fact that $5 \mid u$). Then $\pm 2^{e-1}(t + u\sqrt{5}) = (1 \pm \sqrt{5})^e$; therefore

$$\pm 2^{e-1}u = e + 5\binom{e}{3} + 5^2\binom{e}{5} + \cdots,$$

so $2^{e-1}u \equiv \pm e \pmod 5$ and since $5 \mid u$ then $5 \mid e$. Thus $e = 5f$. Let

$$\frac{m + n\sqrt{5}}{2}\left(\frac{1 \pm \sqrt{5}}{2}\right)^f = \frac{c' + d'\sqrt{5}}{2},$$

where $c' \equiv d' \pmod 2$. Then $a + b\sqrt{5} = \pm((c' + d'\sqrt{5})/2)^5$. By the remark at the beginning, $c', d'$ are even. Let $c = \pm c'/2$, $d = \pm d'/2$. Then $c, d$ satisfy relations (6.1).

(2)   The proof in this case is very similar, so we only indicate the main steps. First we prove that $\gcd((a + b\sqrt{5})/2, (a - b\sqrt{5})/2 = 1$,

hence $(a + b\sqrt{5})/2 = ((m + n\sqrt{5})/2)^2((t + u\sqrt{5})/2)$ with $m \equiv n$ (mod 2), $t \equiv u$ (mod 2), $t^2 - 5u^2 = \pm 4$. Then $5 \mid u$. If $u = 0$ let $c = \pm m$, $d = \pm n$, so relations (6.2) are satisfied. If $u \neq 0$ then $(t + u\sqrt{5})/2 = \pm((1 \pm \sqrt{5})/2)^e$ with $e > 0$ (actually $e > 1$). Then $e = 5f$, and letting

$$\pm \left(\frac{m + n\sqrt{5}}{2}\right)\left(\frac{1 \pm \sqrt{5}}{2}\right)^f = \frac{c + d\sqrt{5}}{2},$$

it follows that $(a + b\sqrt{5})/2 = ((c + d\sqrt{5})/2)^5$ and again the relations (6.2) are satisfied. □

**(6B)**    *The equation*

(6.3)                         $X^5 + Y^5 + Z^5 = 0$

*has no solution in integers all different from 0.*

PROOF. We assume that there exist nonzero integers $x, y, z$ such that $x^5 + y^5 + z^5 = 0$. We may assume that $\gcd(x, y, z) = 1$ and hence $x, y, z$ are also pairwise relatively prime.

*First Case*: 5 does not divide $xyz$.

Then $x, y, z$ are congruent to $\pm 1$ or $\pm 2$ (modulo 5). Since $x^5 \equiv x$ (mod 5), $y^5 \equiv y$ (mod 5), $z^5 \equiv z$ (mod 5) then $x + y + z \equiv x^5 + y^5 + z^5 = 0$ (mod 5).

If $x, y, z$ are pairwise incongruent modulo 5 then $x + y + z \not\equiv 0$ (mod 5). So, for example, $x \equiv y$ (mod 5). Then $-z \equiv x + y \equiv 2x$ (mod 5). Raising to the fifth power, $x^5 \equiv y^5$ (mod $5^2$), $-z^5 \equiv 2^5 x^5$ (mod $5^2$) hence also $-z^5 = x^5 + y^5 \equiv 2x^5$ (mod $5^2$) and therefore $2^5 x^5 \equiv 2x^5$ (mod $5^2$), so $2^5 \equiv 2$ (mod $5^2$), which is not true. This proves the proposition in the first case.

*Second Case.* $5 \mid z$ (for example).

Then $5 \nmid xy$. Since $\gcd(x, y) = 1$ then either $x, y$ are both odd or of different parity.

We first consider the case where $x, y$ are odd. Then $-z^5 = x^5 + y^5$ is even, so $2, 5$ divide $z$. We may write $z = 2^m 5^n z'$, with $m \geq 1$, $n \geq 1$, $z'$ not a multiple of 2 or 5. Replacing $z$ by $z'$, we have nonzero integers $x, y, z$, pairwise relatively prime such that

(6.4)                         $-2^{5m} 5^{5n} z^5 = x^5 + y^5,$

with $x, y, z$ odd and not multiples of 5, $m \geq 1$, $n \geq 1$. Let $x + y = 2p$, $x - y = 2q$ ($p, q$ integers not equal to 0) so $x = p + q$, $y = p - q$, with $\gcd(p, q) = 1$ and $p, q$ not both odd. Then

$$-2^{5m}5^{5n}z^5 = (p + q)^5 + (p - q)^5 = 2p(p^4 + 10p^2q^2 + 5q^4).$$

Since $5 \mid p$ or $5 \mid p^4 + 10p^2q^2 + 5q^4$ it follows that $5 \mid p$ and we write $p = 5r$. So $5 \nmid q$ and $\gcd(r, q) = 1$, $q, r$ having different parity. Thus

$$-2^{5m}5^{5n}z^5 = 2 \times 5^2 r(q^4 + 50q^2r^2 + 125r^4).$$

Let $t = q^4 + 50q^2r^2 + 125r^4 = (q^2 + 25r^2)^2 - 5(10r^2)^2$. We put $u = q^2 + 25r^2$, $v = 10r^2$, so $u, v$ are not 0, $u$ is odd, $10 \mid v$, $\gcd(u, v) = 1$. Then $t$ is odd, $5 \nmid t$, $\gcd(t, r) = 1$, so $5 \mid r$ (since $5n > 2$).

Since $\gcd(2 \times 5^2 r, t) = 1$ then $2 \times 5^2 r$ and $t$ are fifth powers of integers. But $t = u^2 - 5v^2$ with $u \not\equiv v \pmod 2$, $\gcd(u, v) = 1$, $5 \nmid u$, $5 \mid v$. By (6A), there exist nonzero integers, $c, d$ such that

$$\begin{cases} u = c(c^4 + 50c^2d^2 + 125d^4), \\ v = 5d(c^4 + 10c^2d^2 + 5d^4), \end{cases}$$

and $\gcd(c, d) = 1$, $c \not\equiv d \pmod 2$, $5 \nmid c$. From this, it follows that $5 \mid d$ since $5 \mid r$ thence $5^3 \mid v$. We also note that $d > 0$. Multiplying the last relation by $2 \times 5^3$ we have $(2 \times 5^2 r)^2 = 2 \times 5^3 \times 10r^2 = 2 \times 5^4 d(c^4 + 10c^2d^2 + 5d^4)$ and this number is a fifth power (since $2 \times 5^2 r$ is a fifth power).

But $\gcd(2 \times 5^4 d, c^4 + 10c^2d^2 + 5d^4) = 1$ because $c^4 + 10c^2d^2 + 5d^4$ is odd, $5 \nmid c$ and $\gcd(c, d) = 1$. Hence $2 \times 5^4 d$ and $c^4 + 10c^2d^2 + 5d^4 = (c^2 + 5d^2)^2 - 5(2d^2)^2$ are fifth powers. Again $c^2 + 5d^2$, $2d^2$ are not both odd, $\gcd(c^2 + 5d^2, 2d^2) = 1$, $5 \nmid c^2 + 5d^2$, $5 \mid 2d^2$. By (6A), there exist nonzero integers $c', d'$ such that

$$\begin{cases} c^2 + 5d^2 = c'(c'^4 + 50c'^2d'^2 + 125d'^4), \\ 2d^2 = 5d'(c'^4 + 10c'^2d'^2 + 5d'^4), \end{cases}$$

and $\gcd(c', d') = 1$, $c' \not\equiv d' \pmod 2$, $5 \nmid c'$. From this it follows that $5 \mid d'$ because $5^2 \mid d^2$. We also note that $d' > 0$. Multiplying the last relation by $2 \times 5^8$ we have

$$2^2 \times 5^8 d^2 = (2 \times 5^4 d)^2 = 2 \times 5^9 d'(c'^4 + 10c'^2d'^2 + 5d'^4),$$

and this number is a fifth power. Since $\gcd(2 \times 5^9 d', c'^4 + 10c'^2d'^2 + 5d'^4) = 1$ then $2 \times 5^9 d'$, $c'^4 + 10c'^2d'^2 + 5d'^4$ are fifth powers. This is analogous to the previous assertion that $2 \times 5^4 d$, $c^4 + 10c^2d^2 + 5d^4$

were also fifth powers. Moreover, $0 < d' < d$, because $25d'^5 \le 5d'(c'^4 + 10c'^2 d'^2 + 5d'^4) = 2d^2$, so $0 < d' \le 5\sqrt{(2d^2)/25} < d$. If this procedure would continue, we would reach an integer $d''$ such that $0 < d'' < 1$ and this is absurd.

It remains to consider the case where $x$ and $y$ are of different parity. We omit some details of the computations.

Let $x + y = p$, $x - y = q$ so $p, q$ are odd, $\gcd(p, q) = 1$, $2x = p + q$, $2y = p - q$. Then $-2^5 \times 5^{5n} z^5 = (2x)^5 + (2y)^5 = (p + q)^5 + (p - q)^5 = 2p(p^4 + 10p^2 q^2 + 5q^4)$. Since $5 \mid p$ we write $p = 5r$, so $5 \nmid q$, $\gcd(q, r) = 1$, $q, r$ are both odd,

$$-2^5 \times 5^{5n} z^5 = 2 \times 5^2 rt,$$

where $t = q^4 + 50q^2 r^2 + 125r^4 = u^2 - 5v^2$ with $u = q^2 + 25r^2$, $v = 10r^2$. Then $u, v$ are not 0, $u, v$ are even, $u \equiv 2 \pmod 4$, $5 \nmid t$, $\gcd(t, r) = 1$ so $5 \mid r$ (since $5n > 2$). We write $u = 2u'$, $v = 2v'$, $u', v'$ are odd, $\gcd(u', v') = 1$, $5 \nmid u'$, $5 \mid v'$. If $t' = t/4 = u'^2 - 5v'^2$ then $t' \equiv 0 \pmod 4$, and $-5^{5n} z^5 = 5^2 rt'/4$, with $\gcd(5^2 r, t'/4) = 1$. So $5^2 r$ and $t'/4 = (u'^2 - 5v'^2)/4$ are fifth powers. By (6A) there exist nonzero integers $c, d$ such that

$$\begin{cases} u' = c(c^4 + 50c^2 d^2 + 125d^4)/16, \\ v' = 5d(c^4 + 10c^2 d^2 + 5d^4)/16, \end{cases}$$

and $\gcd(c, d) = 1$, $c, d$ are both odd, $5 \nmid c$. Moreover, since $5 \mid r$ then $5^2 \mid v'$ so $5 \mid d$. We note also that $d > 0$.

Multiplying the last relation by $5^3$ we have

$$(5^2 r)^2 = 5^3 v' = \frac{5^4 d}{4} \left[ \left( \frac{c^2 + 5d^2}{2} \right)^2 - 5d^4 \right],$$

where $((c^2 + 5d^2)/2)^2 - 5d^4 \equiv 0 \pmod 4$. Since the two factors in the right-hand side are relatively prime and $(5^2 r)^2$ is a fifth power, then $5^4 d$ and $\frac{1}{4}[((c^2 + 5d^2)/2)^2 - 5(d^2)^2]$ are fifth powers. By (6A), there exist nonzero integers $c', d'$ such that

$$(c^2 + 5d^2)/2 = c'(c'^4 + 50c'^2 d'^2 + 125d'^4)/16,$$
$$d^2 = 5d'(c'^4 + 10c'^2 d'^2 + 5d'^4)/16,$$

with $\gcd(c', d') = 1$, $c', d'$ both odd, $5 \nmid c'$. Moreover, $5 \mid d'$ and

TABLE 3. FLT for the exponent 5.

| Author | Case | Year |
|---|---|---|
| Gauss | both | 1863 (posthumous publication) |
| Schopis | first | 1825 |
| Lebesgue | both | 1843 |
| Lamé | both | 1847 |
| Gambioli | both | 1901 and 1903/4 |
| Werebrusow | both | 1905 |
| Mirimanoff | first | 1909 |
| Rychlik | both | 1910 |
| Hayashi[2] | both | 1911 |
| van der Corput | both | 1915 |
| Terjanian | both | 1987 |

$d' > 0$. Multiplying the last relation by $5^8$ we have

$$5^8 d^2 = (5^4 d)^2 = \frac{5^9 d'}{16}(c'^4 + 10c'^2 d'^2 + 5d'^4)$$

$$= \frac{5^9 d'}{4}\left[\left(\frac{c'^2 + 5d'^2}{2}\right)^2 - 5(d'^2)^2\right].$$

Again $5^9 d'$ and $\frac{1}{4}[((c'^2 + 5d'^2)/2)^2 - 5(d'^2)^2]$ are fifth powers. This is analogous to the previous assertion. Moreover $0 < d' < d$, because $25d'^5 \leq 16d^2$. The continuation of this procedure would lead to a contradiction. ☐

In 1912, Plemelj proved the following extension of the preceding theorem, see also Nagell (1958):

**(6C)**    *The equation*

$$X^5 + Y^5 + Z^5 = 0$$

*has only trivial solutions in integers of the number field $\mathbb{Q}(\sqrt{5})$.*

Other proofs of Fermat's theorem for the exponent 5 are given by the authors in Table 3.

---

[2]This proof may be incorrect, according to private communications.

# Bibliography

1825 Schopis, *Einige Sätze aus der unbestimmten Analytik*, Progr. Gummbinnen, 1825, pp. 12–15.

1828 Dirichlet, G.L., *Mémoire sur l'impossibilité de quelques équations indéterminées du $5^e$ degré*, J. Reine Angew. Math., **3** (1828), 354–375; reprinted in *Werke*, Vol. I, pp. 1–20 and 21–46, G. Reimer Verlag, Berlin, 1889, and also by Chelsea, New York, 1969.

1830 Legendre, A.M., *Théorie des Nombres* ($3^e$ édition), Vol. II, p. 5, Firmin Didot Fréres, Paris, 1830; reprinted by A. Blanchard, Paris, 1955.

1843 Lebesgue, V.A., *Théorèmes nouveaux sur l'équation indéterminée $x^5 + y^5 = az^5$*, J. Math. Pures Appl., (1), **8** (1843), 49–70.

1847 Lamé, G., *Mémoire sur la résolution en nombres complexes de l'équation $A^5 + B^5 + C^5 = 0$*, J. Math. Pures Appl., (1), **12** (1847), 137–171.

1875 Gauss, C.F., *Zur Theorie der complexen Zahlen.* (I) Neue Theorie der Zerlegung der Cuben, *Werke*, Vol. II, pp. 387–391, Königl. Ges. Wiss. Göttingen, 2nd ed., 1875.

1901 Gambioli, D., *Memoria bibliografica sull'ultimo teorema di Fermat*, Period. Mat., **16** (1901), 145–192.

1903/4 Gambioli, D., *Intorno all' ultimo teorema di Fermat*, Il Pitagora, **10** (1903/4), 11–13 and 41–42.

1905 Werebrusow, A.S., *On the equation $x^5 + y^5 = Az^5$* (in Russian), Moskov. Math. Samml., **25** (1905), 466–473.

1909 Mirimanoff, D., *Sur le dernier théorème de Fermat*, Enseign. Math., **11** (1909), 49–51.

1910 Rychlik, K., *On Fermat's last theorem for $n = 5$* (in Bohemian), Časopis Pěst. Mat., **39** (1910), 185–195, 305–317.

1911 Hayashi, T., *On Fermat's last theorem*, Indian Math. Club, Madras, **3** (1911), 111–114; reprinted in Science Reports, Tôhoku Imp. Univ., (1), **1** (1911/12), 51–54.

1912 Plemelj, J., *Die Unlösbarkeit von $x^5 + y^5 + z^5 = 0$ im Körper $k(\sqrt{5})$*, Monatsh. Math., **23** (1912), 305–308.

1915 van der Corput, J.G., *Quelques formes quadratiques et quelques équations indéterminées*, Nieuw Archief Wisk., **11** (1915), 45–75.

1958 Nagell, T., *Sur l'équation* $x^5 + y^5 = z^5$, Arkiv. Mat., **3** (1958), 511–514.

1973/4 Terjanian, G., *L'équation* $x^n + y^n = z^n$ *pour* $n = 5$ *et* $n = 14$, Sém. Th. des Nombres, Bordeaux, 1973/4, exp. no. 5, 6 pp.

1987 Terjanian, G., *Sur une question de V.A. Lebesgue*, Ann. Inst. Fourier, **37** (1987), 19–37.

1999 Ribenboim, P., *Classical Theory of Algebraic Num bers*, Springer-Verlag, New York, 1999.

## I.7. Fermat's Equation of Degree Seven

In 1839, Lamé proved Fermat's theorem for the exponent 7. Lebesgue found a simpler proof in 1840. Genocchi devised in 1874 and 1876 a still simpler proof (using an idea already found in Legendre (1830)), which is reproduced in Nagell's book (1951):

**(7A)**

(1) *If* $x, y, z$ *are the roots of a cubic equation with coefficients in* $\mathbb{Q}$, *and* $x^7 + y^7 + z^7 = 0$ *then either* $xyz = 0$ *or* $x, y, z$ *are proportional (in a certain order) to the cubic roots of* 1, *namely* $1, \zeta = (-1 + \sqrt{-3})/2, \zeta^2 = (-1 - \sqrt{-3})/2$.

(2) *The equation* $x^7 + y^7 + z^7 = 0$ *has only the trivial solution in integers.*

PROOF. (1) Suppose that $x, y, z$ are the roots of $f(X) = X^3 - pX^2 + qX - r$, with $p, q, r \in \mathbb{Q}$. Then

$$\begin{cases} p = x + y + z, \\ q = xy + xz + yz, \\ r = xyz. \end{cases}$$

Case I: $p = 0$.

We use the identity (see Section II.5)

$$(X + Y)^7 - X^7 - Y^7 = 7XY(X + Y)(X^2 + XY + Y^2)^2.$$

If $x + y + z = 0$ and $x^7 + y^7 + z^7 = 0$, then $7xy(x+y)(x^2+xy+y^2)^2 = 0$. Hence either $x = 0$, or $y = 0$, or $z = -(x + y) = 0$, or $xyz \neq 0$, but $x^2 + xy + y^2 = 0$. Therefore $(y/x)^2 + y/x + 1 = 0$ and so

$y/x$ is a cubic root of $1, y/x \neq 1$. Thus $y = x\zeta$ (or $y = x\zeta^2$) and $z = -(x+y) = -x(1+\zeta) = x\zeta^2$, so $x, y, z$ are proportional to $1, \zeta, \zeta^2$ (or to $1, \zeta^2, \zeta$).

Case II: $p \neq 0$.

Let $k \geq 1$ and $s_k = x^k + y^k + z^k$, the sum of the $k$th powers of the roots of $f(X) = 0$. By Newton's formulas:

$$s_1 \qquad\quad = p,$$
$$s_2 \quad\; = s_1 p - 2q,$$
$$s_3 \; = s_2 p - s_1 q + 3r,$$
$$s_4 = s_3 p - s_2 q + s_1 r,$$
$$s_5 = s_4 p - s_3 q + s_2 r,$$
$$s_6 = s_5 p - s_4 q + s_3 r,$$
$$s_7 = s_6 p - s_5 q + s_4 r.$$

Substituting we obtain

$$x^7 + y^7 + z^7 = p^7 - 7p^5 q + 7p^4 r + 14p^3 q^2 - 21p^2 qr - 7pq^3 + 7pr^2 + 7q^2 r.$$

Let $m = pq - r \in \mathbb{Q}$. Then

$$x^7 + y^7 + z^7 = p^7 - 7(pq - r)(p^4 - p^2 q + q^2) + 7(pq - r)^2 p,$$

that is,

$$x^7 + y^7 + z^7 = p^7 - 7m(p^4 - p^2 q + q^2) + 7m^2 p.$$

Since $x^7 + y^7 + z^7 = 0$ then

$$p^7 - 7m(p^4 - p^2 q + q^2) + 7m^2 p = 0.$$

Let $q/p^2 = Q$, $m/p^3 = M$. Then

$$p^7 - 7p^3 M(p^4 - p^4 Q - p^4 Q^2) + 7p^7 M^2 = 0,$$

so

$$M^2 - M(1 - Q + Q^2) + \tfrac{1}{7} = 0.$$

Since $M$ is a rational number, the discriminant $((1 - Q + Q^2)/2)^2 - \tfrac{1}{7}$ is the square of a nonzero rational number. Let $2Q - 1 = s/t$, where $s, t$ are relatively prime integers, $t > 0$. Then $Q = (t + s)/2t$, hence $64t^4[((3t^2 + s^2)/8t^2)^2 - \tfrac{1}{7}] = s^4 + 6t^2 s^2 - t^4/7 = u^2$ where $u$ is a nonzero rational number.

Since $7u^2$ is an integer, then $u$ must be an integer, so 7 divides $t$, and therefore $7 \nmid s$. Let $t = 7^e v$, with $e \geq 1$, $7 \nmid v$. Hence

$$(7.1) \qquad s^4 + 6 \times 7^{2e} v^2 s^2 - 7^{4e-1} v^4 = u^2.$$

From (7.1) we deduce that

$$(7.2) \qquad (s^2 + 3 \times 7^{2e} v^2)^2 - u^2 = 64 \times 7^{4e-1} v^4,$$

hence

$$(7.3) \quad (s^2 + 3 \times 7^{2e} v^2 + u)(s^2 + 3 \times 7^{2e} v^2 - u) = 64 \times 7^{4e-1} v^4.$$

Now we show that $\gcd(s^2 + 3 \times 7^{2e} v^2 + u, \, s^2 + 3 \times 7^{2e} v^2 - u)$ is a power of 2. Indeed, let $p \neq 2$ be any prime such that $p \mid s^2 + 3 \times 7^{2e} v^2 + u$ and $p \mid s^2 + 3 \times 7^{2e} v^2 - u$. Then $p \mid 2u$, so $p \mid u$. Also $p \mid 2(s^2 + 3 \times 7^{2e} v^2)$, so $p \mid s^2 + 3 \times 7^{2e} v^2$. If $p = 7$, since $e \geq 1$ then $7 = p \mid s$, which is a contradiction. If $p \neq 7$, by (7.3) $p \mid 64 \times 7^{4e-1} v^4$, so $p \mid v$, hence $p \mid s$, and this is again a contradiction. Thus the greatest common divisor of the two factors in (5.3) is a power of 2.

Subcase (a): $v$ is odd.

From (5.3) it follows that

$$(7.4) \qquad \begin{cases} s^2 + 3 \times 7^{2e} v^2 \pm u = 7^{4e-1} A a^4, \\ s^2 + 3 \times 7^{2e} v^2 \mp u = B b^4, \end{cases}$$

where $a, b$ are relatively prime integers, $ab = v$ (so $a, b$ are odd) and $A, B$ are even and $AB = 64$.

From (7.4) we deduce that

$$(7.5) \qquad 2s^2 + 6 \times 7^{2e} v^2 = 7^{4e-1} A a^4 + B b^4.$$

Noting that if $x$ is odd then $x^2 \equiv 1 \pmod 8$ and that $7 \equiv -1 \pmod 8$ then

$$s^2 \equiv -3 \times 7^{2e} a^2 b^2 + 7^{4e-1} \frac{A}{2} a^4 + \frac{B}{2} b^4 \equiv -3 - \frac{A}{2} + \frac{B}{2} \pmod 8.$$

We consider the various possibilities for $A, B$:

$$\begin{array}{llll}
A = 32, & B = 2 & \rightarrow & s^2 \equiv -2 \pmod 8, \quad \text{impossible,} \\
A = 16, & B = 4 & \rightarrow & s^2 \equiv -1 \pmod 8, \quad \text{impossible,} \\
A = 8, & B = 8 & \rightarrow & s^2 \equiv -3 \pmod 8, \quad \text{impossible,} \\
A = 4, & B = 16 & \rightarrow & s^2 \equiv 3 \pmod 8, \quad \text{impossible,} \\
A = 2, & B = 32 & \rightarrow & s^2 \equiv 4 \pmod 8.
\end{array}$$

In this last case, (7.5) becomes

$$s^2 + 3 \times 7^{2e} a^2 b^2 = 7^{4e-1} a^4 + 16 b^4,$$

hence multiplying with 64,

$$64 s^2 + 6 \times 7^{2e} \times 32 a^2 b^2 - 64 \times 16 b^4 = 7^{4e-1} \times 64 a^4,$$

so

$$64 s^2 - (32 b^2 - 3 \times 7^{2e} a^2)^2 = 7^{4e-1} \times 64 a^4 - 7^{4e} \times 3^2 a^4 = 7^{4e-1} a^4,$$

and therefore

$$(7.6) \ (8s + 32 b^2 - 3 \times 7^{2e} a^2)(8s - 32 b^2 + 3 \times 7^{2e} a^2) = 7^{4e-1} a^4.$$

We note that $\gcd(8s + 32 b^2 - 3 \times 7^{2e} a^2, 8s - 32 b^2 + 3 \times 7^{2e} a^2) = 1$. Indeed, since $a$ is odd, the above numbers are odd. If $p \mid 8s + 32 b^2 - 3 \times 7^{2e} a^2$ and $p \mid 8s - 32 b^2 + 3 \times 7^{2e} a^2$ then $p \mid 16s$, so $p \mid s$, and similarly $p \mid 32 b^2 - 3 \times 7^{2e} a^2$. But $p \mid 7^{4e-1} a^4$ (by (7.6)). If $p = 7$ then $p = 7 \mid s$, which is impossible. If $p \neq 7$ then $p \mid a$ so $p \mid b$, again a contradiction. This proves the assertion, and therefore

$$(7.7) \qquad \begin{cases} 8s \pm 32 b^2 \mp 3 \times 7^{2e} a^2 = c^4, \\ 8s \mp 32 b^2 \pm 3 \times 7^{2e} a^2 = 7^{4e-1} d^4, \end{cases}$$

where $c, d$ are relatively prime integers, $cd = a$. Hence $c, d$ are odd. From (7.7) we derive the congruence

$$\mp 3 \equiv c^4 \pmod{8},$$

which is impossible. We have therefore shown that subcase (a) is impossible.

Subcase (b): $v$ is even.

Then $s$ is odd (because $t$ is even), hence $u$ is also odd. We write (5.2) as

$$(s^2 + 3 \times 7^{2e} v^2)^2 - u^2 = 4 \times 7^{4e-1} (2v)^4,$$

and noting that both factors of (7.3) are even, and their greatest common divisor is 2, we deduce that

$$(7.8) \qquad \begin{cases} s^2 + 3 \times 7^{2e} v^2 \pm u = 2 \times 7^{4e-1} A^4, \\ s^2 + 3 \times 7^{2e} v^2 \mp u = 2 \times B^4, \end{cases}$$

with $A, B$ relatively prime integers, $AB = 2v$, so either $A$ or $B$ is even. Hence

(7.9)
$$s^2 = -3 \times 7^{2e} v^2 + 7^{4e-1} A^4 + B^4 = -\tfrac{3}{4} \times 7^{2e} A^2 B^2 + 7^{4e-1} A^4 + B^4.$$

If $B$ is even then $A$ is odd and $s^2 \equiv -3v^2 - 1 \pmod 8$, hence $s^2 \equiv -1 \pmod 4$, which is impossible.

If $A$ is even, then $B$ is odd. Let $A = 2A_1$, hence from (5.9)

$$s^2 = -3 \times 7^{2e} A_1^2 B^2 + 7^{4e-1} \times 2^4 A_1^4 + B^4$$

and therefore $1 \equiv s^2 \equiv -3A_1^2 + 1 \pmod 8$; hence $3A_1^2 \equiv 0 \pmod 8$ so $4 \mid A_1$. We write $A_1 = 4A_2$, hence $A = 8A_2$ and (7.9) is rewritten as

$$s^2 = -3 \times 16 \times 7^{2e} A_2^2 B^2 + 7^{4e-1} \times 8^4 \times A_2^4 + B^4,$$

so

$$s^2 - (B^2 - 3 \times 8 \times 7^{2e} A_2^2)^2 = 7^{4e-1} \times 8^4 \times A_2^4 - 3^2 \times 8^2 \times 7^{4e} A_2^4$$
(7.10)
$$= 7^{4e-1} \times 8^2 \times A_2^4.$$

This gives

(7.11) $\quad (s - B^2 + 3 \times 8 \times 7^{2e} A_2^2)(s + B^2 - 3 \times 8 \times 7^{2e} A_2^2)$
$$= 7^{4e-1} \times 4 \times (2A_2)^4.$$

The two factors of (7.11) are even, and it may be seen, as before, that their greatest common divisor is 2. Hence

(7.12) $\quad \begin{cases} s \mp B^2 \pm 3 \times 8 \times 7^{2e} A_2^2 = 2c_2^4, \\ s \pm B^2 \mp 3 \times 8 \times 7^{2e} A_2^2 = 7^{4e-1} \times 2d_2^4, \end{cases}$

where $c_2, d_2$ are relatively prime integers such that $c_2 d_2 = 2A_2$.

From (7.12) we deduce by subtraction that $\mp B^2 \pm 3 \times 8 \times 7^{2e} A_2^2 = c_2^4 - 7^{4e-1} d_2^4$, hence

(7.13) $\qquad \pm B^2 = \pm 6 \times 7^{2e} c_2^2 d_2^2 - c_2^4 + 7^{4e-1} d_2^4.$

Since any nonzero square modulo 7 is congruent to $1, 2$ or $4$, then $B^2 + c_2^4 \not\equiv 0 \pmod 7$. So we must have the negative signs in (7.13):

(7.14) $\qquad B^2 = c_2^4 + 6 \times 7^{2e} c_2^2 d_2^2 - 7^{4e-1} d_2^4.$

This equation is of the same form as (7.1). Moreover,

$$v = \frac{AB}{2} = 4A_2 B \geq 2c_2 d_2 > d_2.$$

TABLE 4. FLT for the exponent 7.

| Author | Cases | Year |
|---|---|---|
| Legendre (Sophie Germain) | first | 1823 (see Chapter II, §3) |
| Genocchi | both | 1864 |
| Pepin | both | 1876 |
| Maillet | both | 1897 |

So, we may proceed by infinite descent and conclude that the subcase (b) is also impossible. Thus, case II is not possible, and this proves (1).

(2) If $x, y, z$ are integers such that $x^7 + y^7 + z^7 = 0$, we consider the polynomial $f(X) = X^3 - pX^2 + qX - r$, which has the roots $x, y, z$. By (1), since $x, y, z$ cannot be proportional to $1, \zeta, \zeta^2$, then $xyz = 0$. $\square$

Other proofs of Fermat's theorem for the exponent 7 are given by the authors in Table 4.

## Bibliography

1823 Legendre, A.M., *Recherches sur quelques objets d'analyse in-déterminée, et particulièrement sur le théorème de Fermat*, Mém. Acad. Roy. Sci. Institut France, **6** (1823), 1–60; reprinted as the "Second Supplément" in 1825, to a printing of *Essai sur la Théorie des Nombres* ($2^e$ édition), Courcier, Paris; reprinted in Sphinx Oedipe, **4** (1909), 97–128.

1839 Lamé, G., *Mémoire sur le dernier théorème de Fermat*, C. R. Acad. Sci. Paris, **9** (1839), 45–46.

1839 Cauchy, A. and Liouville, J., *Rapport sur un mémoire de M. Lamé relatif au dernier théorème de Fermat*, C. R. Acad. Sci. Paris, **9** (1839), 359–363; also appeared in J. Math. Pures Appl., **5** (1840), 211–215 and *Oeuvres Complètes*, Sér. 1, Vol. 4, pp. 494–504, Gauthier-Villars, Paris, 1884.

1840 Lamé, G., *Mémoire d'analyse indéterminée démontrant que l'équation $x^7 + y^7 = z^7$ est impossible en nombres entiers*, J. Math. Pures Appl., **5** (1840), 195–211.

1840 Lebesgue, V.A., *Démonstration de l'impossibilité de résoudre l'équation $x^7 + y^7 + z^7 = 0$ en nombres entiers*, J. Math. Pures

Appl., **5** (1840), 276–279.

1840 Lebesgue, V.A., *Addition à la note sur l'équation $x^7 + y^7 + z^7 = 0$*, J. Math. Pures Appl., **5** (1840), 348–349.

1864 Genocchi, A., *Intorno all'equazioni $x^7 + y^7 + z^7 = 0$*, Annali Mat., **6** (1864), 287–288.

1874 Genocchi, A., *Sur l'impossiblilité de quelques égalités doubles*, C. R. Acad. Sci. Paris, **78** (1874), 433–436.

1876 Genocchi, A., *Généralisation du théorème de Lamé sur l'impossibilité de l'équation $x^7 + y^7 + z^7 = 0$*, C. R. Acad. Sci. Paris, **82** (1876), 910–913.

1876 Pepin, T., *Impossibilité de l'équation $x^7 + y^7 + z^7 = 0$*, C. R. Acad. Sci. Paris, **82** (1876), 676–679 and 743–747.

1897 Maillet, E., *Sur l'équation indéterminée $ax^{\lambda^t} + by^{\lambda^t} = cz^{\lambda^t}$*, Assoc. Française Avanc. Sci., St. Etienne, Sér. II, **26** (1897), 156–168.

1951 Nagell, T., *Introduction to Number Theory*, Wiley, New York, 1951; reprinted by Chelsea, New York, 1962.

## I.8. Other Special Cases

There have been numerous papers devoted to the proof of Fermat's theorem for special exponents, other than 3, 4, 5, 7. The methods used were specific to the exponent in question and in most instances not susceptible of generalization.

We note that according to an oral communication of Terjanian, Hayashi's proof (1911) for the second case and exponent 13 has a mistake.

Brčić-Kostić studied the equation $x^4 + y^2 = z^6$ in 1956 and showed that it has no solution in relatively prime integers; however, it has nontrivial solutions in integers which are not pairwise relatively prime.

Now we indicate the elementary proofs of Breusch (1960) for the exponents 6 and 10. Of course, the theorem for these exponents follows from the truth for the exponents 3 and 5. But the proofs which we present will be entirely independent of the above results. The following preliminary results, by the method of infinite descent, were also proved by Breusch:

TABLE 5. FLT for various exponents.

| Author | Exponent | Case | Year |
|---|---|---|---|
| Kausler | 6 | both | 1806 |
| Sophie Germain | all primes less | first | 1823 |
| (Legendre) | than 100 | | (see Chapter III) |
| Dirichlet | 14 | both | 1832 |
| Lamé[3] | 11, 17, 23, 29, 41 | first | 1847 |
| Matthews | 11, 17 | first | 1885–6 |
| Tafelmacher[4] | 11, 17, 23, 29 | first | 1892 |
| Thue | 6 | both | 1896 |
| Tafelmacher | 6 | both | 1897 |
| Lind | 6 | both | 1909 |
| Mirimanoff | 11, 17 | first | 1909 |
| Kapferer | 6, 10 | both | 1913 |
| Swift | 6 | both | 1914 |
| Kokott | 11 | first | 1915 |
| Fell | 11, 17, 23 | first | 1943 |
| Breusch | 6, 10 | both | 1960 |
| Terjanian | 14 | both | 1974 |

**(8A)**    *There exist no positive integers $x, y, u, v$ such that*

(8.1) $$x^2 + y^2 = u^2 + v^2$$

*and*

(8.2) $$xy = 2uv.$$

PROOF. Assume the contrary, and among all possible solutions, consider the one with minimal positive product $xy$. From this minimal choice it follows that no three of the four integers $x, y, u, v$ can have a common factor greater than 1. Also $\gcd(x, y) = 1$, because if $p$ is any prime dividing $x$ and $y$ then $p$ divides $u$ or $v$, a contradiction. Similarly, $\gcd(u, v) = 1$. Hence one of $x, y$ is even, and the other is odd. For example, let $2 \mid y$, so $x^2 + y^2$ is odd, so $u$ or $v$ is even, say $u$ is. Therefore from (8.2) we deduce that $4 \mid y$; from $x^2 \equiv 1$

---

[3]Lamé's paper (1847) has no proofs.

[4]Tafelmacher's proof (1892) holds only for the first case; see also Dickson, *History of the Theory of Numbers*, Vol. II, pp. 755.

(mod 8) it follows that $x^2 + y^2 \equiv 1$ (mod 8). Again, by (8.1), $u^2 \equiv 0$ (mod 8), so $u \equiv 0$ (mod 4) and $8 \mid y$.

Let $y = 2^{r+1}y'$ (with $r \geq 2, y'$ odd), then $u = 2^r u'$, with $u'$ odd. Thus

$$(8.3) \qquad \begin{array}{l} xy' = u'v \qquad \text{and} \qquad x, y', u'v \text{ are odd,} \\ \gcd(x, y') = \gcd(u'v) = 1. \end{array}$$

Moreover, if

$$\begin{array}{ll} a = \gcd(x, u'), & b = \gcd(x, v), \\ c = \gcd(y', u'), & d = \gcd(y', v), \end{array}$$

then $a, b, c, d$ are pairwise relatively prime and $b, d$ are odd, so $b^2 \equiv d^2 \equiv 1$ (mod 8). So $x = ab$, $y' = cd$, $u' = ac$, $v = bd$. Therefore the original equation becomes

$$a^2 b^2 + 2^{2r+2}c^2 d^2 = 2^{2r}a^2 c^2 + b^2 d^2.$$

Letting $t = 2^r c$, we have

$$(8.4) \qquad 3t^2 d^2 = (a^2 - d^2)(t^2 - b^2),$$

where $t, a, b, d$ are pairwise relatively prime and $4 \mid t$. It follows that $d^2$ divides $t^2 - b^2$ and $t^2 - b^2$ divides $3d^2$.

This holds if and only if one of the following conditions is satisfied:

(I) $t^2 - b^2 = 3d^2$;
(II) $t^2 - b^2 = d^2$;
(III) $t^2 - b^2 = -3d^2$; and
(IV) $t^2 - b^2 = -d^2$.

But $t^2 \equiv 0$ (mod 8) and $b^2 \equiv d^2 \equiv 1$ (mod 8). Therefore the cases I, II, III are not possible. From $t^2 - b^2 = -d^2$ it follows that $3t^2 = d^2 - a^2$, so

$$(8.5) \qquad b^2 = d^2 + t^2 = a^2 + 4t^2.$$

By (1A) there exist integers $m, n > 0$ and integers $r, s > 0$ such that

$$(8.6) \qquad \begin{cases} b = m^2 + n^2, \\ t = 2mn, \end{cases}$$

and

$$(8.7) \qquad \begin{cases} b = r^2 + s^2, \\ 2t = 2rs. \end{cases}$$

From these, we obtain the relations

(8.8) $$r^2 + s^2 = m^2 + n^2$$

and

(8.9) $$rs = 2mn.$$

Since $rs = t = 2^r c \leq 2^r y' < y \leq xy$, we have found a new solution for the original system, contradicting the minimality of the solution $(x, y, u, v)$. $\square$

**(8B)**    *There exist no positive integers $x, y, u, v$ such that*

(8.10) $$x^2 - y^2 = u^2 + v^2$$

*and*

(8.11) $$xy = 2uv.$$

PROOF. We assume the contrary and consider positive integers satisfying the above relations and such that $xy$ is minimal.

Proceeding as before, one of $x, y$ is even and the other is odd. Thus $x^2 - y^2 = u^2 + v^2 \equiv 1 \pmod 4$; therefore necessarily $x$ is odd and $y$ is even. As in the preceding proof, we arrive at equations (8.3), (8.4), (8.5), and (8.6) with $t = 2^r c, r \geq 2$. And we obtain the relation $a^2 b^2 - 2^{2r+2} c^2 d^2 = 2^{2r} a^2 c^2 + b^2 d^2$, hence

(8.12) $$5t^2 d^2 = (a^2 - d^2)(b^2 - t^2).$$

So $d^2$ divides $b^2 - t^2$ and $b^2 - t^2$ divides $5d^2$. By the same reasoning as previously, taking congruences modulo 8, we see that $b^2 - t^2 = d^2$. It follows that $5t^2 = a^2 - d^2$, hence

(8.13)     $b^2 = d^2 + t^2$    and    $b^2 + 4t^2 = a^2$.

By (1A) there exist integers $m, n > 0$ and integers $r, s > 0$ such that

(8.14) $$\begin{cases} b = m^2 + n^2, \\ t = 2mn, \end{cases}$$

and

(8.15) $$\begin{cases} b = r^2 - s^2, \\ 2t = 2rs. \end{cases}$$

We obtain the relations

(8.16) $$r^2 - s^2 = m^2 + n^2$$

and

(8.17) $$rs = 2mn.$$

Since $rs < xy$, we have again reached a contradiction to the minimality of the solution $(x, y, u, v)$.  □

Using these facts, Breusch proved:

**(8C)**   *There exist no positive integers $x, y, z$ such that $x^6 + y^6 = z^6$.*

PROOF. Assume on the contrary that $x, y, z$ are pairwise relatively prime positive integers such that $x^6 + y^6 = z^6$. Then 3 does not divide both $x$ and $y$, so we may assume that $3 \nmid x$. We have

$$x^6 = z^6 - y^6 = (z + y)(z - y)(z^2 + zy + y^2)(z^2 - zy + y^2).$$

The last two factors in the above product must be odd, since $x, y$ are not both even. Moreover, it is easily seen that each factor $z^2 + zy + y^2$ and $z^2 - zy + y^2$ is relatively prime to the three other factors of the right-hand side (because $3 \nmid x$). Hence by the unique factorization of integers,

$$\begin{cases} z^2 + zy + y^2 = b^6, \\ z^2 - zy + y^2 = c^6, \end{cases}$$

with integers $b > c > 0$. Adding and subtracting these relations, we have $2(z^2 + y^2) = b^6 + c^6$ and $2zy = b^6 - c^6$. But $b, c$ are odd, so $b^3 + c^3 = 2m$, $b^3 - c^3 = 2n$ (with $m > 0, n > 0$), so squaring and adding, we get $b^6 + c^6 = 2(m^2 + n^2)$, and multiplying, $b^6 - c^6 = 4mn$. Hence $z^2 + y^2 = m^2 + n^2$, $zy = 2mn$. According to (8A), this is impossible.  □

**(8D)**   *There exist no positive integers $x, y, z$ such that*

$$x^{10} + y^{10} = z^{10}.$$

PROOF. Let $x, y, z$ be positive, pairwise relatively prime integers such that $x^{10} + y^{10} = z^{10}$. Then 5 does not divide both $x$ and $y$, so we may assume $5 \nmid x$. We have

$$
\begin{aligned}
x^{10} &= z^{10} - y^{10} \\
&= (z + y)(z - y)(z^4 + z^3 y + z^2 y^2 + z y^3 + y^4) \\
&\quad \times (z^4 - z^3 y + z^2 y^2 - z y^3 + y^4).
\end{aligned}
$$

The last two factors in the above product must be odd, since $y, z$ are not both even. Moreover, it is easily seen that each of the last two factors is relatively prime to the three other factors on the right-hand side (since $5 \nmid x$). Hence by the unique factorization of integers,

$$
\begin{cases}
z^4 + z^3 y + z^2 y^2 + z y^3 + y^4 = b^{10}, \\
z^4 - z^3 y + z^2 y^2 - z y^3 + y^4 = c^{10},
\end{cases}
$$

with integers $b > c > 0$. Adding and subtracting these relations, we have $2(z^4 + z^2 y^2 + y^4) = b^{10} + c^{10}$, $2(z^3 y + z y^3) = b^{10} - c^{10}$. But $b, c$ are odd, so $b^5 + c^5 = 2m$, $b^5 - c^5 = 2n$ (with $m > 0, n > 0$), so squaring and adding, we get $b^{10} + c^{10} = 2(m^2 + n^2)$, and multiplying, $b^{10} - c^{10} = 4mn$. Hence

$$
\begin{cases}
z^4 + z^2 y^2 + y^4 = m^2 + n^2, \\
z y (z^2 + y^2) = 2mn.
\end{cases}
$$

Now letting $z^2 + y^2 = r$, $zy = s$, then $r^2 - s^2 = m^2 + n^2$, $rs = 2mn$. This is impossible by (8B).  □

Fermat's theorem for the exponent 14 was established by Dirichlet (1832) before Lamé settled the case of exponent 7. Dirichlet also showed:

**(8E)**    *The equation*

$$
X^{14} - Y^{14} = 2^m 7^{1+n} Z^{14}
$$

*(with $m \geq 0, n \geq 0$) has no solution in nonzero integers $x, y, z$ with $x, y$ relatively prime.*

Terjanian proved in 1974:

**(8F)**    *If $a$ is a natural number, $a \neq 0$, multiple of 7, and without prime factor $p \equiv 1 \pmod{7}$, if $x, y, z$ are natural numbers, $x, y \neq 0$,*

$x, y$ *relatively prime, if*

$$x^{14} - y^{14} = az^{14},$$

*then $x = y = 1$ and $z = 0$.*

From this, he deduced a simple proof of Fermat's theorem for the exponent 14.

In 1885, Matthews gave a proof for the first case of Fermat's theorem for the exponents 11 and 17. In 1943, Fell indicated a distinct proof for 11 and claimed that his method also solved the first case for the exponents 17 and 23.

## Bibliography

1806 Kausler, C.F., *Nova demonstratio theorematis nec summam, nec differentiam duorum cubo-cuborum cubu-cubum esse posse*, Novi Acta Acad. Petrop., **15** (1806), 146–155.

1823 Legendre, A.M., *Recherches sur quelques objets d'analyse indéterminée, et particulièrement sur le théorème de Fermat*, Mém. Acad. Roy. Sci. Institut France, **6** (1823), 1–60; reprinted as the "Second Supplément" in 1825, to a printing of *Essai sur la Théorie des Nombres* ($2^e$ édition), Courcier, Paris; reprinted in Sphinx-Oedipe, **4** (1909), 97–128.

1832 Dirichlet, G.L., *Démonstration du théorème de Fermat pour le cas des $14^e$ puissances*, J. Reine Angew. Math., **9** (1832), 390–393; reprinted in *Werke*, Vol. I, pp. 189–194, G. Reimer, Berlin, 1889; reprinted by Chelsea, New York, 1969.

1847 Lamé, G., *Troisième mémoire sure le dernier théorème de Fermat*, C. R. Acad. Sci. Paris, **24** (1847), 888.

1885 Matthews, G.B., *Note in connexion with Fermat's last theorem*, Messenger Math., **15** (1886), 68–74.

1892 Tafelmacher, W.L.A., *Sobre el teorema de Fermat de que la ecuación $x^n + y^n = z^n$ no tiene solución en numeros enteros $x, y, z$, i siendo $n > 2$*, Ann. Univ. Chile, Santiago, **82** (1892), 271–300 and 415–437.

1896 Thue, A., *Über die Auflösbarkeit einiger unbestimmter Gleichungen*, Det Kongel. Norske Videnskabers Selskabs Skrifter, 1896, no. 7, 14 pp.; reprinted in *Selected Mathematical Papers*, pp. 19–30, Universitetsforlaget, Oslo, 1977.

1897 Tafelmacher, W.L.A., *La ecuación $x^3 + y^3 = z^2$: Una demonstración nueva del teorema de Fermat para el caso de las sestas potencias*, Ann. Univ. Chile, Santiago, **97** (1897), 63–80.

1909 Lind, B., *Einige zahlentheoretische Sätze*, Arch. Math. Phys., (3), **15** (1909), 368–369.

1909 Mirimanoff, D., *Sur le dernier théorème de Fermat*, Enseign. Math., **11** (1909), 49–51.

1911 Hayashi, T., *On Fermat's last theorem*, Indian Math. Club, Madras, **3** (1911), 111–114. Reprinted in Sci. Rep., Tôhoku Imp. Univ., (1), **1** (1911/2), 51–54.

1913 Kapferer, H., *Beweis des Fermatschen Satzes für die Exponenten 6 und 10*, Archiv Math. Phys., (3), **21** (1913), 143–146.

1914 Swift, E., *Solution to Problem 206*, Amer. Math. Monthly, **21** (1914), 238–239.

1915 Kokott, P., *Über einen Spezialfall des Fermatschen Satzes*, Arch. Math. Phys., (3), **24** (1915), 90–91.

1916 Swift, E., *Solution to Problem 209*, Amer. Math. Monthly, **23** (1916), 261.

1943 Fell, J., *Elementare Beweise des großen Fermatschen Satzes für einige besondere Fälle*, Deutsche Math., **7** (1943), 184–186.

1956 Brčić-Kostić, M., *Solution of the diophantine equation $x^4 + y^2 = z^6$* (in Serbo-Croatian, Esperanto summary), Bull. Soc. Math. Phys. Serbie, **8** (1956), 125–136.

1960 Breusch, R., *A simple proof of Fermat's last theorem for $n = 6$, $n = 10$*, Math. Mag., **33** (1960), 279–281.

1974 Terjanian, G., *L'équation $x^n + y^n = z^n$ pour $n = 5$ et $n = 14$*, Sém. Th. des Nombres, Bordeaux, 1973/4, exp. no. 5, 6 pp.

1974 Terjanian, G., *L'équation $x^{14} + y^{14} = z^{14}$ en nombres entiers*, Bull. Sci. Math., Sér. 2, **98** (1974), 91–95.

## I.9. Appendix

### Would this Be Fermat's "Marvelous Proof"?[5]

We can see in Euler's *Opera Posthuma Mathematica et Physica*, Petropoli, 1862, Vol. 1, pp. 231–232, in "Fragmenta arithmetica ex adversariis mathematicis deprompta" the following argument, attributed to A.J. Lexell. It represents an attempt to apply the method of infinite descent to Fermat's equation, in a way which could have been Fermat's.

The attempt failed.

Assume that FLT is false for some exponent $n > 2$. We may assume $n$ to be an odd prime (in the argument we only require $n$ to be odd) and that there exist nonzero integers (not necessarily positive) $a, b, c$, such that $a^n + b^n = c^n$ and $c$ is even, $a$ and $b$ are odd, $a \neq b$, and $\gcd(a, b, c) = 1$. Let

$$\begin{cases} x = c^n/2, \\ y = (a^n - b^n)/2, \\ z = abc^{n-2}/2, \end{cases}$$

so $x, y, z$ are integers, $x$ is even. Then

$$\begin{cases} x + y = a^n, \\ x - y = b^n, \end{cases}$$

hence

$$\frac{x^2 - y^2}{4x^2} = \left(\frac{ab}{c^2}\right)^n = \left(\frac{z}{x}\right)^n.$$

It follows that

$$x(x^n - 4z^n) = x^{n-1}\left[x^2 - 4x^2\left(\frac{z}{x}\right)^n\right]$$
$$= x^{n-1}y^2 = (x^{(n-1)/2}y)^2.$$

Let $d = \gcd(x, z)$, so $d = c^{n-2}/2$ because $\gcd(ab, c) = 1$. Let $x = dx'$, $z = dz'$, so $x'$ is even, $\gcd(x', z') = 1$ and $d^{n+1}x'(x'^n - 4z'^n)$ is a square, hence so is $x'(x'^n - 4z'^n)$, with the two factors relatively prime. Thus there exist integers $r, s$ such that

$$\begin{cases} x' = r^2, \\ x'^n - 4z'^n = s^2. \end{cases}$$

---

[5] I am indebted to E. Bombieri who called my attention to what follows.

So
$$r^{2n} - s^2 = (r^n + s)(r^n - s) = 4z'^n.$$
Since $r \equiv s \pmod 2$, then $\gcd(r^n + s, r^n - s) = 2$, so
$$\begin{cases} r^n + s = 2t^2, \\ r^n - s = 2u^n, \end{cases}$$
and adding, we obtain $r^n = t^n + u^n$, which gives a nontrivial solution of Fermat's equation, with $r$ even. If we would have $r < c$ then by descent we would reach a contradiction. However,
$$r^2 = x' = \frac{x}{d} = c^2,$$
so $r = c$ and the descent method is not applicable.

# II
# 4 Interludes

In this chapter we discuss topics which will be required in the subsequent developments. Their importance is not restricted to their applications to Fermat's last theorem.

## II.1. p-Adic Valuations

Let $p$ be a prime, let $a$ be a nonzero integer, and let $v_p(a)$ be the exponent of $p$ in the factorization of $a$ as a product of prime-powers:

$$a = p^{v_p(a)} b, \quad \text{where } p \nmid b,$$

$v_p(a)$ is the *p-adic value* of $a$. By convention we also set $v_p(0) = \infty$.
We note:

$$(1.1) \quad \begin{cases} v_p(ab) = v_p(a) + v_p(b); \\ v_p(a+b) \geq \min\{v_p(a), v_p(b)\}; \\ \text{if } v_p(a) < v_p(b_1), v_p(b_2), \dots, v_p(b_k) \\ \qquad \text{then } v_p(a + b_1 + b_2 + \cdots + b_k) = v_p(a). \end{cases}$$

If $v_p(a) = e \geq 1$ we say that $p^e$ is the exact power of $p$ dividing $a$ and we write $p^e \parallel a$.

More generally, if $\gcd(a, a') = 1$, $a' > 0$, we define $v_p(a/a') = v_p(a) - v_p(a')$. Then for any rational numbers $r, s$, the above properties (1.1) are still satisfied. The mapping $v_p : \mathbb{Q} \to \mathbb{Z} \cup \{\infty\}$ is the *p-adic valuation* of $\mathbb{Q}$.

We say that $r = a/a'$ is *p-integral* if $v_p(r) \geq 0$. Clearly every integer is $p$-integral (for every prime $p$). The set $\mathbb{Z}_p$ of $p$-integral rational numbers is a subring of the field $\mathbb{Q}$ of rational numbers. Explicitly, $r \in \mathbb{Z}_p$ if and only if $r = 0$ or $r = p^k a/b$ with $k \geq 0$, $b > 0$, $\gcd(a, b) = 1$, $p \nmid a$, $p \nmid b$.

If $r, s \in \mathbb{Q}$, $e \geq 1$, we write $r \equiv s \pmod{p^e}$ when $v_p(r - s) \geq e$; we also say that $p^e$ divides $r - s$ (with respect to $\mathbb{Z}_p$). This relation of congruence satisfies the ordinary properties of the congruence of integers modulo a natural number.

It is also clear that a rational number $r$ is in $\mathbb{Z}$ if and only if $r$ is $p$- integral for every prime $p$.

The following two results have numerous applications. For any real numbers $x$ let $[x]$ denote the unique integer such that $[x] \leq x < [x] + 1$. $[x]$ is called the *integral part* of $x$.

In 1808, Legendre determined the exact power $p^m$ of the prime $p$ that divides a factorial $a!$ (so $p^{m+1}$ does not divide $a!$). There is a very nice expression of $m$ in terms of the $p$-adic development of $a$:

$$a = a_k p^k + a_{k-1} p^{k-1} + \cdots + a_1 p + a_0,$$

where $p^k \leq a < p^{k+1}$ and $0 \leq a_i \leq p - 1$ (for $i = 0, 1, \ldots, k$). The integers $a_0, a_1, \ldots, a_k$ are the digits of $a$ in base $p$.

For example, in base 5, we have $328 = 2 \times 5^3 + 3 \times 5^2 + 3$, so the digits of 328 in base 5 are 2, 3, 0, 3. Using the above notation:

**(1A)**    *If $a \geq 1$ then $v_p(a!) = m$ where*

$$m = \sum_{i=1}^{\infty} \left[ \frac{a}{p^i} \right] = \frac{a - (a_0 + a_1 + \cdots + a_k)}{p - 1}.$$

PROOF. By definition $a! = p^m b$, where $p \nmid b$. Let $a = q_1 p + r_1$ with $0 \leq q_1$, $0 \leq r_1 < p$; so $q_1 = [a/p]$. The multiples of $p$, not bigger than $a$, are $p, 2p, \ldots, q_1 p \leq a$. So $p^{q_1}(q_1!) = p^m b'$, where $p \nmid b'$. Thus $q_1 + m_1 = m$, where $p^{m_1}$ is the exact power of $p$ which divides $q_1!$.

Since $q_1 < a$, by induction,

$$m_1 = \left[\frac{q_1}{p}\right] + \left[\frac{q_1}{p^2}\right] + \left[\frac{q_1}{p^3}\right] + \cdots \; ;$$

but

$$\left[\frac{q_1}{p^i}\right] = \left[\frac{[a/p]}{p^i}\right] = \left[\frac{a}{p^{i+1}}\right],$$

as may be easily verified. So

$$m = \left[\frac{a}{p}\right] + \left[\frac{a}{p^2}\right] + \left[\frac{a}{p^3}\right] + \cdots .$$

Now we derive the second expression, involving the *p*-adic digits of $a = a_k p^k + \cdots + a_1 p + a_0$. Then

$$\left[\frac{a}{p}\right] = a_k p^{k-1} + \cdots + a_1,$$

$$\left[\frac{a}{p^2}\right] = a_k p^{k-2} + \cdots + a_2,$$

$$\vdots$$

$$\left[\frac{a}{p^k}\right] = a_k.$$

So

$$\sum_{i=0}^{\infty} \left[\frac{a}{p^i}\right] = a_1 + a_2(p + 1) + a_3(p^2 + p + 1) + \cdots$$

$$+ a_k(p^{k-1} + p^{k-2} + \cdots + p + 1)$$

$$= \frac{1}{p-1}\{a_1(p - 1) + a_2(p^2 - 1) + \cdots + a_k(p^k - 1)\}$$

$$= \frac{1}{p-1}\{a - (a_0 + a_1 + \cdots + a_k)\}. \quad \square$$

In 1852, Kummer used Legendre's result to determine the exact power $p^m$ of $p$ dividing a binomial coefficient

$$\binom{a+b}{a} = \frac{(a+b)!}{a!\, b!},$$

where $a \geq 1$, $b \geq 1$.

**(1B)**    *The exact power of $p$ dividing $\binom{a+b}{a}$ is equal to $\varepsilon_0 + \varepsilon_1 + \cdots + \varepsilon_t$, which is the number of "carry-overs" when performing the addition of $a, b$ written in base $p$.*

PROOF. Let

$$a = a_0 + a_1 p + \cdots + a_t p^t,$$
$$b = b_0 + b_1 p + \cdots + b_t p^t,$$

where $0 \le a_i \le p - 1$, $0 \le b_i \le p - 1$, and either $a_t \ne 0$ or $b_t \ne 0$. Let $S_a = \sum_{i=0}^{t} a_i$, $S_b = \sum_{i=0}^{t} b_i$ be the sums of the $p$-adic digits of $a, b$. Let $c_i$, $0 \le c_i \le p - 1$, and $\varepsilon_i = 0$ or $1$, be defined successively as follows:

$$a_0 + b_0 = \varepsilon_0 p + c_0,$$
$$\varepsilon_0 + a_1 + b_1 = \varepsilon_1 p + c_1,$$
$$\varepsilon_1 + a_2 + b_2 = \varepsilon_2 p + c_2,$$
$$\vdots$$
$$\varepsilon_{t-1} + a_t + b_t = \varepsilon_t p + c_t.$$

Multiplying these equations successively by $1, p, p^2, \ldots$ and adding them,

$$a + b + \varepsilon_0 p + \varepsilon_1 p^2 + \cdots + \varepsilon_{t-1} p^t = \varepsilon_0 p + \varepsilon_1 p^2 + \cdots + \varepsilon_t p^{t+1}$$
$$+ c_0 + c_1 p + \cdots + c_t p^t.$$

So, $a + b = c_0 + c_1 p + \cdots + c_t p^t + \varepsilon_t p^{t-1}$, and this is the expression of $a + b$ in the base $p$. Similarly, by adding those equations,

$$S_a + S_b + (\varepsilon_0 + \varepsilon_1 + \cdots + \varepsilon_{t-1}) = (\varepsilon_0 + \varepsilon_1 + \cdots + \varepsilon_t) p + S_{a+b} - \varepsilon_t.$$

By Legendre's result,

$$(p - 1)m = (a + b) - S_{a+b} - a + S_a - b + S_b$$
$$= (p - 1)(\varepsilon_0 + \varepsilon_1 + \cdots + \varepsilon_t).$$

Hence, the result of Kummer.  □

This theorem of Kummer was rediscovered by Lucas in 1878. In 1991, Frasnay extended the result replacing integers by $p$-adic integers.[1]

---

[1] This result is apparently still unpublished; a preprint was given to the author.

The results of Legendre and Kummer have found many applications in the so-called *p-adic Analysis*.

## Bibliography

1808 Legendre, A.M., *Essai sur la Théorie des Nombres* ($2^e$ édition), Courcier, Paris, 1808.

1852 Kummer, E.E., *Über die Ergänzungssätze zu den allgemeinen Reziprozitätsgesetzen*, J. Reine Angew. Math., **44** (1852), 93–146; reprinted in *Collected Papers*, Vol. I (editor A. Weil), pp. 485–538, Springer-Verlag, New York, 1975.

1878 Lucas, E., *Sur les congruences des nombres eulériens et des coefficients différentiels*, Bull. Soc. Math. France, **6** (1878), 49–54.

## II.2. Cyclotomic Polynomials

Let $n \geq 1$ and $\zeta_n = \cos(2\pi/n) + i\sin(2\pi/n)$. So $\zeta_1 = 1$, $\zeta_2 = -1$, $\zeta_3 = (-1 + i\sqrt{3})/2$, $\zeta_4 = i$, $\zeta_5 = \cos 72° + i\sin 72°$, $\zeta_6 = (1 + i\sqrt{3})/2$, etc. . . .

All the powers of $\zeta_n$ are also $n$th roots of 1, so they are roots of the polynomial $X^n - 1$. If $\omega$ is any $n$th root of 1, the smallest $d \geq 1$ such that $\omega^d = 1$ is called the order of the root of unity $\omega$ and necessarily $d$ divides $n$; we say then that $\omega$ is a primitive $d$th root of $q$.

The powers $\zeta_n^j$ for $j = 1, 2, \ldots, n$ are all distinct with $\zeta_n^n = 1$; so $\zeta_n$ is a primitive root of unity of order $n$. Since there are $n$ $n$th roots of 1, every $n$th-root of 1 is a power $\zeta_n^j$. Moreover, as is easily seen, $\zeta_n^j$ is a primitive $n$th root of 1 if and only if $\gcd(j, n) = 1$. Thus the number of primitive $n$th roots of 1 is equal to $\varphi(n)$, where $\varphi(n)$ denotes the *totient of $n$* and $\varphi$ is Euler's function.

The $n$th *cyclotomic polynomial* is

$$(2.1) \qquad \Phi_n(X) = \prod_{\gcd(j,n)=1} (X - \zeta_n^j)$$

(product for all $j$, $1 \leq j < n$, $\gcd(j, n) = 1$). It is a monic polynomial of degree $\varphi(n)$. Since the polynomial remains invariant by the permutation of its roots, from Galois theory the coefficients are in $\mathbb{Z}$.

By grouping the $n$th roots of 1 according to their order, we obtain

(2.2) $$X^n - 1 = \prod_{d|n} \Phi_d(X).$$

From this, or directly, we obtain

(2.3) $\quad \Phi_p(X) = X^{p-1} + X^{p-2} + \cdots + X + 1,$

(2.4) $\quad \Phi_{p^e}(X) = \dfrac{X^{p^e} - 1}{X^{p^{e-1}} - 1}$

$$= X^{p^{e-1}(p-1)} + X^{p^{e-1}(p-2)} + \cdots + X^{p^{e-1}} + 1,$$

for any prime $p$ and $e \geq 1$.

If $m \mid n$ and $m \neq n$ we have therefore

(2.5) $$X^n - 1 = (X^m - 1)\Phi_n(X) \prod_d \Phi_d(X)$$

(product for all $d$, $1 \leq d < n$, $d \mid n$, $d \nmid m$). Let $\mu$ be the *Möbius function*

(2.6) $\quad \mu(p_1 \cdots p_r) = \begin{cases} 0 \text{ if the primes } p_i \text{ are not distinct,} \\ (-1)^r \text{ otherwise.} \end{cases}$

Then

(2.7) $$\Phi_n(X) = \prod_{d|n} (X^{n/d} - 1)^{\mu(d)}.$$

We note the following properties: If $p$ is a prime and $p$ divides $m$ then

(2.8) $$\Phi_{pm}(X) = \Phi_m(X^p) \quad (\text{when } p \mid m).$$

If $p$ does not divide $m$ and $s \geq 1$ then

(2.9) $\quad \Phi_{p^s m}(X) = \dfrac{\Phi_m(X^{p^s})}{\Phi_m(X^{p^{s-1}})} \quad (\text{when } p \nmid m, \ s \geq 1).$

We now consider the corresponding homogenized polynomials in two indeterminates. Let

(2.10) $$\Phi_n(X, Y) = Y^{\varphi(n)} \Phi_n\left(\frac{X}{Y}\right),$$

then

(2.11) $$X^n - Y^n = \prod_{d|n} \Phi_d(X, Y).$$

If $m \mid n$ and $m \neq n$ then

(2.12) $$X^n - Y^n = (X^m - Y^m)\Phi_n(X, Y) \prod_d \Phi_d(X, Y)$$

(product for all $d$, $1 \leq d < n$, $d \mid n$, $d \nmid m$). We have also

(2.13) $$\Phi_n(X, Y) = \prod_{d|n} \left(X^{n/d} - Y^{n/d}\right)^{\mu(d)},$$

and, as before,

(2.14) $$\Phi_{pm}(X, Y) = \Phi_m(X^p, Y^p) \quad \text{(when } p \mid m),$$

(2.15) $$\Phi_{p^s m}(X, Y) = \frac{\Phi_m\left(X^{p^s}, Y^{p^s}\right)}{\Phi_m\left(X^{p^{s-1}}, Y^{p^{s-1}}\right)} \quad \text{(when } p \nmid m, \ s \geq 1).$$

## II.3. Factors of Binomials

Let $a, b$ be nonzero distinct integers. In this section we consider binomials $a^n \pm b^n$ as well as the integers $(a^n - b^n)/(a - b)$ and we discuss their factors.

**(3A)**    *Let $a, b$ be nonzero distinct integers.*
   (1) *If $p \neq 2$, $p \nmid ab$, and $v_p(a - b) = e \geq 1$, then $v_p\left(a^{p^r} - b^{p^r}\right) = e + r$ for every $r \geq 1$.*
   (2) *If $2 \nmid ab$ and $v_2(a - b) = e \geq 2$, then $v_2\left(a^{2^r} - b^{2^r}\right) = e + r$ for every $r \geq 1$.*
   (3) *If $p$ is any prime and $p \mid a^p - b^p$, then $p^2 \mid a^p - b^p$.*

PROOF. (1)    It suffices to show that $v_p(a^p - b^p) = e + 1$, and then repeat the argument. By hypothesis, $a = b + kp^e$, where $p \nmid k$. Then

$$a^p = b^p + \binom{p}{1}b^{p-1}kp^e + \binom{p}{2}b^{p-2}k^2p^{2e} + \cdots + k^p p^{pe}.$$

Since $p$ divides $\binom{p}{j}$ for $j = 1, \dots, p - 1$, then

$$v_p \left[ \binom{p}{j} b^{p-j} k^j p^{je} \right] \geq 1 + je.$$

From $v_p(k^p p^{pe}) = pe$, it follows that $v_p(a^p - b^p) = e + 1$.

(2)    As in (1), it suffices to show that $v_2(a^2 - b^2) = e + 1$. By hypothesis, $a = b + 2^e k$ with $e \geq 2$ and $a, b, k$ odd. Then $a^2 = b^2 + 2^{e+1} k + 2^{2e} k^2$; since $e + 1 < 2e$, then $v_2(a^2 - b^2) = e + 1$.

(3)    By hypothesis, $a \equiv a^p \equiv b^p \equiv b \pmod{p}$; raising to the $p$th power, $a^p \equiv b^p \pmod{p^2}$, so $p^2 \mid a^p - b^p$.    □

If $n \geq 1$ and $a, b$ are distinct nonzero integers, let

(3.1)    $$Q_n(a, b) = \frac{a^n - b^n}{a - b} = \sum_{k=0}^{n-1} a^k b^{n-1-k}.$$

By convention we define $Q_0(a, b) = 0$. We note the following expression for $Q_n(a, b)$ $(n \geq 1)$:

(3.2)    $$Q_n(a, b) = \frac{[(a - b) + b]^n - b^n}{a - b}$$

$$= (a - b)^{n-1} + \binom{n}{1}(a - b)^{n-2} b + \cdots$$

$$+ \binom{n}{n-2}(a - b)b^{n-2} + nb^{n-1}$$

$$= (a - b)e + nb^{n-1},$$

where $e \in \mathbb{Z}$. Also, if $n = p$ is a prime number, then

(3.3)    $$Q_p(a, b) = (a - b)^{p-1} + pf,$$

where $f \in \mathbb{Z}$.

We shall now indicate some properties of the integers $Q_n(a, b)$. Jacquemet (before 1729) proved (5) below when $n = p$ is an odd prime. Euler proved in 1738: if $p$ is an odd prime, $a > 1$, then $\gcd(Q_p(a, \pm 1), a \pm 1) = 1$ or $p$; if $p$ divides $a \pm 1$ then $v_p(Q_p(a, \pm 1)) = 1$; moreover, if $p \neq 3$, $a \neq 2$, then $Q_p(a, \pm 1)$ is odd and greater than 1.

In 1769 Lagrange proved (4) below, as well as (6) when $n = p$ is a prime number. In 1837, Kummer proved (4) when $n = p$ is a prime number. In 1888, Sylvester proved a special case of (3). In 1897, F.

Lucas proved special cases of (5) and (6). More recently, properties (4), (5), (6), and (7) have been established by Inkeri (1946) and Vivanti (1947). In the present form, most of the proposition is in Möller's paper (1955).

**(3B)**    *Let $n > m \geq 1$, and let $a, b$ be nonzero distinct integers.*
*Then*

   (1) *If $n = mq + r$ with $r \geq 0$ then*

$$Q_n(a, b) = a^r Q_q(a^m, b^m) Q_m(a, b) + b^{mq} Q_r(a, b).$$

   *If $n = mq - r$ with $r \geq 0$ then*

$$Q_n(a, b) = \left[a^{m-r} Q_{q-1}(a^m, b^m) + b^{n-m}\right] Q_m(a, b)$$
$$- a^{m-r} b^{n-m} Q_r(a, b).$$

*Assuming $a, b$ relatively prime, we have:*

   (2) *If $d = \gcd(n, m)$ then $Q_d(a, b) = \gcd\left(Q_n(a, b), Q_m(a, b)\right)$.*
   (3) *$\prod_{p|n} Q_p(a, b)$ divides $Q_n(a, b)$.*
   (4) *$\gcd(Q_n(a, b), a - b) = \gcd(n, a - b)$.*
   (5) *If $p \mid a - b$, $p \nmid n$ then $p \nmid Q_n(a, b)$.*
   (6) *If $p$ is an odd prime dividing $a - b$ then $v_p\left(Q_n(a, b)\right) = v_p(n)$.*
   (7) *If $4 \mid a - b$ then $v_2\left(Q_n(a, b)\right) = v_2(n)$.*
   *If $2 \mid a - b$ but $4 \nmid a - b$ then $v_2\left(Q_n(a, b)\right) \geq v_2(n)$.*
   (8) *If $n$ is odd then $Q_n(a, b)$ is odd.*
   (9) *If $n$ is odd and $e > 0$ then $\gcd(Q_n(a, b), a^{2^e n} + b^{2^e n}) = 1$.*
  (10) *If every prime factor of $n$ divides $a - b$ then $n(a - b)$ divides $a^n - b^n$.*

PROOF. (1)    Let $n = mq + r$. Then

$$Q_n(a, b) = \frac{a^{mq+r} - b^{mq+r}}{a - b}$$
$$= \frac{a^{mq+r} - a^r b^{mq} + a^r b^{mq} - b^{mq+r}}{a - b}$$
$$= a^r \times \frac{a^{mq} - b^{mq}}{a - b} + \frac{a^r - b^r}{a - b} \times b^{mq}$$
$$= a^r \times \frac{a^{mq} - b^{mq}}{a^m - b^m} \times \frac{a^m - b^m}{a - b} + \frac{a^r - b^r}{a - b} \times b^{mq}$$
$$= a^r Q_q(a^m, b^m) Q_m(a, b) + Q_r(a, b) b^{mq}.$$

Now let $n = mq - r$. Then

$$\left[a^{m-r}Q_{q-1}(a^m, b^m) + b^{n-m}\right]Q_m(a,b) - a^{m-r}b^{n-m}Q_r(a,b)$$

$$= a^{m-r}\frac{a^{(q-1)m} - b^{(q-1)m}}{a^m - b^m} \times \frac{a^m - b^m}{a - b} + b^{n-m} \times \frac{a^m - b^m}{a - b}$$

$$\quad - a^{m-r}b^{n-m}\frac{a^r - b^r}{a - b}$$

$$= \frac{a^n - a^{m-r}b^{(q-1)m} + a^m b^{n-m} - b^n - a^m b^{n-m} + a^{m-r}b^{n-m+r}}{a - b}$$

$$= \frac{a^n - b^n}{a - b} = Q_n(a,b).$$

(2)  Since $a, b$ are relatively prime then for every $k \geq 1$, $a$ and $b$ are also relatively prime to $Q_k(a,b) = a^{k-1} + a^{k-2}b + \cdots + ab^{k-2} + b^{k-1}$. By (1), $Q_d(a,b)$ divides $Q_n(a,b)$ and $Q_m(a,b)$. Let $r, s$ be positive integers such that $d = sm - rn$ (or $d = rn - sm$). So $sm = rn + d$ and by (1),

$$Q_s(a^m, b^m)Q_m(a,b) = Q_{sm}(a,b) = a^d Q_r(a^n, b^n)Q_n(a,b) + b^{rn}Q_d(a,b).$$

If $t$ divides $Q_m(a,b)$ and $Q_n(a,b)$ then $t$ divides $b^{rn}Q_d(a,b)$; but $Q_m(a,b)$ and $b$ are relatively prime, hence $t \mid Q_d(a,b)$, showing the statement.

(3)    By (2), the integers $Q_p(a,b)$ (for primes $p$ dividing $n$) are pairwise relatively prime. By (1) if $p|n$ then $Q_p(a,b)$ divides $Q_n(a,b)$. Hence $\prod_{p|n} Q_p(a,b)$ divides $Q_n(a,b)$.

(4) Since $\gcd(a,b) = 1$, it follows from (3.2) that $\gcd(Q_n(a,b), a - b) = \gcd(n, a - b)$.

(5)    This is an obvious consequence of (4).

(6)    Let $n = p^r m$, $p \nmid m$, $r \geq 0$ so $v_p(n) = r$. Let $a_1 = a^m$, $b_1 = b^m$. Since $p \mid a - b$, by (5) $p$ does not divide $Q_m(a,b) = (a^m - b^m)/(a - b) = (a_1 - b_1)/(a - b)$, so $v_p(a_1 - b_1) = v_p(a - b) \geq 1$. Hence $v_p\left(a_1^{p^r} - b_1^{p^r}\right) = v_p(a_1 - b_1) + r$. Thus

$$v_p\left(Q_n(a,b)\right) = v_p\left(\frac{a_1^{p^r} - b_1^{p^r}}{a_1 - b_1}\right) + v_p\left(\frac{a_1 - b_1}{a - b}\right) = r = v_p(n).$$

(7)    Let $n = 2^r m$, $2 \nmid m$, $r \geq 0$, so $v_2(n) = r$. Let $a_1 = a^m$, $b_1 = b^m$. As in (6), $v_2(a_1 - b_1) = v_2(a - b) = e \geq 1$. If $e \geq 2$ then

$v_2 \left( a_1^{2^r} - b_1^{2^r} \right) = e + r$, so

$$v_2 \left( Q_n(a, b) \right) = v_2 \left( \frac{a_1^{2^r} - b_1^{2^r}}{a_1 - b_1} \right) + v_2 \left( \frac{a_1 - b_1}{a - b} \right) = r = v_2(n).$$

However, if $e = 1$ then $v_2 \left( a_1^{2^r} - b_1^{2^r} \right) \geq r + 1$ (and it may be greater than $r + 1$), so we can only conclude that $v_2 \left( Q_n(a, b) \right) \geq v_2(n)$.

(8)    If $a \not\equiv b \pmod 2$ then $a^n \not\equiv b^n \pmod 2$ so $Q_n(a, b)$ is odd. If $a \equiv b \pmod 2$ since $a, b$ are relatively prime, they are odd. It follows that $Q_n(a, b) = a^{n-1} + a^{n-2}b + \cdots + ab^{n-2} + b^{n-1}$ is the sum of an odd number of odd summands, so it is odd.

(9)    Let $p$ be a prime, $r \geq 1$, and $p^r \mid Q_n(a, b)$, $p^r \mid a^{2^e n} + b^{2^e n}$. By (7), $p \neq 2$. Since $p^r \mid a^n - b^n$ then $a^{2^e n} \equiv b^{2^e n} \pmod{p^r}$ so $p^r \mid 2a^{2^e n}$. Therefore $p$ divides $a$, hence also $b$, which is not possible.

(10)    Let $p$ be any prime factor of $n$; by hypothesis $p \mid a - b$, hence by (6) or (7), $v_p(n) \leq v_p \left( Q_n(a, b) \right)$, therefore $v_p(n(a - b)) \leq v_p(Q_n(a, b)(a - b)) = v_p(a^n - b^n)$. Since $p$ is arbitrary, this shows that $n(a - b)$ divides $a^n - b^n$. $\square$

We indicate now the following complementary result proved by Inkeri in 1946:

**(3C)**    Let $p$ be an odd prime, $n \geq 1$, and let $a, b$ be nonzero relatively prime integers such that $a \neq b$. Then:

(1) $Q_{p^n}(a, b) = \prod_{m=1}^{n} Q_p(a^{p^{m-1}}, b^{p^{m-1}})$.
(2) If $p$ does not divide $a^{p^n} - b^{p^n}$ then the integers $a - b$, $Q_p(a, b)$, $Q_p(a^p, b^p), \ldots, Q_p(a^{p^{n-1}}, b^{p^{n-1}})$ are pairwise relatively prime.
(3) If $p \mid a^{p^n} - b^{p^n}$, if $i, j$ are integers such that $1 \leq i < j \leq n$ then

$$\gcd \left( Q_p \left( a^{p^{i-1}}, b^{p^{i-1}} \right), Q_p \left( a^{p^{j-1}}, b^{p^{j-1}} \right) \right) = p,$$

$$\gcd \left( a - b, Q_p \left( a^{p^{i-1}}, b^{p^{i-1}} \right) \right) = p.$$

(4) If $v_p \left( a^{p^n} - b^{p^n} \right) = e \geq 1$ then $e \geq n + 1$ and $v_p(a - b) = e - n$.

PROOF. (1)

$$Q_{p^n}(a, b) = \prod_{m=1}^{n} \frac{a^{p^m} - b^{p^m}}{a^{p^{m-1}} - b^{p^{m-1}}} = \prod_{m=1}^{n} Q_p(a^{p^{m-1}}, b^{p^{m-1}}).$$

(2)   We have $a^{p^n} - b^{p^n} = q_0 q_1 \cdots q_n$ where $q_0 = a - b$,

$$q_j = \frac{a^{p^j} - b^{p^j}}{a^{p^{j-1}} - b^{p^{j-1}}} = Q_p \left( a^{p^{j-1}}, b^{p^{j-1}} \right).$$

By hypothesis, $p \nmid q_j$ (for $j = 0, 1, \ldots, n$). If $l$ is a prime and $l$ divides $q_i, q_j$ (with $0 \le i < j \le n$), then $l \ne p$ and $l \mid a^{p^i} - b^{p^i}$. Since

$$Q_{p^{j-i}} \left( a^{p^i}, b^{p^i} \right) = \frac{a^{p^j} - b^{p^j}}{a^{p^i} - b^{p^i}} = q_{i+1} q_{i+2} \cdots q_j,$$

then $l$ divides $Q_{p^{j-1}}(a^{p^i}, b^{p^i})$. By (3B)(4), we have $l = p$, a contradiction.

(3)   Assume that $l$ is a prime, $e \ge 1$, and $l^e$ divides $q_i$ and $q_j$ $(0 \le i < j \le n)$. Then $l^e$ divides $a^{p^i} - b^{p^i}$ and also $a^{p^{j-1}} - b^{p^{j-1}} = (a^{p^i} - b^{p^i}) q_{i+1} \cdots q_{j-1}$. Since $l^e$ divides $Q_p(a^{p^{j-1}}, b^{p^{j-1}}) = q_j$, by (3B)(4) $l^e$ divides $p$, so $l^e = p$.

(4)   If $p$ divides $a^{p^n} - b^{p^n}$ then $p \nmid a$, $p \nmid b$. From

$$a^{p^{n-1}} \equiv \left( a^{p^{n-1}} \right)^p \equiv \left( b^{p^{n-1}} \right)^p \equiv b^{p^{n-1}} \pmod{p},$$

it follows that $p \mid a^{p^{n-1}} - b^{p^{n-1}}$. By (3B)(6), $v_p(Q_p(a^{p^{n-1}}, b^{p^{n-1}})) = 1$, hence $v_p(a^{p^{n-1}} - b^{p^{n-1}}) = e - 1$, with $e - 1 \ge 1$. Repeating this argument, $v_p(a - b) = e - n$ with $e \ge n + 1$.   $\square$

We shall need later the following estimates:

**(3D)**   *Let $n$ be odd, $n \ge 3$, and let $a, b$ be nonzero distinct integers such that $a + b \ge 1$. Then $Q_n(a, b) \ge n$. The equality holds exactly when $n = 3$, $a = 2$, $b = -1$ or $n = 3$, $a = -1$, $b = 2$.*

PROOF. Since $Q_n(a, b) = Q_n(b, a)$ and $a \ne b$, we may assume without loss of generality that $a > b$. If $b \ge 1$ then

$$Q_n(a, b) = a^{n-1} + a^{n-2}b + \cdots + ab^{n-2} + b^{n-1} \ge 1 + 1 + \cdots + 1 = n.$$

If $b \leq -1$ then $a \geq 1 - b \geq 2$, $b^2 \geq 1$ and

$$
\begin{aligned}
Q_n(a, b) &= (a + b)(a^{n-2} + a^{n-4}b^2 + \cdots + ab^{n-3}) + b^{n-1} \\
&\geq 2^{n-2} + 2^{n-4} + \cdots + 2 + 1 \\
&= 2\left(4^{(n-3)/2} + 4^{(n-5)/2} + \cdots + 1\right) + 1 \\
&= 2 \cdot \frac{4^{(n-1)/2} - 1}{4 - 1} + 1 = \tfrac{2}{3}\left(2^{n-1} - 1\right) + 1 \\
&= \frac{2^n + 1}{3} \geq n,
\end{aligned}
$$

when $n \geq 3$. If we have $Q_n(a, b) = n$ then we must have equalities all through, and this requires that $n = 3$, $a = 2$, $b = -1$. By symmetry, we may have $Q_n(a, b) = n$ also when $n = 3$, $a = -1$, $b = 2$.  □

We note explicitly that if $p$ is an odd prime, if $a > b \geq 1$ then $Q_p(a, b) > p$. This fact is a special case of the above result, but may also be seen directly, since $Q_p(a, b) = a^{p-1} + a^{p-2}b + \cdots + ab^{p-2} + b^{p-1} > p$.

Before proceeding, we note: If $1 \leq b < a$, by (2.13), $\Phi_n(a, b) > 0$. If $n$ is odd, $n \geq 3$, and $a, b$ are not both even, then $\Phi_n(a, b)$ is odd. Also, if $n$ is odd, $n \geq 3$, and $a, b$ are not both even, then $\Phi_{2n}(a, b) = \Phi_n(-a, b)$ is odd.

Let $a, b$ be nonzero distinct relatively prime integers, let $n \geq 1$. We say that the prime $p$ is a *primitive factor* of $a^n - b^n$ (respectively, $a^n + b^n$) if $p$ divides $a^n - b^n$ (respectively, $a^n + b^n$) but $p$ does not divide $a^m - b^m$ (respectively, $a^m + b^m$) for every $m$, $1 \leq m < n$. We first note that if 2 is a primitive factor of $a^n \pm b^n$ then $n = 1$, because $a \equiv b \pmod 2$.

Next we observe:

**(3E)**    *If $n \geq 2$ then $p$ is a primitive factor of $a^n + b^n$ if and only if $p$ is a primitive factor of $a^{2n} - b^{2n}$.*

PROOF. If $p$ is a primitive factor of $a^{2n} - b^{2n}$ then $p \nmid a^n - b^n$ so $p \mid a^n + b^n$; moreover if $p \mid a^k + b^k$ where $1 \leq k \leq n$ then $p \mid a^{2k} - b^{2k}$, so $2k = 2n$, showing that $p$ is a primitive factor of $a^n + b^n$.

Conversely, if $p$ is a primitive factor of $a^n + b^n$ (with $n \geq 2$) then $p \neq 2$ and $p \mid a^{2n} - b^{2n}$. If $1 \leq k < 2n$ and $p \mid a^k - b^k$, let $k = 2^e m$, with $e \geq 1$ and $m$ odd. Then

$$a^k - b^k = \left(a^{2^{e-1}m} - b^{2^{e-1}m}\right)\left(a^{2^{e-1}m} + b^{2^{e-1}m}\right).$$

Since $2^{e-1}m < n$, by hypothesis $p \nmid a^{2^{e-1}m} + b^{2^{e-1}m}$, so $p \mid a^{2^{e-1}m} - b^{2^{e-1}m}$. Repeating this argument, $p$ divides $a^m - b^m$, where $m$ is odd. Thus $a^m \equiv b^m \pmod{p}$ and also $a^n \equiv -b^n \pmod{p}$. Then $a^{mn} \equiv b^{mn} \pmod{p}$, $a^{nm} \equiv -b^{nm} \pmod{p}$ hence $p \mid b$ (because $p \neq 2$), so $p \mid a$, contrary to the hypothesis. Thus $p$ is a primitive factor of $a^{2n} - b^{2n}$. □

The first result is very easy to prove!

**(3F)**    Let $a, b$ be nonzero relatively prime integers, let $n \geq 1$, and let $p$ be a prime. The following statements are equivalent:

(1) $p$ is a primitive factor of $a^n - b^n$.
(2) $p \mid a^n - b^n$, but if $1 \leq m < n$, $m \mid n$, then $p \nmid a^m - b^m$.
(3) $p \mid \Phi_n(a, b)$, but if $1 \leq m < n$, then $p \nmid \Phi_m(a, b)$.
(4) $p \mid \Phi_n(a, b)$, but if $1 \leq m < n$, $m \mid n$, then $p \nmid \Phi_m(a, b)$.
(5) $p \nmid b$ and if $b'$ is such that $bb' \equiv 1 \pmod{p}$ then the order of $ab'$ modulo $p$ is equal to $n$.

PROOF. The implications $(1) \to (2)$ and $(3) \to (4)$ are trivial. The equivalences $(1) \leftrightarrow (3)$ and $(2) \leftrightarrow (4)$ follow at once from the expression $a^n - b^n = \prod_{d|n} \Phi_d(a, b)$.

Noting that $(ab')^d \equiv 1 \pmod{p}$ holds if and only if $p \mid a^d - b^d$, then (5) is obviously equivalent to (1) and to (2). □

Let $p$ be a prime not dividing $n$. Then the congruence $X^n - 1 \equiv 0 \pmod{p}$ does not have double roots, since the derivative $nX^{n-1}$ has only the root 0 modulo $p$. It follows from (2.5) that if $a$ is an integer, $a \neq 1$, and $p \nmid a$ if $\Phi_{p-1}(a) \equiv D \pmod{p}$ then $\Phi_m(a) \not\equiv 0 \pmod{p}$ for every $m$, $1 \leq m < p - 1$, $m$ dividing $p - 1$.

From this observation and from (3F) we deduce: If $a \neq 1$ and $p$ does not divide $a$, then $a$ is a primitive root modulo $p$ (i.e., the order of $a$ modulo $p$ is $p - 1$) if and only if $\Phi_{p-1}(a) \equiv 0 \pmod{p}$.

The following proposition appears in the paper of Birkhoff and Vandiver (1904) and once more in Inkeri's paper (1946) for the case where $n$ is an odd prime. The inclusion $E_1 \subseteq E_2$ was first shown by Legendre (1830).

**(3G)**    *Let $n \geq 2$. With the preceding hypothesis, the following sets of primes coincide:*

    $E_1$: *the set of primitive factors of $a^n - b^n$;*

    $E_2$: *the set of primes $p$ such that $p \equiv 1 \pmod{n}$ and $p \mid \Phi_n(a,b)$; and*

    $E_3$: *the set of primes $p$ such that $p \nmid n$ and $p \mid \Phi_n(a,b)$.*

PROOF. We show that $E_1 \subseteq E_2$. By hypothesis, $p \mid a^n - b^n$ but $p \nmid a^m - b^m$ for all $m$, $1 \leq m < n$. By (3F), if $bb' \equiv 1 \pmod{p}$ then the order of $ab'$ modulo $p$ is $n$. Hence $n \mid p - 1$. Moreover, if $1 \leq m < n$, $m \mid n$, then $\Phi_m(a,b)$ divides $a^m - b^m$; since $p \nmid a^m - b^m$ then $p \nmid \Phi_m(a,b)$. But $p$ divides $a^n - b^n = \prod_{m \mid n} \Phi_m(a,b)$, hence $p$ divides $\Phi_n(a,b)$.

Obviously $E_2 \subseteq E_3$, because if $n \mid p - 1$ then $n < p$, hence $p \nmid n$.

Now we show that $E_3 \subseteq E_1$. Clearly $p \mid a^n - b^n$. Suppose $p$ is not a primitive factor of $a^n - b^n$. Then there exists $m$, $1 \leq m < n$, $m \mid n$, such that $p \mid a^m - b^m$. From

$$a^n - b^n = \Phi_n(a,b)(a^m - b^m) \prod_{d \neq n,\, d \mid n,\, d \nmid m} \Phi_d(a,b),$$

by hypothesis $p$ divides $(a^n - b^n)/(a^m - b^m)$. We write $n = md$, $a^m = a_1$, $b^m = b_1$, hence

$$\frac{a^n - b^n}{a^m - b^m} = \frac{a_1^d - b_1^d}{a_1 - b_1} = \sum_{i=0}^{d-1} a_1^{d-1-i} b_1^i \equiv d a_1^{d-1} \pmod{p}.$$

But $p \mid a_1 - b_1$, so $p \nmid a_1$, hence $p \mid d$, so $p \mid n$, which is a contradiction, concluding the proof.  $\square$

Next we prove:

**(3H)**    *Let $a, b$ be nonzero distinct relatively prime integers, and $n \geq 1$. Let $p$ be a primitive factor of $a^d - b^d$, let $v_p(a^d - b^d) = r \geq 1$ and assume that $r \geq 2$ if $p = 2$. Then*

    (1) $v_p(\Phi_d(a,b)) = r$;

    (2) *if $t \geq 1$ then $v_p(\Phi_{dp^t}(a,b)) = 1$; and*

    (3) *if $t \geq 0$, $k > 1$, $p \nmid k$ then $v_p(\Phi_{kdp^t}(a,b)) = 0$.*

PROOF. (1)  By (3F), $p \nmid \Phi_l(a,b)$ for all $l$, $1 \leq l < d$. It follows that $v_p(\Phi_d(a,b)) = v_p(a^d - b^d)$.

(2)  We have, by (2.12),

$$a^{dp^t} - b^{dp^t} = \Phi_{dp^t}(a,b) \cdot \left(a^{dp^{t-1}} - b^{dp^{t-1}}\right) \prod_{e \mid d,\, 1 \le e < d} \Phi_{ep^t}(a,b).$$

We note that $p \nmid \Phi_{ep^t}(a,b)$ when $e < d$, otherwise $p \mid a^{ep^t} - b^{ep^t}$; since $p$ is a primitive factor of $a^d - b^d$ then by (3F)(5), $d$ divides $ep^t$; but $p \nmid d$ so $d \mid e$, a contradiction. But by (3B)(6) and (3B)(7),

$$v_p\left(a^{db^t} - b^{db^t}\right) = r + t,$$

$$v_p\left(a^{db^{t-1}} - b^{dp^{t-1}}\right) = r + t - 1.$$

Therefore $v_p(\Phi_{dp^t}(a,b)) = 1$.

(3)  We have

$$a^{dkp^t} - b^{dkp^t} = \Phi_{dkp^t}(a,b) \prod_{e \mid dkp^t,\, e \nmid dp^t,\, e < dkp^t} \Phi_e(a,b) \cdot \left(a^{dp^t} - b^{dp^t}\right).$$

By (3B)(6) and (3B)(7), $v_p(a^{dkp^t} - b^{dkp^t}) = r + t$, $v_p(a^{dp^t} - b^{dp^t}) = r + t$, hence $v_p(\Phi_{dkp^t}(a,b)) = 0$.  □

For every integer $n \ge 2$, let $P[n]$ denote the largest prime factor of $n$.

**(3I)**  Let $a > b \ge 1$, let $\gcd(a,b) = 1$, and let $n \ge 2$. Let $p$ be a primitive factor of $a^f - b^f$ such that $p \mid \Phi_n(a,b)$. Then:
   (1)  There exists $j \ge 0$ such that $n = fp^j$ with $p \nmid f$.
   (2)  If $j > 0$, then $p = P[n]$.
   (3)  If $j > 0$ and $p^2 \mid \Phi_n(a,b)$, then $n = p = 2$.
   (4)  $\gcd(\Phi_n(a,b), n) = 1$ or $P[n]$.

PROOF.  (1)  By (2.11), $\Phi_n(a,b)$ divides $a^n - b^n$; then $p \mid a^n - b^n$, hence $f \mid n$ by (3F). Since $p \mid a^{p-1} - b^{p-1}$, again $f \mid p - 1$, so $f < p$. Let $n = fp^j w$ with $j \ge 0$, $p \nmid fw$. Write $r = fp^j$. By (3.2),

$$\frac{a^n - b^n}{a^r - b^r} \equiv w b^{w-1} \pmod{a^r - b^r}.$$

Since $p \mid a^r - b^r$ (because $f \mid r$), then

$$\frac{a^n - b^n}{a^r - b^r} \equiv w b^{w-1} \pmod{p}.$$

If $n < m$ then by (2.12), $\Phi_n(a,b)$ divides $(a^n - b^n)/(a^r - b^r)$. Since $p \nmid b$, (because $\gcd(a,b) = 1$), then $p \mid w$, which is absurd. So $n = fp^j$.

(2)    From $f < p$, if $j > 0$, then $p = P[n]$.

(3)    Let $j > 1$ and $s = fp^{j-1}$, so $n = ps$. Then

$$\frac{a^n - b^n}{a^s - b^s} = \frac{[(a^s - b^s) + b^s]^p - b^{sp}}{a^s - b^s}$$

$$= pb^{s(p-1)} + \binom{p}{2}(a^s - b^s)b^{s(p-2)} + \binom{p}{3}(a^s - b^s)^2 b^{s(p-3)}$$

$$+ \cdots + (a^s - b^s)^{p-1}.$$

If $p \geq 3$, since $p \mid a^s - b^s$, then

$$\frac{a^n - b^n}{a^s - b^s} \equiv p \pmod{p^2}.$$

On the other hand, by (2.12), $\Phi_n(a,b)$ divides $(a^n - b^n)/(a^s - b^s)$, hence $p^2 \mid \Phi_n(a,b)$. Thus, if $p^2 \mid \Phi_n(a,b)$, then necessarily $p = 2$. So $f \leq p - 1$ implies $f = 1$ and $n = 2^j$, so $\Phi_n(a,b) \not\equiv 0 \pmod{r}$, which is absurd. This shows that $j = 1$ and $n = 2$.

(4)    Assume that there exists a prime $p$ dividing $\gcd(\Phi_n(a,b), n)$. By (1) and (2), $p = P[n]$. By (3), if $p^2 \mid \gcd(\Phi_n(a,b), n)$, then $n = p = 2$, so $p^2 \nmid n$. This shows the assertion. $\square$

The following very interesting theorem was proved by Bang (1886) in a particular case. In 1892, Zsigmondy proved the stronger version presented here. It was rediscovered by Birkhoff and Vandiver (1904) and by various other mathematicians, like Dickson (1905), Carmichael (1913), Kanold (1950), Artin (1955), Hering (1974), Lüneburg (1981) and maybe others.

**(3J)**    *Let $a > b \geq 1$, $\gcd(a,b) = 1$, $n \geq 1$.*

(1) *$a^n - b^n$ has a primitive factor, with the following exceptions:*
    (a) *$n = 1$, $a - b = 1$;*
    (b) *$n = 2$, $a + b$ a power of 2; and*
    (c) *$n = 6$, $a = 2$, $b = 1$.*

(2) *$a^n + b^n$ has a primitive factor, with the following exception:*
    *$n = 3$, $a = 2$, $b = 1$.*

PROOF. (1)   It is clear that in cases (a), (b), (c), $a^n - b^n$ does not have a primitive factor. If $n = 1$ and $a - b$ does not have a primitive factor, then $a - b = 1$.

Let $n = 2$ and assume that $a^2 - b^2$ does not have a primitive factor. From $a^2 - b^2 = (a + b)(a - b)$ and $\gcd(a + b, a - b) = 1$ or 2, if $p$ is an odd prime dividing $a + b$, then $p$ divides $a^2 - b^2$. But $p$ is not a primitive factor, so $p \mid a - b$, hence $p$ divides $a$ and $b$, which is absurd. This shows that $a + b$ is a power of 2.

Now let $n \geq 3$ and assume again that $a^n - b^n$ does not have a primitive factor. Let $p = P[n]$ and $v_p(\Phi_n(a, b)) = j \geq 0$. Define

$$\Phi_n^*(a, b) = \frac{\Phi_n(a, b)}{p^j}.$$

(1°) Assume that $\Phi_n^*(a, b) = 1$. Let $\zeta_1, \zeta_2, \ldots, \zeta_{\varphi(n)}$ be the primitive $n$th roots of 1. From

$$|a - \zeta_i b| = \left|\frac{a}{b} - \zeta_i\right| > b\left(\frac{a}{b} - 1\right) = a - b$$

and a previous remark,

$$\Phi_n(a, b) = |\Phi_n(a, b)| = \prod_{i=1}^{\varphi(n)} |a - \zeta_i b| > (a - b)^{\varphi(n)} \geq 1 = \Phi_n^*(a, b).$$

So $j \geq 1$ and $p \mid \Phi_n(a, b)$, hence $p$ divides $a^n - b^n$; so $p$ is a primitive factor of $a^f - b^f$, where $f$ divides $n$. By (3I), $\gcd(n, \Phi(a, b)) = p$ and also $p^2 \nmid \Phi_n(a, b)$.

In conclusion, $\Phi_n(a, b) = p$, because $\Phi_n^*(a, b) = 1$. Moreover, from $p \mid n$, it follows that $p - 1$ divides $\varphi(n)$. This implies in turn that $p = \Phi_n(a, b) \geq (a - b)^{\varphi(n)} \geq (a - b)^{p-1}$, hence $a - b = 1$.

If $p^2 \mid n$ let $n = pm$, then $p - 1 \leq \varphi(m)$ and by (2.14)

$$p = \Phi_n(a, b) = \Phi_m(a^p - b^p) > (a^p - b^p)^{\varphi(m)} \geq (a^p - b^p)^{p-1},$$

because $p \mid m$. Thus $a^p - b^p = 1$, which is not compatible with $a - b = 1$.

Thus, from (3I), $n = pf$, $p \nmid f$, where $p$ is a primitive factor of $a^f - b^f$. Note also that $f \mid p - 1$, so $f < p$. From $\varphi(n) = (p - 1)\varphi(f)$ it follows that

$$p(a^p - b^p) > p(a^f - b^f) \geq \Phi_n(a, b)\Phi_f(a, b) = \Phi_f(a^p - b^p) > (a^p - b^p)^{\varphi(f)},$$

using (2.12). Therefore $p > (a^p - b^p)^{\varphi(f)-1}$, hence necessarily $\varphi(f) = 1$, thus $f = 1$ or $f = 2$, so $n = p$ or $n = 2p$.

If $n = p$, then $p = \Phi_p(a, b) = a^{p-1} + a^{p-2}b + \cdots + ab^{p-2} + b^{p-1} = (a^p - b^p)/(a - b) = a^p - b^p$, and this is absurd because $a - b = 1$. If $n = 2p$, from $3 \leq p$ it follows from (2.12) that

$$p = \Phi_{2p}(a, b) = \frac{a^p + b^p}{a + b}.$$

By (3D), necessarily $a = 2$, $b = 1$, and $p = 3$, so $n = 6$.

(2°) Assume that $a^n - b^n$ does not have a primitive factor. If suffices to show that $\Phi_n^*(a, b) = 1$ and the result follows from (1°). Let $p$ be a prime dividing $\Phi_n(a, b)$, so $p \mid a^n - b^n$. Then there exists $f$, dividing $n$, $1 \leq f < n$, such that $p$ is a primitive factor of $a^f - b^f$. By (3I), $p = P[n]$ and $\Phi_n(a, b) = p^j$ with $j \geq 1$. Hence $\Phi_n^*(a, b) = 1$.

(2)   If $n = 3$, $a = 2$, $b = 1$, then $a^n + b^n = 2^3 + 1$ has no primitive factor. Conversely, if $n = 1$ and $a + b \geq 2$, so there is a primitive factor.

If $n = 2$ and $a^2 + b^2$ does not have a primitive factor, then $a^2 + b^2 = 2^k$ (with $k \geq 2$). Indeed, if $p$ is an odd prime dividing $a^2 + b^2$, then $p \mid a + b$, so $p \mid a^2 - b^2$, hence $p \mid 2a^2$; it follows that $p \mid a$ and also $p \mid b$, which is absurd. From $a^2 + b^2 = 2^k$ ($k \geq 2$), $\gcd(a, b) = 1$, it follows that $a, b$ are odd, hence $a^2 + b^2 \equiv 2 \pmod 4$, which is a contradiction, proving that $a^2 + b^2$ has a primitive factor.

If $n \geq 3$, it follows from (1) that $a^{2n} - b^{2n}$ has a primitive factor $p$ with the only exception $n = 3$, $a = 2$, $b = 1$. If $p = 2$ then $a, b$ are odd, so $2 \mid a + b$, which is not compatible with 2 being a primitive factor of $a^n - b^n$.

By (3E), $a^n + b^n$ has a primitive factor, with the exception indicated.   □

It follows from this theorem and (3F) that if $a \geq 2$, then each number in the sequence

$$\Phi_3(a), \Phi_4(a), \Phi_5(a), \Phi_6(a), \Phi_7(a), \ldots$$

(with $\Phi_6(a)$ deleted when $a = 2$) has a prime factor which is not a factor of any of the preceding numbers.

The following results are also of interest:

**(3K)**    Let $1 \leq m < n$, and $a > b \geq 1$, with $\gcd(a, b) = 1$. If $\gcd(\Phi_m(a, b), \Phi_n(a, b)) \neq 1$, then $P[n] = \gcd(\Phi_m(a, b), \Phi_n(a, b))$.

PROOF. If $n = 2$, then $m = 1$. If $\gcd(a - b, a + b) \neq 1$, then $\gcd(a - b, a + b) = 2$.

Now assume $n \geq 3$. Let $p$ be a prime and let $e \geq 1$ be such that $p^e \mid \Phi_m(a, b)$, $p^e \mid \Phi_n(a, b)$. Then $p \mid a^m - b^m$, $p \mid a^n - b^n$, so $p$ is not a primitive factor of $a^n - b^n$. By (3G) $p \mid n$ and by (3I), $p = P[n]$, $\Phi_n(a, b) = pc$, $p \nmid c$, so $e = 1$. Since $p$ was an arbitrary common divisor of $\Phi_m(a, b)$, $\Phi_n(a, b)$, this proves that $P[n] = \gcd(\Phi_m(a, b), \Phi_n(a, b))$. $\square$

**(3L)**     *Let $p$ be any prime, let $0 \leq i < j$, and let $a \geq b > 1$, $\gcd(a, b) = 1$. Then*

$$\gcd(\Phi_{p^i}(a, b), \ \Phi_{p^j}(a, b)) = \begin{cases} 1 & \text{if } p \nmid a - b, \\ p & \text{if } p \mid a - b. \end{cases}$$

PROOF. By (3K), if $d = \gcd(\Phi_{p^i}(a, b), \Phi_{p^j}(a, b)) \neq 1$ then $d = p$.

Assume first that $p \neq 2$. If $p \mid a - b$, then $a^{p^{j-1}} \equiv a \equiv b \equiv b^{p^{j-1}}$ (mod $p$) so by (3B), $p$ divides $\Phi_{p^j}(a, b) = (a^{p^j} - b^{p^j})/(a^{p^{j-1}} - b^{p^{j-1}})$. Similarly, $p$ divides $\Phi_{p^i}(a, b)$. Finally, if $p \nmid a - b$, then $a^{p^j} \equiv a \not\equiv b \equiv b^{p^j}$ (mod $p$), so $p \nmid a^{p^j} - b^{p^j}$ and a fortiori $p \nmid \Phi_{p^j}(a, b)$. Thus, $\gcd(\Phi_{p^i}(a, b), \Phi_{p^j}(a, b)) = 1$.

If $p = 2$, then $\Phi_1(a, b) = a - b$ and $\Phi_{2^k}(a, b) = a^{2^{k-1}} + b^{2^{k-1}}$ (for $k \geq 1$). So if $a \equiv b$ (mod 2), then 2 divides $\gcd(\Phi_{2^i}(a, b), \Phi_{2^j}(a, b))$, and conversely. $\square$

The following corollary of (3L) will be useful:

**(3M)**     *If $a > b \geq 1$ are integers and $n \geq 2$ then $P[a^n - b^n] > n$ and $P[a^n + b^n] > 2n$.*

PROOF. We may assume without loss of generality that $\gcd(a, b) = 1$. Indeed, let $d = \gcd(a, b)$ and let $a = da_1$, $b = db_1$, so $a_1 > b_1 \geq 1$ and $\gcd(a_1, b_1) = 1$. Moreover, $a^n \pm b^n = d^n(a_1^n \pm b_1^n)$, hence $P[a_1^n \pm b_1^n] \leq P[a^n \pm b^n]$; so it suffices to show that $n < P[a_1^n - b_1^n]$ and $2n < P[a_1^n + b_1^n]$. Thus, we assume $\gcd(a, b) = 1$.

(1) If $a = 2$, $b = 1$, $n = 6$ then $a^n - b^n = 2^6 - 1 = 63 = 3^2 \times 7$ and $P[2^6 - 1] = 7 > 6$. In the other cases, by (3L) let $p$ be a primitive factor of $a^n - b^n$. By (3G), $p \equiv 1$ (mod $n$) so $p = 1 + kn$, hence $P[a^n - b^n] \geq p > n$.

(2)   By (3L), let $p$ be a primitive factor of $a^{2n} - b^{2n}$. By (3G), $p \equiv 1 \pmod{2n}$, so $p = 1 + 2kn > 2n$. By a previous remark $p$ is also a primitive factor of $a^n + b^n$. Hence $P[a^n + b^n] \geq p > 2n$.   $\square$

## Bibliography

1729 (or before) Jacquemet, C., Manuscript in the Bibliothèque Nationale de Paris (see Dickson, *History of the Theory of Numbers*, Vol. II, p. 731).

1738 Euler, L., *Theorematum quorundam arithmeticorum demonstrationes*, Comm. Acad. Sci. Petrop., **10**, 1738 (1747), 125–146; reprinted in *Opera Omnia*, Vol. II, *Comm. Arithmeticae*, Vol. I, pp. 38–58.

1769 Lagrange, J.L., *Résolution des équations numériques*, Mém. Acad. Roy. Sci. Belles-Lettres Berlin, **23** (1769); reprinted in *Oeuvres*, Vol. II, pp. 527–532 and 539–578, Gauthier-Villars, Paris, 1868.

1830 Legendre, A.M., *Théorie des Nombres* ($3^e$ édition), Vol. I, pp. 226–229, Firmin Didot Frères, Paris, 1830; reprinted by A. Blanchard, Paris, 1955.

1837 Kummer, E.E., *De aequatione $x^{2\lambda} + y^{2\lambda} = z^{2\lambda}$ per numeros integros resolvenda*, J. Reine Angew. Math., **17** (1837), 203–209; reprinted in *Collected Papers*, Vol. I, pp. 135–141, Springer-Verlag, New York, 1975.

1886 Bang, A.S., *Taltheoretiske Undersøgelser*, Tidskrift Mat., København, Ser. 5, **4** (1886), 70–80 and 130–137.

1888 Sylvester, J.J., *On the divisors of the sum of a geometrical series whose first term is unity and common ratio any positive or negative integer*, Nature, **37** (1888), 417–418; reprinted in *Mathematical Papers*, Vol. IV, pp. 625–629, Cambridge University Press, Cambridge, 1912.

1892 Zsigmondy, K., *Zur Theorie der Potenzreste*, Monatsh. Math., **3** (1892), 265–284.

1897 Lucas, F., *Note relative à la théorie des nombres*, Bull. Soc. Math. France, **25** (1897), 33–35.

1904 Birkhoff, G.D. and Vandiver, H.S., *On the integral divisors of $a^n - b^n$*, Ann. of Math., (2), **5** (1904), 173–180.

1905 Dickson, L.E., *On the cyclotomic function*, Amer. Math. Monthly, **12** (1905), 86–89; reprinted in *Collected Mathe-*

*matical Papers* (editor A.A. Albert ), Vol. 3, pp. 136–139, Chelsea, New York, 1975.

1913 Carmichael, R.D., *On the numerical factors of arithmetical forms $\alpha^n \pm \beta^n$*, Ann. of Math., (2), **15** (1913), 30–70.

1946 Inkeri, K., *Untersuchungen über die Fermatsche Vermutung*, Ann. Acad. Sci. Fenn., Ser. A1, Nr. 33, 1946, 60 pp.

1947 Vivanti, G., *Un teorema di aritmetica e la sua relazione colla ipotesi di Fermat*, Rend. Istit. Lombardo, Milano, Cl. Sci. Mat., **11** (80) (1947), 239–246.

1950 Kanold, H.J., *Sätze über Kreisteilungspolynome und ihre Anwendungen auf einige zahlentheoretische Probleme, II*, J. Reine Angew. Math., **187** (1950), 169–182.

1955 Artin, E., *The orders of linear groups*, Comm. Pure Appl. Math., **8** (1955), 355–366.

1955 Möller, K., *Untere Schranke für die Anzahl der Primzahlen, aus denen $x, y, z$ der Fermatschen Gleichung $x^n + y^n = z^n$ bestehen muss.*, Math. Nachr., **14** (1955), 25–28.

1962 Schinzel, A., *On primitive prime factors of $a^n - b^n$*, Proc. Cambridge Philos. Soc., **58** (1962), 555–562.

1964 Kapferer, H., *Verifizierung des symmetrisches Teils der Fermatschen Vermutung für unendlich viele paarweise teilerfremde Exponenten E*, J. Reine Angew. Math., **214/5** (1964), 360–372.

1966 Leopoldt, H.-W., *Lösung einer Aufgabe von Kostrikhin*, J. Reine Angew. Math., **221** (1966), 160–161.

1974 Hering, C., *Transitive linear groups and linear groups which contain irreducible subgroups of prime order*, Geom. Dedicata, **2** (1974), 425–460.

1981 Lüneburg, H., *Ein einfacher Beweis für den Satz von Zsigmondy über primitive Primteiler von $A^n - 1$*. In *Geometries and Groups* (editors M. Aigner and D. Jungnickel), Lecture Notes in Math., **803**, pp. 219–222, Springer-Verlag, New York, 1981.

## II.4. The Resultant and Discriminant of Polynomials

Let

$$F(X,Y) = a_0 X^n + a_1 X^{n-1}Y + \cdots + a_n Y^n \quad \text{(with } a_0 \neq 0\text{)},$$
$$G(X,Y) = b_0 X^m + b_1 X^{m-1}Y + \cdots + b_m Y^m \quad \text{(with } b_0 \neq 0\text{)},$$

where the coefficients $a_i$, $b_j$ belong to an integral domain $A$.

We shall define the resultant of $F, G$, denoted by $R(F,G)$ or also by $\text{Res}(F,G)$. First, if $m = 0$ we define $R(F, b_0) = b_0^n$, while if $n = 0$ we define $R(a_0, G) = a_0^m$. In particular, $R(a_0, b_0) = 1$. Next, if $m \neq 0$ and $n \neq 0$ we define $R(f,g)$ to be the determinant of the following matrix with $m + n$ rows and columns:

$$\begin{pmatrix} a_0 & a_1 & \cdots & \cdots & a_n & 0 & 0 \\ 0 & a_0 & a_1 & \cdots & \cdots & a_n & 0 \\ \vdots & \vdots & \vdots & \vdots & \vdots & \vdots & \vdots \\ 0 & 0 & \cdots & a_0 & a_1 & \cdots & a_n \\ 0 & 0 & \cdots & b_0 & b_1 & \cdots & b_m \\ \vdots & \vdots & \vdots & \vdots & \vdots & \vdots & \vdots \\ 0 & b_0 & b_1 & \cdots & b_m & \cdots & 0 \\ b_0 & b_1 & \cdots & b_m & \cdots & 0 & 0 \end{pmatrix};$$

note that there are $m$ rows containing $a_0, \ldots, a_n$ as entries, followed by $n$ rows with $b_0, \ldots, b_m$.

$R(F,G)$ is called the *resultant* of $F(X,Y)$ and $G(X,Y)$. $R(F,G)$ is a polynomial with coefficients in $\mathbb{Z}$ of degree $m$ in the coefficients $a_i$ and of degree $n$ in the coefficients $b_j$.

The resultant of $\partial F(X,Y)/\partial X$, $\partial F(X,Y)/\partial Y$ is called the *discriminant* of $F(X,Y)$:

$$\text{Discr}(F) = R\left(\frac{\partial F}{\partial X}, \frac{\partial F}{\partial Y}\right).$$

We recall the following well-known properties (see Bôcher (1907) or Cohn (1974)):

**(4A)**    Let $F(X,Y)$, $G(X,Y)$ be binary forms of degrees, respectively, $n \geq 1$, $m \geq 1$. Then:

   (1) $F(X,Y)$ has a (nonconstant) factor proportional to a factor of $G(X,Y)$ if and only if $R(F,G) = 0$.

(2) *If $n \geq 2$ then $F(X,Y)$ has a multiple linear factor if and only if $\mathrm{Discr}(F) = 0$.*

(3) *If $F(X,Y) = \prod_{i=1}^{n}(\alpha_i' X - \alpha_i Y)$ and $G(X,Y) = \prod_{j=1}^{m}(\beta_j' X - \beta_j Y)$ (with $\alpha_i$, $\alpha_i'$, $\beta_j$, $\beta_j'$ elements of a field containing the coefficients of $F, G$ and $\alpha_i' \neq 0$ for each $i$ and $\beta_j' \neq 0$ for each $j$) then*

$$R(F,G) = \prod_{i=1}^{n} G(\alpha_i, \, \alpha_i') = (-1)^{mn} \prod_{j=1}^{m} F(\beta_j, \beta_j')$$

$$= \prod_{i,j}(\beta_j' \alpha_i - \beta_j \alpha_i') = (-1)^{mn} \prod_{i,j}(a_i' \beta_j - a_i \beta_j').$$

*In particular, $R(G, F) = (-1)^{mn} R(F, G)$.*

(4) *If $H(X,Y)$ is also a binary form of degree $l$, then $R(FG, H) = R(F, H)R(G, H)$ and $R(H, FG) = R(H, F)R(H, G)$.*

(5) *If $n \geq m$ and $H(X,Y)$ is a form of degree $\deg(H) = n - m$, then $R(F - HG, G) = R(F, G)$. Similarly, if $m \geq n$ and $K(X,Y)$ is a form of degree $\deg(K) = m - n$ then $R(F, G - KF) = R(F, G)$.*

Now let $f(X), g(X)$ be any nonzero polynomials of degrees $n, m$, respectively. Let $F(X,Y) = Y^n f(X/Y)$, $G(X,Y) = Y^m g(X/Y)$ so $F(X,Y)$, $G(X,Y)$ are binary forms of degrees $n, m$, respectively.

The *resultant* of $f, g$ is by definition $R(f, g) = R(F, G)$. The *discriminant* of $f$ is, by definition: $\mathrm{Discr}(f) = R(f, f')$.

If $n \geq 1$, $m \geq 1$ and $F(X,Y) = \prod_{i=1}^{n}(\alpha_i' X - \alpha_i Y)$, $G(X,Y) = \prod_{j=1}^{m}(\beta_j' X - \beta_j Y)$ (as in (1A)), with $a_i' \neq 0$, $\beta_j' \neq 0$, then $\alpha_i/\alpha_i'$ are the roots of $f(X)$, $\beta_j/\beta_j'$ are the roots of $g(X)$.

If $f(X) = a_0 X^n + a_1 X^{n-1} + \cdots + a_n$ (with $n \geq 1$, $a_0 \neq 0$), let $F(X,Y) = Y^n f(X/Y) = \prod_{i=1}^{n}(\alpha_i' X - \alpha_i Y)$.

The discriminants of the polynomial $f(X)$ and of the binary form $F(X,Y)$ are related as follows:

(4.1) $$\mathrm{Discr}(F) = \frac{n^{n-2}}{a_0} \mathrm{Discr}(f).$$

Indeed, the derivative of $f(X)$ is

$$f'(X) = n a_0 X^{n-1} + (n-1) a_1 X^{n-2} + \cdots + a_{n-1};$$

the corresponding binary form is $G(X,Y) = Y^{n-1} f'(X/Y)$. On the other hand, $\partial F / \partial X = Y^{n-1} f'(X/Y) = G(X,Y)$, $\partial F / \partial Y =$

$$nY^{n-1}f(X/Y) - XY^{n-2}f'(X/Y) = (1/Y)[nF(X,Y) - XG(X,Y)].$$

So, we have, on the one hand

$$R\left(\frac{\partial F}{\partial X}, Y\frac{\partial F}{\partial Y}\right) = R\left(\frac{\partial F}{\partial X}, Y\right) R\left(\frac{\partial F}{\partial X}, \frac{\partial F}{\partial Y}\right) = na_0 \operatorname{Discr}(F).$$

On the other hand,

$$R\left(\frac{\partial F}{\partial X}, Y\frac{\partial F}{\partial Y}\right) = R(G, nF - XG) = R(G, nF)$$

$$= R(G, n) \cdot R(G, F) = n^{n-1}(-1)^{n(n-1)}R(F, G)$$
$$= n^{n-1}(-1)^{n(n-1)}R(f, f') = n^{n-1}\operatorname{Discr}(f).$$

For the convenience of the reader we write explicitly the properties of the resultant and the discriminant for polynomials in one indeterminate.

**(4B)**    *For polynomials $f, g, h, k$, with $\deg(f) = n$, $\deg(g) = m$:*

(1) $R(g, f) = (-1)^{mn}R(f, g)$.
(2) *If $n \le m$ and $\deg(h) \le m - n$ then $R(f, g) = R(f, g + fh)$.*
(3) $R(hk, g) = R(h, g) \cdot R(k, g)$,
    $R(g, hk) = R(g, h) \cdot R(g, k)$.
(4) $R(f^s, g) = [R(f, g)]^s$ *for every integer $s \ge 1$.*
(5) $R((X - a)^s, g) = [g(a)]^s$ *where $a \in A$, $s \ge 1$.*
(6) *If $f = a_0 \prod_{i=1}^{n}(X - \alpha_i)$ and $g = b_0 \prod_{j=1}^{m}(X - \beta_i)$ then*

$$R(f, g) = a_0^m b_0^n \prod_{i=1}^{n}\prod_{j=1}^{m}(\alpha_i - \beta_j)$$

$$= a_0^m \prod_{i=1}^{n} g(\alpha_i)$$

$$= (-1)^{mn}b_0^m \prod_{j=1}^{n} f(\beta_j).$$

(7) *If $f = a_0 \prod_{i=1}^{n}(X - \alpha_i)$, then*

$$\operatorname{Discr}(f) = (-1)^{n(n-1)/2}a_0^{2n-1}\prod_{i<j}(\alpha_i - \alpha_j)^2.$$

(8) *If $f = hk$, $\deg(h) = r$, and $\deg(k) = s$, then*

$$\operatorname{Discr}(f) = (-1)^{rs}\operatorname{Discr}(h)\operatorname{Discr}(k)[R(h, k)]^2.$$

**(4C)**

(1) *If $f, g \in A[X]$ are nonconstant and $R(f, g) \neq 0$, then $f, g$ are relatively prime.*

(2) *If $A = K$ is a field, and if $f, g \in K[X]$ are relatively prime, then $R(f, g) \neq 0$.*

PROOF. (1)    Assume that $f, g$ have a common nonconstant factor $h \in A[X]$. So $f = h f_1$ and $g = h g_1$. By (4B)(3),

$$R(f, g) = R(h, h) \cdot R(h, g_1) \cdot R(f_1, h) \cdot R(f_1, g_1) = 0.$$

(2)    Assume that $f, g \in K[X]$ are relatively prime. By Bézout's theorem, there exist $f_1, g_1 \in K[X]$ such that $g_1 f + f_1 g = 1$; in particular, $\deg(g_1 f) = \deg(f_1 g)$. By (4B)(3),

$$R(g_1 f, f_1 g) = R(g_1, f_1) \cdot R(g_1, g) \cdot R(f, f_1) \cdot R(f, g).$$

If $R(f, g) = 0$, then $R(g_1 f, f_1 g) = 0$. However, by (4B)(2),

$$R(g_1 f, f_1 g) = R(g_1 f, 1 - g_1 f) = R(g_1 f, 1) = 1,$$

which is a contradiction.    □

## Bibliography

1907 Bôcher, M., *Introduction to Higher Algebra*, Macmillan, New York, 1907; reprinted in 1947.

1974 Cohn, P.M., *Algebra*, Vol. I, Wiley, New York, 1974.

# III
# Algebraic Restrictions
# on Hypothetical Solutions

Assume that $n > 3$ and $x, y, z$ are nonzero pairwise relatively prime integers such that

$$x^n + y^n = z^n.$$

In this chapter we derive algebraic relations which must be satisfied by $x, y, z, n$. In some cases, these lead to a contradiction showing that for the exponent in question, Fermat's equation has only the trivial solution.

## III.1. The Relations of Barlow

Let $p$ be an odd prime and suppose that there exist nonzero pairwise relatively prime integers $x, y, z$ such that $x^p + y^p + z^p = 0$. To begin, we observe that $x + y + z \neq 0$. Indeed, $x, y, z$ cannot be all positive (nor all negative), so we assume, for example, that $x > 0$, $y > 0$ and $z < 0$. Then $(x+y)^p > x^p + y^p = -z^p$ since $x + y > -z$, thus $x + y + z \neq 0$.

We shall indicate relations which the integers $x, y, z$ must satisfy.

The first such results were proved by Barlow (1810, 1811) and discovered independently by Abel in 1823, who stated them without proof, in a letter to Holmboe. The results below were given with complete proofs by Legendre, as early as 1823, and were known to Sophie Germain. Later, they were rediscovered by Lindemann (1901, 1907) and appeared in papers by Catalan (1886), Tafelmacher (1892), Fleck (1909), Lind (1910), Bachmann (1919), James (1938), Racliş (1944), etc.

**(1A)** *If there exist nonzero integers $x, y, z$ such that $x^p + y^p + z^p = 0$, $\gcd(x, y, z) = 1$ and $p$ does not divide $z$, then there exist relatively prime integers $t, t_1$, not multiples of $p$, such that*

$$x + y = t^p, \qquad \frac{x^p + y^p}{x + y} = t_1^p, \qquad z = -tt_1.$$

*Moreover, $t_1$ is odd and $t_1 > 1$.*

PROOF. From the hypothesis, $x, y, z$ are pairwise relatively prime. Consider the integer

(1.1)
$$Q_p(x, -y) = \frac{x^p + y^p}{x + y} = x^{p-1} - x^{p-2}y + \cdots - xy^{p-2} + y^{p-1}.$$

Since $x + y + z \equiv x^p + y^p + z^p \equiv 0 \pmod{p}$ and $p \nmid z$ then $p \nmid x + y$. By Chapter II, (3B), $\gcd(x + y, Q_p(x, -y)) = 1$.

From $(-z)^p = x^p + y^p = (x + y)Q_p(x, -y)$ we conclude that $x + y, Q_p(x, -y)$ are $p$th powers, i.e., there exist integers $t, t_1$ such that $x + y = t^p$, $Q_p(x, -y) = t_1^p$, so $-z = tt_1$ and $\gcd(t, t_1) = 1$.

We show that $t_1$ is odd. From (1.1) we see that $Q_p(x, -y)$ is the sum of an odd number of terms; among these terms, $x^{p-1}$ or $y^{p-1}$ is odd (because $x, y$ are not both even). Thus $Q_p(x, -y)$ must be odd, hence $t_1$ is also odd. Finally, from (1E) since $x > y$ (or $y > x$) then $x - y \geq 1$ (or $y - x \geq 1$), hence $t_1^p = Q_p(x, -y) = Q_p(y, -x) \geq p$, so $t_1 > 0$ and in fact $t_1 > 1$. $\square$

If $x, y, z$ satisfy $x^p + y^p + z^p = 0$, if $p$ does not divide $x, y, z$ and if $\gcd(x, y, z) = 1$ then by the previous result there exist integers $r, s, t, r_1, s_1, t_1$, not multiples of $p$, such that

(1.2)
$$\begin{cases} x + y = t^p, & (x^p + y^p)/(x + y) = t_1^p, & z = -tt_1, \\ y + z = r^p, & (y^p + z^p)/(y + z) = r_1^p, & x = -rr_1, \\ z + x = s^p, & (z^p + x^p)/(z + x) = s_1^p, & y = -ss_1. \end{cases}$$

Moreover, $r, s, t, r_1, s_1, t_1$ are pairwise relatively prime, $r_1, s_1, t_1$ are odd and greater than 1. We note that $r^p + s^p + t^p = 2(x+y+z) \neq 0$. By addition and subtraction, it follows that

$$(1.3) \quad \begin{cases} x = -r^p + (r^p + s^p + t^p)/2 = (-r^p + s^p + t^p)/2, \\ y = -s^p + (r^p + s^p + t^p)/2 = (r^p - s^p + t^p)/2, \\ z = -t^p + (r^p + s^p + t^p)/2 = (r^p + s^p - t^p)/2. \end{cases}$$

We have the following complement of (1A) known to Sophie Germain and reproduced by Legendre (1823). Proofs were also given by Fleck (1909), Lind (1910), Pomey (1923) and again by Spunar (1928), James (1938), Pérez-Cacho (1958), and Draeger (1959) (in a different form).

**(1B)**  *If $p$ is an odd prime not dividing $z$, every prime divisor $q$ of $t_1$ is congruent to 1 modulo $2p$. In particular, $t_1 \equiv 1 \pmod{2p}$. If, moreover, $p$ does not divide $xyz$, then every prime divisor of $r_1 s_1 t_1$ is congruent to 1 modulo $2p^2$. In particular, $r_1 \equiv 1 \pmod{2p^2}$, $s_1 \equiv 1 \pmod{2p^2}$ and $t_1 \equiv 1 \pmod{2p^2}$.*

PROOF. Let $q$ be a prime dividing $t_1$. Then $q$ divides $x^p + y^p$ but $q$ does not divide $x + y = t^p$ because $\gcd(t, t_1) = 1$. By Chapter II, (3G), $q \equiv 1 \pmod{p}$. Since $q - 1$ is even then $q \equiv 1 \pmod{2p}$.

Now we suppose that $p \nmid xyz$ and that $q$ is a prime dividing $r_1$, hence $q$ divides $x$. Therefore $q$ does not divide $yz$. We note also that $\gcd(r, r_1) = 1$, so $q$ does not divide $y + z$.

Thus we have $y \equiv t^p \pmod{q}$, $z \equiv s^p \pmod{q}$, hence $t^p + s^p \equiv y + z \not\equiv 0 \pmod{q}$ and $t^{p^2} + s^{p^2} \equiv y^p + z^p \equiv -x^p \equiv 0 \pmod{q}$. So $q$ is a primitive factor of $t^{p^2} + s^{p^2}$. By Chapter II, (3E), $q$ is a primitive factor of $t^{2p^2} - s^{2p^2}$, hence by Chapter II, (3G), $q \equiv 1 \pmod{2p^2}$. In particular, $r_1 \equiv 1 \pmod{2p^2}$.

The proof is similar for the prime factors of $s_1$ and $t_1$.  $\square$

Now we give the relations which must be satisfied by would-be solutions in the second case; these facts (including $n \geq 2$) were known to Sophie Germain and were given by Legendre (1823).

**(1C)**  *Let $x, y, z$ be nonzero integers such that $p$ divides $z$, $x^p + y^p + z^p = 0$ and $\gcd(x, y, z) = 1$. Then there exist an integer $n \geq 2$ and pairwise relatively prime integers $r, s, t, r_1, s_1, t_1$, not multiples* *of $p$, such that $r_1, s_1, t_1$ are odd and greater than 1 and satisfy the relations*

(1.4)

$$\begin{cases} x + y = p^{p^{n-1}}t^p, & (x^p + y^p)/(x + y) = pt_1^p, & z = -p^n t t_1, \\ y + z = r^p, & (y^p + z^p)/(y + z) = r_1^p, & x = -r r_1, \\ z + x = s^p, & (z^p + x^p)/(z + x) = s_1^p, & y = -s s_1. \end{cases}$$

*Moreover, if $q$ is any prime dividing $t_1$ then $q \equiv 1 \pmod{p^2}$; in particular, $t_1 \equiv 1 \pmod{2p^2}$.*

PROOF. $x, y, z$ are pairwise relatively prime. If $p$ divides $z$ then $p$ does not divide $x$ nor $y$, and from $x + y + z \equiv x^p + y^p + z^p \equiv 0 \pmod{p}$ it follows that $x + y \equiv -z \equiv 0 \pmod{p}$.

Let $m \geq 2$ be such that $x + y = p^{m-1}t'$, where $p$ does not divide $t'$. Let $Q_p(x, -y) = (x^p + y^p)/(x + y)$. Since $p$ divides $u = x + y$ then by Chapter II, (3B)(6), $v_p(Q_p(x, -y)) = v_p(p) = 1$. So $Q_p(x, -y) = pt_1'$, $p \nmid t_1'$. By Chapter II, (3B)(4), $\gcd(x + y, Q_p(x, -y)) = p$, so $\gcd(t', t_1') = 1$.

Since $-z^p = x^p + y^p = p^m t' t_1'$ then by unique factorization $p$ divides $m$, and $t'$, $t_1'$ are $p$th powers of integers. We may write $m = pn$,

$$x + y = p^{pn-1}t^p,$$
$$\frac{x^p + y^p}{x + y} = pt_1^p,$$
$$z = -p^n t t_1,$$

where $n \geq 1$, $t, t_1 \in \mathbb{Z}$, $\gcd(t, t_1) = 1$, $p$ does not divide $t$ nor $t_1$.

Since $p \nmid x$ and $p \nmid y$, by (1A) there exist integers $r, r_1, s, s_1$, not multiples of $p$, such that

$$y + z = r^p, \quad (y^p + z^p)/(y + z) = r_1^p, \quad x = -r r_1,$$
$$z + x = s^p, \quad (z^p + x^p)/(z + x) = s_1^p, \quad y = -s s_1,$$

and $\gcd(t, r, s) = \gcd(t_1, r_1, s_1) = 1$, $\gcd(r, r_1) = \gcd(s, s_1) = 1$. The proofs that $r_1, s_1, t_1$ are odd and that $r_1 > 1$, $s_1 > 1$ are the same as in (1A).

Now we show that $t_1 > 1$. By Chapter II, (3D), $pt_1^p = (x^p + y^p)/(x + y) = Q_p(x, -y) = Q_p(y, -x) \geq p$, hence $t_1 \geq 1$ and it suffices to show that $t_1 \neq 1$. If $t_1 = 1$ then again by the same result, assuming for example $x > y$, this would imply: $p = 3$, $x = 2$, $y = 1$, hence $2^3 + 1^3 + z^3 = 0$ which is impossible.

Now we show that if $q$ is any prime factor of $t_1$ then $q \equiv 1$ (mod $p^2$). We have $z \equiv 0$ (mod $q$) hence $y \equiv r^p$ (mod $q$), $x \equiv s^p$ (mod $q$) and $0 \equiv -x^p + y^p + z^p \equiv r^{p^2} + s^{p^2}$ (mod $q$). On the other hand, $q$ does not divide $r^p + s^p$ (otherwise $q$ divides $x + y$ hence $q$ divides $t$, contrary to the fact that $\gcd(t, t_1) = 1$). It follows from Chapter II, (3D), that $q \equiv 1$ (mod $p^2$). This implies that $t_1 \equiv 1$ (mod $2p^2$).

It remains to show that $n \geq 2$. In fact, $r^p + s^p = 2z + (x + y) = -2p^n t t_1 + p^{pn-1} t^p \equiv 0$ (mod $p$). By Chapter II, (3H), $r^p + s^p \equiv 0$ (mod $p^2$). Since $p \nmid t t_1$ it follows that $2z = (r^p + s^p) - p^{pn-1} t^p \equiv 0$ (mod $p^2$), so $n \geq 2$. $\square$

From (1.1) we deduce that $r^p + s^p + p^{pn-1} t^p = 2(x + y + z) \neq 0$.

The fact that the exact power of $p$ dividing $x + y$ is $p^{pn-1}$ has been proved again and again (even in 1955 by Stone) by authors unaware that this result has been known for a long time.

We write the relations analogous to (1.3), assuming that $p \mid z$:

(1.5)
$$\begin{cases} x = -r^p + (r^p + s^p + p^{pn-1} t^p)/2 = (-r^p + s^p + p^{pn-1} t^p)/2, \\ y = -s^p + (r^p + s^p + p^{pn-1} t^p)/2 = (r^p - s^p + p^{pn-1} t^p)/2, \\ z = -p^{pn-1} t^p + (r^p + s^p + p^{pn-1} t^p)/2 = (r^p + s^p - p^{pn-1} t^p)/2. \end{cases}$$

In the case of a squarefree exponent it is still possible to indicate some relations which are reminiscent of the Barlow relations; see Stewart (1977):

**(1D)**   *If $n > 2$ is a square-free integer, if $x, y, z$ are nonzero pairwise relatively prime integers such that $x^n + y^n = z^n$ (respectively, if $n$ is odd and $x^n - y^n = z^n$), then $z - y = 2^u d^{n-1} a^n$ (respectively, $z + y = 2^u d^{n-1} a^n$) where $a, d$ are natural numbers, $u$ is equal to $0$ or $1$, and $2^u$ and $d$ divide $n$.*

PROOF. We first consider the case where $x^n + y^n = z^n$ and we write $z - y = a' a^n$ where $a, a' \geq 1$ and for every prime $p$ the $p$-adic value of $a'$ is $v_p(a') < n$.

If $p$ is a prime dividing $a'$ then $p$ divides $n$. Otherwise, $p \nmid n$ and by Chapter II, (3B)(5), $p \nmid Q_n(z, y) = (z^n - y^n)/(z - y)$. So

$v_p(a') + nv_p(a) = v_p(z - y) = v_p(z^n - y^n) = v_p(x^n) = nv_p(y)$. Hence $n$ divides $v_p(a')$ and $n \leq v_p(a')$, which contradicts the hypothesis. This shows that $p \mid n$, so $v_p(n) = 1$, because $n$ is square free.

Now we determine $v_p(a')$ when $p$ divides $a'$. First, let $p$ be an odd prime. By Chapter II, (3B)(6), $v_p(Q_n(z, y)) = v_p(n) = 1$. So $v_p(a') + nv_p(a) + 1 = v_p(z-y) + v_p(Q_n(z, y)) = v_p(z^n - y^n) = v_p(x^n) = nv_p(x)$; hence $v_p(a') \equiv -1 \pmod{n}$ and therefore $v_p(a') = n - 1$.

Now let $p = 2$. If $4 \mid z - y$, by the result already quoted, Part (7), we have $v_2(Q_n(z, y)) = v_2(n) = 1$, so as above we conclude that $v_2(a') = n - 1$. If $2 \mid z - y$ but $4 \nmid z - y$ then $v_2(a') + nv_2(a) = v_2(z - y) = 1$, so $v_2(a') = 1$. There remains the possibility that $2$ does not divide $z - y$.

Putting these facts together, we may write $z - y = 2^u d^{n-1} a^n$, where $u = 0$ or $1$, $2^u$ divides $n$, and $d^{n-1}$ divides $n$. Now let $n$ be odd and $x^n - y^n = z^n$. Then $x^n + (-y)^n = z^n$ and by the first part of the proof $z + y$ has the form indicated.  □

In particular, if $n$ is a square-free integer, $n > 2$, if $x, y, z$ are nonzero pairwise relatively prime integers such that $x^n + y^n = z^n$, then

$$\begin{cases} z - x = 2^{u_1} d_1^{n-1} a_1^n, \\ z - y = 2^{u_2} d_2^{n-1} a_2^n, \end{cases}$$

and, moreover, if $n$ is odd then

$$x + y = 2^{u_3} d_3^{n-1} a_3^n$$

where $a_1, a_2, a_3, d_1, d_2, d_3$ are natural numbers, $u_1, u_2, u_3$ are equal to $0$ or $1$, and $2^{u_1}, 2^{u_2}, 2^{u_3}, d_1, d_2, d_3$ divide $n$.

## Bibliography

1810 Barlow, P., *Demonstration of a curious numerical proposition*, J. Nat. Phil. Chem. Arts, **27** (1810), 193–205 (this paper is referred to in Dickson's *History of the Theory of Numbers*, Vol. II, p. 733).

1811 Barlow, P., *An Elementary Investigation of Theory of Numbers* (pp. 153–169), J. Johnson, St. Paul's Church-yard, London, 1811.

1823 Abel, N., Extraits de quelques lettres à Holmboe, Copenhague, l'an $\sqrt[3]{6064321219}$ (en comptant la fraction décimale),

*Oeuvres Complètes*, Vol. II, 2nd ed., pp. 254–255, Grondahl, Christiania, 1881.

1823 Legendre, A.M., *Recherches sur quelques objets d'analyse indéterminée, et particulièrement sur le théorème de Fermat*, Mém. Acad. Sci., Institut France, **6** (1823), 1–60; appeared as "Second Supplément" in 1825, to a printing of *Essai sur la Théorie des Nombres* ($2^e$ édition), Courcier, Paris; reprinted in Sphinx-Oedipe, **4** (1909), 97–128.

1886 Catalan, E., *Sur le dernier théorème de Fermat* (Mélanges Mathématiques, CCXV), Mém. Soc. Roy. Sci. Liège Sér. 2, **13** (1886), 387–397.

1892 Tafelmacher, W.L.A., *Sobre el teorema de Fermat de que la ecuación $x^n + y^n = z^n$ no tiene solución en numeros enteros $x, y, z$ i siendo $n > 2$*, Ann. Univ. Chile, Santiago, **82** (1892), 271–300 and 415–437.

1901 Lindemann, F., *Über den Fermatschen Satz betreffend die Unmöglichkeit der Gleichung $x^n = y^n + z^n$*, Sitzungsber. Akad. Wiss. München, Math., **31** (1901), 185–202; corrigenda p. 495.

1907 Lindemann, F., *Über das sogenannte letzte Fermatsche Theorem*, Sitzungsber. Akad. Wiss. München, Math., **37** (1907), 287–352.

1909 Fleck, A., *Miszellen zum großen Fermatschen Problem*, Sitzungsber. Berliner Math. Ges., **8** (1909), 133–148.

1910 Bachmann, P., *Niedere Zahlentheorie*, Teubner, Leipzig, 1910; reprinted by Chelsea, New York, 1966.

1910 Lind, B., *Über das letzte Fermatsche Theorem*, Abhandl. zur Geschichte d. Math. Wiss., no. 26, 1910, 23–65.

1919 Bachmann, P., *Das Fermatproblem in seiner bisherigen Entwicklung*, W. de Gruyter, Berlin, 1919; reprinted by Springer-Verlag, Berlin, 1976.

1923 Pomey, L., *Sur le dernier théorème de Fermat*, C. R. Acad. Sci. Paris, **177** (1923), 1187–1190.

1928 Spunar, V.M., *On Fermat's last theorem*, J. Washington Acad. Sci., **18** (1928), 385–395.

1938 James, G., *A higher upper limit to the parameters in Fermat's equation*, Amer. Math. Monthly, **45** (1938), 439–445.

1944 Racliş, N., *Démonstration du grand théorème de Fermat pour des grandes valeurs de l'exposant*, Bull. École Polytechnique

Bucarest, **15** (1944), 3–19.

1955 Stone, D.E., *On Fermat's last theorem*, Math. Mag., **28** (1955), 295–296.

1958 Pérez-Cacho, L., *Sobre algunas cuestiones de la teoria de números*, Rev. Mat. Hisp.-Amer., (4), **18** (1958), 10–27 and 113–124.

1958/9 Draeger, M., *Das Fermat-Problem*, Wiss. Z. Techn. Hochschule Dresden, **8** (1958/9), 941–946.

1977 Stewart, C.L., *A note on the Fermat equation*, Mathematika, **24** (1977), 130–132.

## III.2. Secondary Relations for Hypothetical Solutions

In §1 we have seen that if $x, y, z$ are nonzero pairwise relatively prime integers, $p$ an odd prime and if $x^p + y^p + z^p = 0$ then the Barlow relations must be satisfied; in particular, there exist integers $r, s, t, r_1, s_1, t_1$ satisfying certain properties. In this section, we give further properties which must be satisfied by these integers.

If $m, n$ are nonzero integers, $\gcd(m, n) = 1$ and $n$ is odd, let $(m/n)$ denote the Jacobi symbol.

The following consequence of Barlow's relations was first indicated by Pierre in 1943.

**(2A)** *Let $x, y, z$ be nonzero relatively prime integers such that $x^p + y^p + z^p = 0$.*

*(1)    If $p \nmid xyz$ then*

$$\left(\frac{r_1}{s_1 t_1}\right) = \left(\frac{s_1}{t_1 r_1}\right) = \left(\frac{t_1}{r_1 s_1}\right) = +1.$$

*(2)    If $p \mid z$ then*

$$\left(\frac{r_1}{p s_1 t_1}\right) = \left(\frac{s_1}{p t_1 r_1}\right) = \left(\frac{p t_1}{r_1 s_1}\right) = +1.$$

PROOF. (1)    To begin, we note that the polynomial $pXY(X + Y)$ divides $(X + Y)^p - (X^p + Y^p)$ (in $\mathbb{Z}[X, Y]$) (see also Section VII.2).

By (1A) we have $(x+y)^{p-1} - t_1^p \equiv 0 \pmod{pxy}$. Then $t_1^p \equiv (x+y)^{p-1} \pmod{r_1 s_1}$, hence

$$\left(\frac{t_1}{r_1 s_1}\right) = \left(\frac{t_1^p}{r_1 s_1}\right) = +1.$$

By symmetry,

$$\left(\frac{r_1}{s_1 t_1}\right) = \left(\frac{s_1^p}{r_1 t_1}\right) = +1.$$

(2)    If $p \mid z$, by (1C) we have $(x+y)^{p-1} - pt_1^p \equiv 0 \pmod{pxy}$. Hence $pt_1^p \equiv (x+y)^{p-1} \pmod{r_1 s_1}$. Therefore,

$$\left(\frac{pt_1}{r_1 s_1}\right) = \left(\frac{pt_1^p}{r_1 s_1}\right) = +1.$$

As in the proof of (1), $(y+z)^{p-1} - r_1^p \equiv 0 \pmod{pxy}$. Therefore

$$r_1^p \equiv (y+z)^{p-1} \pmod{ps_1 t_1},$$

hence

$$\left(\frac{r_1}{ps_1 t_1}\right) = \left(\frac{r_1^p}{ps_1 t_1}\right) = +1.$$

Similarly,

$$\left(\frac{s_1}{pr_1 t_1}\right) = +1. \quad \square$$

## Bibliography

1943 Pierre, C., *Sur le théorème de Fermat $a^n + b^n = c^n$*, C. R. Acad. Sci. Paris, **217** (1943), 37–39.

# IV
# Germain's Theorem

In this chapter we give the beautiful theorem of Sophie Germain for the first case of Fermat's theorem.

## IV.1. Sophie Germain's Theorem

Sophie Germain, a French mathematician, contemporary of Cauchy and Legendre, proved a very beautiful theorem of an entirely new kind which established "d'un trait de plume" (in Legendre's expression) the first case of Fermat's theorem for every prime $p < 100$. Her method is still being explored by other mathematicians. Legendre developed S. Germain's ideas in his paper of 1823.

We begin with an easy observation; (3) was also given by Bang (1935).

**(1A)**  *Let $q$ be a prime and $n \geq 3$ be an odd integer. The following statements are equivalent:*

(1) *There exist integers $a, b, c$, not multiples of $q$, such that $a^n + b^n + c^n \equiv 0 \pmod{q}$.*

(2) *There exist integers $d, e$, not multiples of $q$, such that $d^n \equiv e^n + 1 \pmod{q}$.*

*Moreover, if $q - 1 = 2kn$, the above statements are equivalent to:*

(3) *There exist roots $u, u'$ of the congruence $X^{2k} - 1 \equiv 0 \pmod{q}$ such that $u' \equiv u + 1 \pmod{q}$.*

PROOF. $(1) \rightarrow (2)$   Since $q \nmid c$ there exist integers $d, e \in \mathbb{Z}$ such that

$$\begin{cases} dc \equiv -a \pmod{q}, \\ ec \equiv b \pmod{q}. \end{cases}$$

Then $q \nmid de$, $(dc)^n \equiv (ec)^n + c^n \pmod{q}$, so $d^n \equiv e^n + 1 \pmod{q}$.

$(2) \rightarrow (1)$   This is trivial.

Now we assume that $q - 1 = 2kn$.

$(2) \rightarrow (3)$   Let $u = e^n$, $u' = d^n$, then $u^{2k} \equiv e^{q-1} \equiv 1 \pmod{q}$ and similarly $(u')^{2k} \equiv d^{q-1} \equiv 1 \pmod{q}$, with $u' \equiv u + 1 \pmod{q}$.

$(3) \rightarrow (2)$   Let $h$ be a primitive root modulo $q$. Let $u = h^m$, so $h^{2km} \equiv u^{2k} \equiv 1 \pmod{q}$, hence $q - 1 = 2kn$ divides $2km$, so $n$ divides $m$. Thus $u \equiv e^n \pmod{q}$. Similarly $u' \equiv d^n \pmod{q}$ and $d^n \equiv e^n + 1 \pmod{q}$.   □

Now we give Legendre's version of Sophie Germain's theorem:[1]

**(1B)**   *Let $p, q$ be distinct odd primes and assume that the following conditions are satisfied:*

(1) *If $a, b, c$ are integers such that $a^p + b^p + c^p \equiv 0 \pmod{q}$ then $q \mid abc$.*

(2) *$p$ is not congruent modulo $q$ to the $p$th power of an integer.*

*Then the first case of Fermat's theorem is true for the exponent $p$.*

PROOF. Let $x, y, z$ be pairwise relatively prime integers, not multiples of $p$, such that $x^p + y^p + z^p = 0$. Then $x^p + y^p + z^p \equiv 0 \pmod{q}$

---

[1]See footnote, p. 13 of Legendre's paper of 1823, where he wrote: "This proof which, one has to agree, is very ingenious, is due to Mlle. Sophie Germain, who cultivates with success the physical and mathematical sciences, as witnesses the prize she has been awarded by the Academy for her paper on vibrations of elastic blades. She is also the author of the proposition in art. 13 as well the one which concerns the particular form of the prime divisors of $\alpha$, given in art. 11." [Here, these correspond to propositions ((2B), (2C)).]

and by hypothesis (1), $q \mid xyz$. We may assume, for example, that $q \mid x$, hence $q \nmid yz$.

Since $p \nmid xyz$ there exist integers $r, s, t, r_1, s_1, t_1$ satisfying the relations (1.2) and (1.3) of Chapter III. Since $q \mid x$ then $-r^p + s^p + t^p \equiv 0$ (mod $q$). By hypothesis (1), $q$ divides one of the integers $r, s, t$. Since $s$ divides $y$, $t$ divides $z$, and $q$ does not divide $yz$ and $q$ does not divide $st$, so $q$ divides $r$. But $t_1^p = (x^p + y^p)/(x + y) \equiv y^{p-1}$ (mod $q$), because $q \mid x$.

Since $q$ divides $r$, so $y \equiv -z$ (mod $q$). Hence

$$r_1^p = \frac{y^p + z^p}{y + z} = y^{p-1} + y^{p-2}(-z) + \cdots + (-z)^{p-1}$$
$$\equiv py^{p-1} \equiv pt_1^p \pmod{q}.$$

Since $t_1 \not\equiv 0$ (mod $q$) there exists an integer $t'$ such that $t't_1 \equiv 1$ (mod $q$), hence $p \equiv (t'r_1)^p$ (mod $q$), which contradicts the second assumption.  □

Before proceeding, we comment on the above conditions.

In the next section, we shall introduce the Wendt determinant, which will serve to test the existence of integers $x, y, z$, not multiples of $q$, such that $x^p + y^p + z^p \equiv 0$ (mod $q$).

**(1C)**  *If $p$ and $q$ are odd primes and $q - 1 = 2pk$, $k$ a natural number, then condition (2) of (1B) is equivalent to each of the following:*

(2′) $(2k)^{2k} \not\equiv 1$ (mod $q$); *and*
(2″) $p^{2k} \not\equiv 1$ (mod $q$).

PROOF. We show first that (2) implies (2′). Let $h$ be a primitive root modulo $q$ and let $p \equiv h^s$ (mod $q$). If $(2k)^{2k} \equiv 1$ (mod $q$) then $h^{2ks} \equiv p^{2k} \equiv (2k)^{2k}p^{2k} \equiv (2kp)^{2k} \equiv (q - 1)^{2k} \equiv 1$ (mod $q$); hence $q - 1 = 2kp$ divides $2ks$ so $p \mid s$ and $p \equiv a^p$ (mod $q$) with $a \equiv h^{s/p}$ (mod $q$).

Now we show that (2′) implies (2″). If $p^{2k} \equiv 1$ (mod $q$) then $(2k)^{2k} \equiv (2k)^{2k}p^{2k} \equiv (q - 1)^{2k} \equiv 1$ (mod $p$).

Finally, we prove that (2″) implies (2). If there exists an $a$ such that $p \equiv a^p$ (mod $q$) then $p^{2k} \equiv a^{2kp} \equiv a^{q-1} \equiv 1$ (mod $q$).  □

In Table 6 (see Legendre, 1823), we indicate, for each $p < 100$, the choice of $q$, of a primitive root $h$ modulo $q$ and the set $R$ of residues of $p$th powers, modulo $q$. The computations, which are quite lengthy, are done using the primitive root modulo $q$. They establish condition (2).

For larger values of $p$ the computations become forbidding. However, a glance at the table reveals that in each case, $q$ has been chosen to equal $2p+1$, or $4p+1$, or $8p+1$, or $10p+1$, or $14p+1$, or $16p+1$. Indeed, the following corollaries of Sophie Germain's theorem hold (Legendre, 1823):

**(1D)**    *If $p$ is an odd prime and $q = 2p + 1$ is also a prime, then the first case of Fermat's conjecture is true for $p$.*

PROOF. We show that $q$ satisfies the assumptions of (1B).

If $x, y, z$ are integers not multiples of $q$ and $x^p + y^p + z^p \equiv 0$ (mod $q$), then from $p = (q - 1)/2$ we have $x^p \equiv \pm 1$ (mod $q$), $y^p \equiv \pm 1$ (mod $q$), $z^p \equiv \pm 1$ (mod $q$). Hence $0 \equiv x^p + y^p + z^p \equiv \pm 1 \pm 1 \pm 1 \not\equiv 0$ (mod $q$), a contradiction.

Similarly, if $p \equiv a^p$ (mod $q$) then condition (2') is not satisfied, so $2p + 1 = q$ divides $2^2 - 1 = 3$, which is absurd.    □

Legendre extended this criterion (1823):

**(1E)**    *If $p$ is a prime, $p > 3$ and $q = 4p+1$, or $q = 8p+1$, or $q = 10p + 1$, or $q = 14p + 1$, or $q = 16p + 1$ is also a prime, then the first case of Fermat's theorem is true for the exponent $p$.*

PROOF. We show that in each case $q$ satisfies the assumptions of (1B).

*Case 1:* Let $q = 4p + 1$.

If $p \equiv a^p$ (mod $q$) then by condition (2') above, $4^4 \equiv 1$ (mod $q$) so $4p + 1 = q$ divides $255 = 3 \times 5 \times 17$, which is absurd.

For the first condition, let $w$ be a primitive fourth root of 1, modulo $q$. So $\{1, w, w^2, w^3\}$ are the roots of $X^4 - 1 \equiv 0$ (mod $q$), and $w^2 \equiv -1$ (mod $q$), $w^3 \equiv -w$ (mod $q$). If the first condition is not verified, then by (1A) there exist $i \neq j$, $0 \leq i, j \leq 3$, such that $w^j \equiv w^i + 1$ (mod $q$). Apart from trivial cases, this leads to one of the following possibilities: $w \equiv \pm 2$ (mod $q$) or $2w \equiv \pm 1$ (mod $q$). Raising to the square it follows that $q = 5$, which is absurd.

*Case 2:* Let $q = 8p + 1$.

If $p \equiv a^p$ (mod $q$), then proceeding as before, $8p + 1 = q$ divides

TABLE 6

| $p$ | $q$ | $h$ | $R$ |
|---|---|---|---|
| 3 | $7 = 2 \times 3 + 1$ | 3 | $\pm 1$ |
| 5 | $11 = 2 \times 5 + 1$ | 2 | $\pm 1$ |
| 7 | $29 = 4 \times 7 + 1$ | 2 | $\pm 1, \pm 12$ |
| 11 | $23 = 2 \times 11 + 1$ | 5 | $\pm 1$ |
| 13 | $53 = 4 \times 13 + 1$ | 2 | $\pm 1, \pm 23$ |
| 17 | $137 = 8 \times 17 + 1$ | 3 | $\pm 1, \pm 10, \pm 37, \pm 41$ |
| 19 | $191 = 10 \times 19 + 1$ | 19 | $\pm 1, \pm 7, \pm 39, \pm 49, \pm 82$ |
| 23 | $47 = 2 \times 23 + 1$ | 5 | $\pm 1$ |
| 29 | $59 = 2 \times 29 + 1$ | 2 | $\pm 1$ |
| 31 | $311 = 10 \times 31 + 1$ | 17 | $\pm 1, \pm 6, \pm 36, \pm 52, \pm 95$ |
| 37 | $149 = 4 \times 37 + 1$ | 2 | $\pm 1, \pm 44$ |
| 41 | $83 = 2 \times 41 + 1$ | 2 | $\pm 1$ |
| 43 | $173 = 4 \times 43 + 1$ | 2 | $\pm 1, \pm 80$ |
| 47 | $659 = 14 \times 47 + 1$ | 2 | $\pm 1, \pm 12, \pm 55, \pm 144, \pm 249$ $\pm 270, \pm 307$ |
| 53 | $107 = 2 \times 53 + 1$ | 2 | $\pm 1$ |
| 59 | $827 = 14 \times 59 + 1$ | 2 | $\pm 1, \pm 20, \pm 124, \pm 270, \pm 337$ $\pm 389, \pm 400$ |
| 61 | $977 = 16 \times 61 + 1$ | 3 | $\pm 1, \pm 52, \pm 80, \pm 227, \pm 252$ $\pm 357, \pm 403, \pm 439$ |
| 67 | $269 = 4 \times 67 + 1$ | 2 | $\pm 1, \pm 82$ |
| 71 | $569 = 8 \times 71 + 1$ | 3 | $\pm 1, \pm 76, \pm 86, \pm 277$ |
| 73 | $293 = 4 \times 73 + 1$ | 2 | $\pm 1, \pm 138$ |
| 79 | $317 = 4 \times 79 + 1$ | 2 | $\pm 1, \pm 114$ |
| 83 | $167 = 2 \times 83 + 1$ | 5 | $\pm 1$ |
| 89 | $179 = 2 \times 89 + 1$ | 2 | $\pm 1$ |
| 97 | $389 = 4 \times 97 + 1$ | 2 | $\pm 1, \pm 115$ |

$8^8 - 1$, so it divides $8^4 - 1 = 4095 = 3^2 \times 5 \times 7 \times 13$ or $8^4 + 1 = 4097 = 17 \times 241$, which is an absurdity.

If the first condition is not verified, if $w$ is a primitive eighth root of 1 modulo $q$, there exist $i \neq j$, $0 \leq i, j \leq 7$, such that $w^j \equiv w^i + 1$ (mod $q$). Since $w^4 \equiv -1$ (mod $q$) we may consider the congruences $w^j \equiv \pm w^i \pm 1$ (mod $q$), with $0 \leq i \leq 3$. The primitive eighth roots of 1 modulo $q$ being $\pm w$, $\pm w^3$, and $w^3 \equiv -w^{-1}$ (mod $q$), we are reduced to study the following possibilities:

(i) $w^j \equiv \pm 2$ (mod $q$) ($j = 1, 2$);
(ii) $w^2 \equiv w \pm 1$ (mod $q$); and
(iii) $2w \equiv \pm 1$ (mod $q$).

We discuss the various cases.

(i)    Raising to the fourth power leads to $q \mid 15$ or $q = 17$, an absurdity.

(ii)    Raising to the square,

$$-1 \equiv w^2 \pm 2w + 1 \equiv \begin{cases} w + 1 + 2w + 1 \equiv 3w + 2 \ (\text{mod } q), \\ w - 1 - 2w + 1 \equiv -w \ (\text{mod } q). \end{cases}$$

Hence $w \equiv \pm 1$ (mod $q$), an absurdity.

(iii)    Raising to the fourth power, $q = 17$, absurd.

*Case* 3: Let $q = 16p + 1$.

If $p \equiv a^p$ (mod $q$), then with the same method, $16p + 1 = q$ divides $16^{16} - 1 = (16^8 + 1)(16^4 + 1)(16^2 + 1) \times 17 \times 15$. Clearly $16p + 1$ does not divide 15, 17, 257. If $16p + 1$ divides $16^4 + 1 = 65537$, which is a prime (in fact a Fermat prime, $2^{16} + 1$), then $p = 16^3$, which is absurd. If $16p + 1$ divides $16^8 + 1 = 2^{32} + 1 = 641 \times 6700417$ (decomposition into primes given by Euler), then $p = 40$ or $418776$, again absurd.

If the first condition is not verified, if $w$ is a primitive sixteenth root of 1 modulo $q$, since $\pm w$, $\pm w^3$, $\pm w^5$, $\pm w^7$ are all the primitive sixteenth roots of 1 modulo $q$, and since $w^8 \equiv -1$ (mod $q$) there exist $i, j$, $0 \leq i \leq j \leq 7$, such that $w^j \equiv \pm w^i \pm 1$ (mod $q$). This leads to one of the following congruences, with all possible sign combinations:

(i) $w^j \equiv \pm 2$ (mod $q$) ($j = 1, 2, \ldots, 7$);
(ii) $w^2 \equiv w \pm 1$ (mod $q$);
(iii) $w^3 \equiv \pm w + 1$ (mod $q$);
(iv) $w^4 \equiv w \pm 1$ (mod $q$); and
(v) $w^4 \equiv \pm w^2 \pm 1$ (mod $q$).

We discuss the various possibilities.

(i)    Raising to the eighth power leads to $q \mid 257$ or $q \mid 255 = 3 \times 5 \times 17$, an absurdity.

(ii)    Raising to the square leads to $w^4 \equiv -w \pmod q$, which is impossible, or $w^4 \equiv 3w + 2 \pmod q$. Squaring again, this leads to $-1 \equiv 9w^2 + 12w + 4 \pmod q$, and substituting $w^2 \equiv w + 1 \pmod q$ we obtain $3w \equiv -2 \pmod q$. Hence $w^4 \equiv 3w + 2 \equiv 0 \pmod q$, which is impossible.

(iii)    Raising to the cube:

$$-w \equiv \pm w^3 + 3w^2 \pm 3w + 1 \pmod q$$

and substituting,

$$-w \equiv w \pm 1 + 3w^2 \pm 3w + 1 \pmod q.$$

According to the choice of the sign, we have $3w^2 + 5w + 2 \equiv 0 \pmod q$ or $3w^2 - w \equiv 0 \pmod q$.

In the first case, multiplying with $w$ and substituting, we obtain $5w^2 + 5w + 3 \equiv 0 \pmod q$; subtracting, $2w^2 \equiv -1 \pmod q$ and therefore $-2^4 \equiv 1 \pmod q$, so $q \mid 17$, impossible. If $3w \equiv 1 \pmod q$, then $3w^3 \equiv -3w + 3 \equiv 2 \pmod q$ so raising to the cube, $-3^3 w \equiv 8 \pmod q$, hence $q$ divides 17, an absurdity.

(iv)    Squaring: $-1 \equiv w^2 \pm 2w + 1 \pmod q$, so $w^2 \equiv \mp 2w - 2 \pmod q$, hence $w \pm 1 \equiv w^4 \equiv 4w^2 \pm 8w + 4 \pmod q$. This gives, according to the choice of sign,

$$4w^2 \equiv \begin{cases} -7w - 3 \\ 9w - 5 \end{cases} \pmod q$$

and subtracting,

$$0 \equiv \begin{cases} w + 5 \\ w + 3 \end{cases} \pmod q.$$

So $w^4 \equiv w \pm 1 \equiv -4 \pmod q$ and raising to the square, $-1 \equiv 16 \pmod q$ so $q \mid 17$, impossible.

(v)    Raising to the square, $-1 \equiv w^4 \pm 2w^2 + 1 \pmod q$ and substituting $w^4 \equiv \pm w^2 \pm 1 \pmod q$, in all cases we obtain $w^2 \equiv \pm 1 \pmod q$, which is impossible.

*Case* 4: Let $q = 10p + 1$.

If $p \equiv a^p \pmod q$, by the above method $10p + 1 = q$ divides $(10^5 + 1)(10^5 - 1)$. If $q$ divides $10^5 + 1 = 100\,001 = 11 \times 9091$ (this last number is a prime) then $p = 909$, an absurdity. If $q$ divides $10^5 - 1 = 99\,999 = 3^2 \times 41 \times 271$ then $p = 4$ or $27$, an absurdity.

If the first condition is not verified, if $w$ is a primitive tenth root of 1 modulo $q$, then there exist $i, j$, $0 \leq i \leq j < 4$, such that $w^j \equiv \pm w^i \pm 1 \pmod{q}$. The above conditions lead to either one of the following congruences:

(i) $w^j \equiv \pm 2 \pmod{q}$ $(j = 1, 2, 3, 4)$;
(ii) $w^2 \equiv w \pm 1 \pmod{q}$; and
(iii) $w^4 \equiv w \pm 1 \pmod{q}$.

We discuss the various possibilities.

(i)  Raising to the fifth power leads to $q \mid 31$ or $q \mid 33 = 3 \times 11$, impossible.

(ii)  We have

$$w^4 \equiv w^2 \pm 2w + 1 \equiv w \pm 1 \pm 2w + 1 \equiv \begin{cases} 3w + 2 \\ -4 \end{cases} \pmod{q}.$$

The second case is not possible. In the first case multiplying with $w$ and substituting $-1 \equiv 3w^2 + 2w = 3w + 3 + 2w \pmod{q}$ so $5w \equiv -4 \pmod{q}$. Multiplying with the previous congruence, $-5 \equiv -12w - 8 \pmod{q}$ so $12w \equiv -3 \pmod{q}$ and $4w \equiv -1 \pmod{q}$ hence subtracting, $w \equiv -3 \pmod{q}$ and therefore $q \mid 11$, absurd.

(iii)  Multiplying with $w$: $-1 \equiv w^2 \pm w \pmod{q}$, so $w^2 \equiv \mp w - 1 \pmod{q}$ and this was considered in case (ii).

*Case* 5: Let $q = 14p + 1$.

If $p \equiv a^p \pmod{q}$, we see in the same way that $q$ divides $14^7 + 1$ or $14^7 - 1$. But $14^7 + 1 = 105\,413\,505 = 3 \times 5 \times 7\,027\,567$ (this last number is a prime). Then $p = 501\,969$, an absurdity since this number is a multiple of 3. Also, $14^7 - 1 = 105\,413\,503 = 13 \times 8\,108\,731$ (this last number is a prime). Then $p = 579\,195$, which is absurd.

If the first condition is not verified, if $w$ is a primitive fourteenth root of 1 modulo $q$, then there exist $i \neq j$, $0 \leq i, j \leq 13$, such that $w^j \equiv w^i + 1 \pmod{q}$. Since $w^7 \equiv -1 \pmod{q}$, the above conditions lead to either one of the following congruences:

(i) $w^j \equiv \pm 2 \pmod{q}$ $(j = 1, 2, \dots, 6)$;
(ii) $w^2 \equiv w \pm 1 \pmod{q}$; and
(iii) $w^3 \equiv \pm w + 1 \pmod{q}$.

We discuss the various cases.

(i)   $-1 \equiv \pm 2^7 \pmod{q}$, so $q \mid 127$ or $129 = 3 \times 43$, which gives $p = 3$, and this was excluded.

(ii)   $w^2 \equiv w \pm 1 \pmod{q}$ gives

$$w^4 \equiv w^2 \pm 2w + 1 \equiv w \pm 1 \pm 2w + 1 \equiv \begin{cases} 3w + 2 \\ -w \end{cases} \pmod{q},$$

the second case being impossible. Squaring,

$$-w \equiv 9w^2 + 12w + 4 \equiv 9w + 9 + 12w + 4 \pmod{q},$$

so $22w \equiv -13 \pmod{q}$. Then $22w^2 - 22 \equiv -13 \pmod{q}$ so $22w^2 \equiv 9 \pmod{q}$. Then $-13w \equiv 9 \pmod{q}$ and from this we obtain $9w \equiv -4 \pmod{q}$, $-4w \equiv 5 \pmod{q}$, $5w \equiv 1 \pmod{q}$; so $25 \equiv -20w \equiv -4 \pmod{q}$, hence $q \mid 29$, which is impossible.

(iii)   $w^3 \equiv \pm w + 1 \pmod{q}$ gives, to the cube:

$$-w^2 \equiv \pm w^3 + 3w^2 \pm 3w + 1$$
$$\equiv \pm w + 1 + 3w^2 \pm 3w + 1 \equiv 3w^2 \pm 4w + 2 \pmod{q},$$

hence $2w^2 \pm 2w + 1 \equiv 0 \pmod{q}$ so $2(\pm w + 1) \pm 2w^2 + w \equiv 0 \pmod{q}$ and

$$\begin{cases} 2w^2 + 3w + 2 \equiv 0 \pmod{q}, \\ -2w^2 - w + 2 \equiv 0 \pmod{q}, \end{cases}$$

hence

$$\begin{cases} w + 1 \equiv 0 \pmod{q}, \\ -3w + 3 \equiv 0 \pmod{q}, \end{cases}$$

and this gives $w \equiv \mp 1 \pmod{q}$, which is impossible.   □

With this criterion, Legendre had actually shown that the first case of Fermat's theorem holds for every prime exponent $p < 197$. Indeed, for each such prime $p$, there exists a prime $q = 2kp + 1$, with $2k \in \{2, 4, 8, 10, 14, 16\}$. On the other hand, $38 \times 197 + 1 = 7487$ is a prime, but $2k \times 197 + 1$ is not a prime if $2k < 38$, $6 \nmid 2k$.

The limitation in Legendre's results was due to the size of the numbers involved. For example, to test whether $p = 197$ is not a $p$th power modulo $q = 7487$, would lead to find whether 7487 divides $38^{19} \pm 1$. Maillet extended Legendre's result in 1897, pushing the limit up to $p = 211$.

Mirimanoff used a method involving Bernoulli numbers, in 1905, to extend the results to 257.

In 1908, Dickson published two papers in which he explored Legendre's ideas and with a more careful analysis involving congruences, he showed that the first case of Fermat's theorem holds for every prime exponent $p < 7000$ (with the exception of $p = 6857$, which he did not take the trouble to examine). See also Maillet's comments (1908).

More progress along this line was made by Krasner (1940), Dénes (1951), and Rivoire (1968).

For the primes $p$ such that $6p + 1$ or $12p + 1$ is also a prime, the method of proof breaks down and does not lead to any conclusion. It should be noted that in 1974 Gandhi announced without proof that if $p$ and $6p+1$ are primes then the first case holds for $p$; since no proof has ever been published, there is reason to doubt of the justification of the statement. In this connection, we quote the paper by Granville and Powell (1988).

An interesting, but very difficult question is whether there exist infinitely many primes $p$ such that $2p + 1$ (or $4p + 1$, or $8p + 1$, etc...) is also a prime. We discuss this problem in the Appendix to this chapter.

We still note here the following result of Vandiver (1926) :

**(1F)**    *Let $p$ and $q = 2kp + 1$ be odd primes (with $k \geq 1$). If $2k = 2^v p^h$, where $h \geq 0$ and $p$ does not divide $v$, and if $2$ is not a pth power modulo $q$, then condition (2) above is satisfied.*

PROOF. We show (2′). If $(2k)^{2k} \equiv 1 \pmod{q}$ or equivalently $p^{2k} \equiv 1 \pmod{p}$, then $2^{2hv} \equiv 2^{2kv}p^{2kh} = (2^v p^h)^{2k} = (2k)^{2k} \equiv 1 \pmod{q}$. Since $p$ does not divide $v$, there exist integers $a, b$ such that $av = 1 + bp$. Then

$$1 \equiv 2^{2kva} \equiv 2^{(1+bp)2k} \equiv 2^{2k}(2^{2kp})^b = 2^{2k}(2^{q-1})^b \equiv 2^{2k} \pmod{q}.$$

If $g$ is a primitive root modulo $q$ and $2 \equiv g^s \pmod{q}$, then $1 \equiv 2^{2k} \equiv g^{2ks} \pmod{q}$. So $q - 1 = 2kp$ divides $2ks$, hence $s = ps'$ and $2 \equiv (g^{s'})^p \pmod{q}$, which is a contradiction. $\square$

Using (1B), Vandiver deduced in 1926 the following result which however had been proved by Wendt in 1894, using his form of Sophie Germain's theorem:

**(1G)**    *If $p$ and $q = 2kp + 1$ are odd primes, with $2k = 2^v p^h$, $h \geq 0$ and $v$ not divisible by $p$, if the congruence $X^p + Y^p + Z^p \equiv 0 \pmod{q}$ has only the trivial solution, then the first case of Fermat's theorem holds for the exponent $p$.*

PROOF. By (1B) and (1F) it suffices to show that 2 is not a $p$th power modulo $q$. If $2 \equiv a^p \pmod{q}$ then $a^p + (-1)^p + (-1)^p \equiv 2 + (-1) + (-1) \equiv 0 \pmod{q}$ contrary to the hypothesis. ☐

Sophie Germain's theorem, corollaries and variations were redis-covered by several authors. In 1953, Thébault proved:

**(1H)**    *If $m \geq 2$ is an integer such that $2m + 1$ is a prime, if there exist pairwise relatively prime nonzero integers $x, y, z$ such that $x^m + y^m = z^m$ then $2m + 1$ divides $xyz$.*

PROOF. If $2m + 1$ does not divide $x$ then by Fermat's little theorem $x^{2m} \equiv 1 \pmod{2m + 1}$ hence $x^m \equiv \pm 1 \pmod{2m + 1}$.
Similarly

$$y^m \equiv \pm 1 \pmod{2m + 1},$$
$$z^m \equiv \pm 1 \pmod{2m + 1},$$

hence $0 = x^m + y^m - z^m = (\pm 1) + (\pm 1) - (\pm 1) \pmod{2m + 1}$, which is impossible. ☐

This same result (even with the further hypothesis that $m$ be prime) is proved again by Stone in 1963 and Gandhi in 1966, in the same journal!
Gandhi showed, also in 1965, a result similar to Thébault's:

**(1I)**    *If $m \geq 2$ is an integer such that $4m + 1$ is a prime, if $x, y, z$ are nonzero pairwise relatively prime integers such that $x^m + y^m = z^m$ then $4m + 1$ divides $xyz$.*

PROOF. If $m = 3$, the statement is trivially true, by Chapter I, §4.
Let $m > 3$ and assume that $4m + 1$ does not divide $xyz$. From $x^m + y^m = z^m$ it follows that $x^{2m} + y^{2m} + 2x^m y^m = z^{2m}$. Since $4m + 1$ is a prime not dividing $x$ then by Fermat's little theorem $x^{4m} \equiv 1 \pmod{4m + 1}$, so $x^{2m} \equiv \pm 1 \pmod{4m + 1}$. Similarly

$$y^{2m} \equiv \pm 1 \pmod{4m + 1}$$

and

$$z^{2m} \equiv \pm 1 \pmod{4m+1}.$$

Hence $\pm 1 \pm 1 + 2x^m y^m \equiv \pm 1 \pmod{4m+1}$ so $2x^m y^m \equiv \pm 1$ or $\pm 3$ $\pmod{4m+1}$ and $\pm 4 \equiv 4x^{2m} y^{2m} \equiv 1$ or $9 \pmod{4m+1}$. This implies that $4m+1 = 3, 5$, or $13$, hence $m = 3$, which is a contradiction. $\square$

This same result (with the further hypothesis that $m$ is prime) is proved again by Gandhi in 1966 and 1970 and by Christilles in 1967.
   Perisastri (1969) proved:

**(1J)**   *If $p > 51$ is a prime such that $8p+1$ is also a prime, if $x, y, z$ are nonzero pairwise relatively prime integers such that $x^p + y^p = z^p$ then $8p+1$ divides $xyz$.*

**(1K)**   *If $m \geq 3$ is an integer such that $3m+1$ is a prime, if $x, y, z$ are pairwise nonzero relatively prime integers such that $x^m + y^m = z^m$, then $3m+1$ divides $xyz$.*

Krishnasastri and Perisastri proved in 1965:

**(1L)**   *If $p$ is an odd prime, if $x, y, z$ are integers such that $x^p + y^p = z^p$ and $p$ does not divide $xz$, then there exists an integer $k \geq 1$ such that $1 + kp$ divides $z$.*

Combining (1C) with Sophie Germain's theorem, we have (see Stone (1963), Perisastri (1968)):

**(1M)**   *Let $p$ and $2p+1$ be odd primes. If $x, y, z$ are nonzero, pairwise relatively prime integers such that $x^p + y^p + z^p = 0$, then $p^2$ divides one (and only one) of the integers $x, y, z$.*

PROOF. By Sophie Germain's theorem we may assume, for example, that $p$ divides $z$. By (1C), $p^2$ divides $z$. $\square$

Pomey obtained in 1923 and 1925, with similar methods, several sufficient conditions for the first case of Fermat's theorem for the prime exponent $p$:

**(1N)**    *Let p be an odd prime and assume that either one of the following conditions is satisfied:*

  (a) $p \equiv 1 \pmod 4$ *and* $2p + 1$ *divides* $2^p + 1$.
  (b) $p \equiv 3 \pmod 4$ *and* $2p + 1$ *divides* $2^p - 1$.
  (c) $4p + 1$ *divides* $2^{2p} + 1$.
  (d) $4p + 1 \equiv 5 \pmod{12}$ *and* $4p + 1$ *divides* $3^{3p} + 1$.
  (e) $8p + 1$ *divides* $2^{4p} - 1$.
  (f) $10p + 1$ *divides* $5^{5p} - 1$.

*Then the first case of Fermat's theorem is true for the exponent p.*

All the above results do not suffice to conclude that there exist infinitely many prime exponents $p$ for which the first case of Fermat's theorem is true. This was first proved, with analytical methods, in 1985 by Adleman and Heath-Brown and and by Fouvry.

Earlier, in 1897, studying the class group of the cyclotomic field, Maillet showed that for every odd prime $p$ there exists an exponent $e$ (depending on $p$) such that the first case of Fermat's theorem is true for the exponent $p^e$. In particular, this implied the existence of an infinite set of pairwise relatively prime exponents for which the first case is true. This last statement was proved again by Kapferer in 1964. His proof was not elementary, since it used the theorems of Furtwängler, as generalized by Moriya (requiring class field theory). In 1978, Powell discovered independently the following very simple proof:

**(1O)**

  (1) *If $p$ is any odd prime, $n = p(p-1)/2 = 2^u m$ where $u \geq 0$, $m$ is odd, if $x, y, z$ are nonzero integers such that $x^n + y^n + z^n = 0$ then $\gcd(m, xyz) \neq 1$.*
  (2) *There exists an infinite set of pairwise relatively prime exponents for which the first case of Fermat's theorem is true.*

PROOF. (1) If $p = 3$ then $n = 3$ and the hypothesis is not satisfied. Let $p > 3$. Suppose that $\gcd(m, xyz) = 1$. Then $p \nmid xyz$ so $x^{(p-1)/2} \equiv \pm 1 \pmod p$ and $x^n \equiv \pm 1 \pmod p$. Similarly $y^n \equiv \pm 1 \pmod p$, $z^n \equiv \pm 1 \pmod p$, hence $x^n + y^n + z^n \not\equiv 0 \pmod p$ and a fortiori, $x^n + y^n + z^n \neq 0$.

  (2)   Assume that $n_1, \ldots, n_k$ are pairwise relatively prime exponents for which the first case of Fermat's theorem is true. Consider

the arithmetic progression $\{-1 + 4n_1 n_2 \cdots n_k t \mid t = 0, 1, 2, \ldots\}$. By Dirichlet's theorem on primes in arithmetic progressions there exists an odd prime $p$ such that $p \equiv -1 \pmod{4n_1 n_2 \cdots n_k}$. Let $n_{k+1} = p(p-1)/2$, so $n_{k+1}$ is odd. Since $\gcd(p(p-1)/2, (p+1)/2) = 1$ then $\gcd(n_{k+1}, n_1 \cdots n_k) = 1$. By (1), the first case is true for the exponent $n_{k+1}$, and this suffices to complete the proof. $\square$

## Bibliography

1823 Legendre, A.M., *Recherches sur quelques objets d'analyse in-déterminée, et particulièrement sur le théorème de Fermat*, Mém. Acad. Roy. Sci. Institut France, **6** (1823), 1–60; reprinted as "Second Supplément" in 1825, to a printing of *Essai sur la Théorie des Nombres* ($2^e$ édition), Courcier, Paris; reprinted in Sphinx-Oedipe, **4** (1909), 97–128.

1879 Germain, S., *Oeuvres Philosophiques* (editor H. Stupuy), pp. 298–302 and 363–364, P. Ritti, Paris, 1879.

1894 Wendt, E., *Arithmetische Studien über den letzten Fermat-schen Satz, welcher aussagt daß die Gleichung $a^n = b^n + c^n$ für $n > 2$ in ganzen Zahlen nicht auflösbar ist*, J. Reine Angew. Math., **113** (1894), 335–346.

1897 Maillet, E., *Sur l'équation indéterminée $ax^{\lambda^t} + by^{\lambda^t} = cz^{\lambda}$*, Assoc. Française Avanc. Sci., St. Etienne, **26** (1897), 156–168.

1905 Mirimanoff, D., *L'équation indéterminée $x^l + y^l + z^l = 0$ et le critérium de Kummer*, J. Reine Angew. Math., **128** (1905), 45–68.

1908 Dickson, L.E., *On the last theorem of Fermat*, Messenger Math., **38** (1908), 14–32.

1908 Dickson, L.E., *On the last theorem of Fermat* (second paper), Quart. J. Pure Appl. Math., **40** (1908), 27–45.

1908 Maillet, E., *Question 612 de Worms de Romilly*, L'Interm. Math., **15** (1908), 247–248.

1910 Bachmann, P., *Niedere Zahlentheorie*, Teubner, Leipzig, 1910; reprinted by Chelsea, New York, 1966.

1923 Pomey, L., *Sur le dernier théorème de Fermat*, C. R. Acad. Sci. Paris, **177** (1923), 1187–1190.

1925 Pomey, L., *Sur le dernier théorème de Fermat*, J. Math. Pures Appl., (9), **4** (1925), 1–22.

1926 Vandiver, H.S., *Note on trinomial congruences and the first case of Fermat's last theorem*, Ann. of Math., **27** (1926), 54–56.

1935 Bang, A.S., *Om tal af Formen $a^m + b^m - c^m$*, Mat. Tidsskrift, **B** (1935), 49–59.

1940 Krasner, M., *À propos du critère de Sophie Germain–Furtwängler pour le premier cas du théorème de Fermat*, Mathematica (Cluj), **16** (1940), 109–114.

1951 Dénes, P., *An extension of Legendre's criterion in connection with the first case of Fermat's last theorem*, Publ. Math. Debrecen, **2** (1951), 115–120.

1953 Thébault, V., *A note on number theory*, Amer. Math. Monthly, **60** (1953), 322–323.

1963 Stone, D.E., *On Fermat's last theorem*, Amer. Math. Monthly, **70** (1963), 976–977.

1964 Kapferer, H., *Verifizierung des symmetrischen Teils der Fermatschen Vermutung für unendlich viele paarweise teilerfremde Exponenten E*, J. Reine Angew. Math., **214/5** (1964), 360–372.

1965 Gandhi, J.M., *A note on Fermat's last theorem*, Math. Notae, **20** (1965), 107–108.

1965 Krishnasastri, M.S.R. and Perisastri, M., *On some diophantine equations*, Math. Student, **33** (1965), 73–76.

1966 Grosswald, E., *Topics from the Theory of Numbers*, Macmillan, New York, 1966.

1966 Gandhi, J.M., *A note on Fermat's last theorem*, Amer. Math. Monthly, **73** (1966), 1106–1107.

1967 Christilles, W.E., *A note concerning Fermat's conjecture*, Amer. Math. Monthly, **74** (1967), 292–294.

1968 Perisastri, M., *A note on Fermat's last theorem*, Amer. Math. Monthly, **75** (1968), 170.

1968 Rivoire, P., *Dernier Théorème de Fermat et Groupe de Classes dans Certains Corps Quadratiques Imaginaires*, Thèse, Université Clermont-Ferrand, 1968, 59 pp.; reprinted in Ann. Sci. Univ. Clermont-Ferrand II, **68** (1979), 1–35.

1969 Perisastri, M., *On Fermat's last theorem*, Amer. Math. Monthly, **76** (1969), 671–675.

1970 Gandhi, J.M., *On Fermat's last theorem*, An. Ştiinţ. Univ. "Al. I. Cuza" Iaşi, Secţ. I (N.S), **16** (1970), 241–248.

1974 Gandhi, J.M., *On Fermat's last theorem*, Notices Amer. Math. Soc., **21** (1974), A-53.

1978 Powell, B., *Proof of a special case of Fermat's last theorem*, Amer. Math. Monthly, **85** (1978), 750–751.

1985 Adleman, L.M. and Heath-Brown, D.R., *The first case of Fermat's last theorem*, Invent. Math., **79** (1985), 409–416.

1985 Fouvry, E., *Théorème de Brun–Titchmarsh. Application au théorème de Fermat*, Invent. Math., **79** (1985), 383–407.

1988 Granville, A. and Powell, B., *On Sophie Germain's type criteria for Fermat's last theorem*, Acta Arith., **50** (1988), 265–277.

## IV.2. Wendt's Theorem

Wendt indicated in 1894 a determinantal criterion for the existence of a nontrivial solution of Fermat's congruence

$$(2.1) \qquad X^p + Y^p + Z^p \equiv 0 \pmod{q},$$

where $p, q$ are distinct odd primes.

To begin, we wish to exclude from our discussion the following trivial case; it also holds without assuming the exponent in (2.1) to be a prime:

**(2A)**    *If $q$ is an odd prime, if $n \geq 1$ is such that $\gcd(n, q-1) = 1$ then there exist integers $x, y, z$, not multiples of $q$, such that $x^n + y^n + z^n \equiv 0 \pmod{q}$.*

PROOF. By hypothesis, $\gcd(n, q-1) = 1$, so there exist integers $a, b$ such that $an + b(q-1) = 1$. Let $x_0, y_0, z_0$ be integers, not multiples of $q$, such that $x_0 + y_0 + z_0 \equiv 0 \pmod{q}$. Then

$$\left\{ \begin{array}{l} x_0^{an} \equiv x_0 \pmod{q}, \\ y_0^{an} \equiv y_0 \pmod{q}, \\ z_0^{an} \equiv z_0 \pmod{q}, \end{array} \right.$$

so $(x_0^a)^n + (y_0^a)^n + (z_0^a)^n \equiv 0 \pmod{q}$.    $\square$

In particular, if $n = p$ is a prime not dividing $q - 1$ then (2.1) has a nontrivial solution.

Wendt's criterion is expressed in terms of the circulant of binomial coefficients. More generally, let $n \geq 1$ and let $\xi_i = \cos 2\pi i/n + \sqrt{-1} \sin 2\pi i/n$ (for $i = 0, 1, \ldots, n-1$) be the $n$ $n$th roots of 1; so $\xi_0 = 1$. The *circulant* of the $n$-tuple $(a_0, a_1, \ldots, a_{n-1})$ of complex numbers $a_i$ is, by definition, the determinant of the matrix

(2.2)
$$
C = \begin{pmatrix}
a_0 & a_1 & \cdots & a_{n-1} \\
a_{n-1} & a_0 & \cdots & a_{n-2} \\
\vdots & \vdots & \ddots & \vdots \\
a_1 & a_2 & \cdots & a_0
\end{pmatrix}.
$$

We denote it by $\mathrm{Circ}(a_0, a_1, \ldots,, a_{n-1})$. The circulant is expressed in terms of $n$th roots of 1 and equally as the resultant of two polynomials (see Chapter II, §4). Spottiswoode (1853) and also Stern (1871) and Muir (1920) proved:

LEMMA 2.1. *Let* $a_0, a_1, \ldots, a_{n-1} \in K$, *let* $G(X) = a_0 + a_1 X + \cdots + a_{n-1}X^{n-1}$ *and let* $\xi_0 = 1, \xi_1, \ldots, \xi_{n-1}$ *be the* $n$th *roots of* 1. *The circulant of* $a_0, a_1, \ldots, a_{n-1}$ *is equal to*

$$
\mathrm{Circ}(a_0, a_1, \ldots, a_{n-1}) = \prod_{i=0}^{n-1} G(\xi_i) = \mathrm{Res}(G(X), X^n - 1)
$$

(*where* Res *denotes the resultant*).

PROOF. Let

$$
A = \begin{pmatrix}
0 & 1 & 0 & \cdots & 0 \\
0 & 0 & 1 & \cdots & 0 \\
\vdots & \vdots & \vdots & \ddots & \vdots \\
1 & 0 & 0 & \cdots & 0
\end{pmatrix}
$$

($n \times n$ matrix), so $I, A, A^2, \ldots, A^{n-1}$ are distinct and $A^n = I$. As is easily seen,

$$
C = a_0 I + a_1 A + a_2 A^2 + \cdots + a_{n-1} A^{n-1}.
$$

The characteristic polynomial of $A$ is $\det(XA - I) = X^n - 1$. Since it has distinct roots $\xi_0 = 1, \xi_1, \ldots, \xi_{n-1}$, then $A$ is diagonalizable, that is, there exists an invertible matrix $U$ (with complex entries) such that

$$UAU^{-1} = \begin{pmatrix} \xi_0 & 0 & \cdots & 0 \\ 0 & \xi_1 & \cdots & 0 \\ \vdots & \vdots & \ddots & \vdots \\ 0 & 0 & \cdots & \xi_{n-1} \end{pmatrix}.$$

Hence

$$UCU^{-1} = \begin{pmatrix} G(\xi_0) & 0 & \cdots & 0 \\ 0 & G(\xi_1) & \cdots & 0 \\ \vdots & \vdots & \ddots & \vdots \\ 0 & 0 & \cdots & G(\xi_{n-1}) \end{pmatrix},$$

where $G(X) = a_0 + a_1 X + \cdots + a_{n-1} X^{n-1}$. So $\mathrm{Circ}(a_0, a_1, \ldots, a_{n-1}) = \det(C) = \det(UCU^{-1}) = \prod_{i=0}^{n-1} G(\xi_i)$. By Chapter II, (4B), we also have $\mathrm{Circ}(a_0, a_1, \ldots, a_n) = \mathrm{Res}(G(X), X^n - 1)$. $\square$

The following result of Wendt is about the circulant of binomial coefficients. Accordingly, for every $n \geq 1$ we define the *Wendt determinant* to be

(2.3)     $$W_n = \mathrm{Circ}\left(1, \binom{n}{1}, \binom{n}{2}, \ldots, \binom{n}{n-1}\right).$$

If $G(X) = 1 + \binom{n}{1}X + \binom{n}{2}X^2 + \cdots + \binom{n}{n-1}X^{n-1} = (1+X)^n - X^n$ then by the lemma, $W_n = \mathrm{Res}(G(X), X^n - 1) = \prod_{n=0}^{n-1}[(1 + \xi_i)^n - 1]$. Now we give Wendt's criterion (1894); see also Matthews (1895), Bang (1935), and an expository presentation by Rivoire (1968):

**(2B)**    *Let $p$ be an odd prime and assume that $q = 2kp + 1$ (with $k \geq 1$) is also a prime. Then there exist integers $x, y, z$, not multiples of $q$, such that $x^p + y^p + z^p \equiv 0 \pmod{q}$ if and only if $q$ divides $W_{2k}$.*

PROOF. By (1A), Fermat's congruence $X^p + Y^p + Z^p \equiv 0 \pmod{q}$ has a nontrivial solution if and only if the system of congruences

$$\begin{cases} X^{2k} \equiv 1 \pmod{q}, \\ (X+1)^{2k} \equiv 1 \pmod{q}, \end{cases}$$

has a nontrivial solution, or equivalently, the system of congruences

$$\begin{cases} X^{2k} - 1 \equiv 0 \pmod{q}, \\ (X+1)^{2k} - X^{2k} \equiv 0 \pmod{q}, \end{cases}$$

has a nontrivial solution. This holds exactly when the resultant of the polynomials $X^{2k} - 1$ and $G(X) = (X+1)^{2k} - X^{2k} = 1 + \binom{2k}{1}X + \binom{2k}{2}X^2 + \cdots + \binom{2k}{2k-1}X^{2k-1}$ is congruent to 0 modulo $q$; by Lemma 2.1 this means that $W_{2k} \equiv 0 \pmod{q}$. $\square$

Now we turn our attention to the computation of Wendt's determinant

$$(2.4) \qquad\qquad W_n = \prod_{j=0}^{n-1} [(1+\xi_j)^n - 1].$$

The following result was stated without proof by Wendt (1894), and proved thereafter also by Matthews (1895), E. Lehmer (1935), Bang (1935), and Frame (1980):

**(2C)**    $W_n = 0$ *if and only if* 6 *divides* $n$.

PROOF. Assume that 6 divides $n$, and let $\xi = \xi_1 = \cos 2\pi/n + \sqrt{-1}\sin 2\pi/n$. Let $l = n/3$, so $\omega = \xi^l$ is a primitive cubic root of 1. Hence $1 + \omega + \omega^2 = 0$, therefore $1 + \xi^l = -\xi^{2l}$ and $(1+\xi^l)^n = 1$. We conclude that $W_n = 0$.

Conversely, if $W_n = 0$ there exists $j$ such that $(1 + \xi_j)^n = 1$, so $\xi_j$, $1 + \xi_j$ are $n$th roots of 1 and since the triangle with vertices $0$, $1$, $1 + \xi_j$,

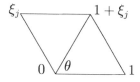

is equilateral, then $\theta = 2\pi/6$ (or $\theta = -2\pi/6$) and $1+\xi_j$ is a primitive sixth root of 1. But $(1 + \xi_j)^n = 1$ hence 6 divides $n$. $\square$

As a corollary:

**(2D)**    *If $p$ and $6mp + 1 = q$ are primes then the congruence* $X^p + Y^p + Z^p \equiv 0 \pmod{q}$ *has a nontrivial solution.*

PROOF. This follows at once from (2B) and (2C). $\square$

In view of (2C), it is customary to modify the definition of the Wendt determinant when 6 divides $n$, by putting

$$W_n = \prod_{i=0}^{n-1} G(\xi_i),$$

where

(2.5) $$G(X) = \frac{(X+1)^n - X^n}{X^2 + X + 1}.$$

In 1935, E. Lehmer indicated, without proof:

**(2E)**    *If $d$ divides $n$ then $W_d$ divides $W_n$.*

PROOF. We may assume that $W_d \neq 0$, i.e., $6 \nmid d$. We note that since $d \mid n$ then each $d$th root of 1 is also an $n$th root of 1. From $W_d \neq 0$, then

$$\frac{W_n}{W_d} = \prod_{\xi_j^d = 1} \frac{(1+\xi_j)^n - 1}{(1+\xi_j)^d - 1} \times \prod_{\xi_i^n = 1, \xi_i^d \neq 1} [(1+\xi_i)^n - 1].$$

If $n = de$ then

$$\frac{(1+\xi_j)^n - 1}{(1+\xi_j)^d - 1} = (1+\xi_j)^{d(e-1)} + \cdots + (1+\xi_j)^d + 1.$$

Therefore $W_n/W_d$ is an algebraic integer, but also a rational number, hence a rational integer.    □

The next property was also indicated by E. Lehmer, without proof. A proof (not the one below) appears in Frame's paper (1980); the weaker statement $2^n - 1$ divides $W_n$ was proved by Bang (1935):

**(2F)**    *If $n \geq 1$ then $W_n = (-1)^{n-1}(2^n - 1)u^2$, where $u$ is an integer.*

PROOF.

$$W_n = \prod_{j=0}^{n-1} [(1 + \xi^j)^n - 1],$$

where $\xi = \cos 2\pi/n + \sqrt{-1} \sin 2\pi/n$. Thus $W_n$ has the factor $2^n - 1$ (when $j = 0$) and if $n$ is even, also the factor $-1$ (when $j = n/2$).

So

$$W_n = (-1)^{n-1}(2^n - 1) \prod_{j \neq 0, n/2} [(1 + \xi^j)^n - 1].$$

Let $u = \prod_{0 < j < n/2}[(1 + \xi^j)^n - 1]$. We note that for every $j \neq 0, n/2$ we have $(1 + \xi^{-j})^n - 1 = [\xi^{-j}(1 + \xi^j)]^n - 1 = (1 + \xi^j)^n - 1$ so $u$ is real and $W_n = (-1)^{n-1}(2^n - 1)u^2$, and it remains to show that $u \in \mathbb{Z}$. Let $\sigma$ be any automorphism of the field $Q(\xi)$, so $\sigma(\xi) = \xi^l$ where $1 \leq l < n$ and $\gcd(l, n) = 1$. Hence $\sigma[(1 + \xi^j)^n - 1] = (1 + \xi^k)^n - 1$ where $0 < k < n/2$ and $jl \equiv \pm k \pmod{n}$ because $(1 + \xi^{-k})^n - 1 = (1 + \xi^k)^n - 1$. If $0 < j, j' < n/2$, let $k, k'$ be such that $0 < k, k' < n/2$ and $jl \equiv \pm k \pmod{n}$, $j'l \equiv \pm k' \pmod{n}$. We note that if $j \neq j'$ then $k \neq k'$, because if $jl \equiv \pm j'l \pmod{n}$ then $j \equiv \pm j' \pmod{n}$ and this would imply $j = j'$. Therefore, different factors of $u$ have distinct images by $\sigma$, so $\sigma(u) = u$ hence $u \in \mathbb{Q}$, being invariant by the automorphism of $\mathbb{Q}(\xi)$. But $u$ is also an algebraic integer, hence $u \in \mathbb{Z}$.  □

For $n$ even, Frame proved (1980):

**(2G)**     If $n = 2m$ and $3 \nmid n$ then $W_n = -3((2^n - 1)/3)^3 u^6$ where $u$ is an integer. In particular, if $p$ is a prime, $p \equiv 5 \pmod 6$, then $W_{p-1} = -3((2^{p-1} - 1)/3)^3 u^6$, where $u$ is an integer.

PROOF. Let $n = 2m$. Since $3 \nmid n$, $\rho = \xi^3$ is also a primitive $n$th root of 1. From $\rho^m = -1$ by (2.4) we have

$$W_n = \prod_{j=0}^{2m-1} \prod_{k=0}^{2m-1} (1 + \rho^{m+j} + \rho^{m+k}).$$

If $j = m$ we have $\prod_{k=0}^{2m-1}(2 + \rho^{m+k}) = 2^{2m} - 1$, similarly if $k = m$ we have $\prod_{k=0}^{2m-1}(2 + \rho^{m+j}) = 2^{2m} - 1$, while if $j = k$ we have $\prod_{j=0}^{2m-1}(1 - 2\rho^j) = 1 - 2^{2m}$. Discounting the repetition of factors with equal summands, and noting that for $j = k = m$ we have the factor 3, then $W_n = -3((2^n - 1)/3)^3 v$ where

$$v = \prod{}' (1 + \rho^{m+j} + \rho^{m+k})$$

($\prod'$ indicates the product for all $(j, k)$ such that $0 \leq j, k \leq 2m - 1$, $j \neq m$, $k \neq m$, and $j \neq k$).

The geometric mean of 1, $\rho^{m+j}$, $\rho^{m+k}$ is $\xi^{j+k}$ (because $\rho = \xi^3$); dividing each factor in the above product by the geometric mean of the summands, we get

$$v = \prod{}'\xi^{j+k}\prod{}'(\xi^{-j-k} + \xi^{m+2j-k} + \xi^{m+2k-j}).$$

The first product is equal to

$$\prod_{j\neq m}\prod_{k\neq m, k\neq j} \xi^j\xi^k = \prod_{j\neq m} \xi^{j(2m-2)} \prod_{k\neq m, k\neq j} \xi^k$$

$$= \prod_{j\neq m}(-1)\xi^{j(2m-3)} \prod_{j=0}^{2m-1} \xi^k$$

$$= \prod_{j\neq m}(-1)\xi^{j(2m-3)}\xi^{(2m-1)2m/2}$$

$$= \prod_{j\neq m} \xi^{j(2m-3)}$$

$$= (-1)^{2m-3} \prod_{j=0}^{2m-1} \xi^{j(2m-3)}$$

$$= -\xi^{(2m-3)(2m-1)2m/2}$$

$$= (-1)^{(2m-3)(2m-1)+1} = 1.$$

The second product is equal to $\prod''(\xi^e + \xi^f + \xi^g)$ (where $\prod''$ is the product for all $(e,f,g)$ such that $0 \leq e,f,g \leq 2m - 1$, $e,f,g$ are distinct and $e + f + g \equiv 0 \pmod{2m}$). Indeed, since $j \neq k$, $j \neq m$, $k \neq m$, and $3 \nmid m$, letting $e, f, g$ be such that $0 \leq e, f, g \leq 2m-1$, and

$$\begin{cases} e \equiv -j - k \pmod{2m}, \\ f \equiv m + 2j - k \pmod{2m}, \\ g \equiv m + 2k - j \pmod{2m}, \end{cases}$$

then $e, f, g$ are distinct and $e + f + g \equiv 0 \pmod{2m}$.

Conversely, for every triple $(e, f, g)$ as indicated, let $j, k$ be such that $0 \leq j, k \leq 2m - 1$, and

$$j \equiv f - e - m \pmod{2m},$$
$$k \equiv g - e - m \pmod{2m},$$

so $j \neq m$, $k \neq m$, and $j \neq k$.

Let $u = \prod''(\xi^e + \xi^f + \xi^g)$ (where $\prod''$ is the product for all $(e, f, g)$ such that $0 \leq e < f < g \leq 2m - 1$ and $e + f + g \equiv 0 \pmod{2m}$).

We show that $u$ is invariant by every automorphism $\sigma$ of $\mathbb{Q}(\xi)$. If $\sigma(\xi) = \xi^l$, with $\gcd(l,n) = 1$, $1 \le l < n$, then $\sigma(\xi^e + \xi^f + \xi^g) = \xi^{el} + \xi^{fl} + \xi^{gl}$ which is a factor of $u$. If $(e',f',g') \ne (e,f,g)$ then $el$, $fl$, $gl$ cannot be congruent modulo $2m$ to $e'l$, $f'l$, $g'l$, respectively, or to these numbers in any order. So different factors of $u$ have different images by $\sigma$, hence $\sigma(u) = u$, showing that the algebraic integer $u$ is rational, hence $u \in \mathbb{Z}$. But $v = u^6$ because each factor $\xi^e + \xi^f + \xi^g$ of $v$ is equal to a factor $\xi^{e'} + \xi^{f'} + \xi^{g'}$ of $u$, where $(e',f',g')$ is obtained by permutation of $(e,f,g)$.

Hence $W_n = -3((2^n - 1)/3)^3 u^6$.    □

Before the next result we need to establish a lemma.

LEMMA 2.2. *Let $n$ be such that $2n + 1 = p$ is a prime, let $\xi = \cos 2\pi/n + \sqrt{-1}\sin 2\pi/n$, and let $P$ be any prime ideal of the cyclotomic field $\mathbb{Q}(\xi)$ which divides $p$, i.e., $P \cap \mathbb{Z} = \mathbb{Z}p$. Then there exists a primitive root $s$ modulo $p$ such that $\xi \equiv s^2 \pmod{P}$.*

PROOF. Indeed, let $g$ be any primitive root modulo $p$, i.e., $g$ modulo $p$ has order $p - 1 = 2n$, so $g^2$ modulo $p$ has order $n$, and the set of elements modulo $p$ with order $n$ is $\{g^{2j} \pmod{p} \mid 1 \le j < n, \gcd(j,n) = 1\}$. If $1 \le j, k < n$, $\gcd(j,n) = \gcd(k,n) = 1$, and $j \ne k$ then $g^{2j} \not\equiv g^{2k} \pmod{p}$. By Chapter II, (3F), $\Phi_n(g^{2j}) \equiv 0 \pmod{p}$ for all such exponents $j$. On the other hand, $\Phi_n(\xi^j) = 0$ so $\Phi_n(\xi^j) \equiv 0 \pmod{P}$ for all $j$, $1 \le j < n$, $\gcd(j,n) = 1$. Hence there exists $j$ such that $\xi \equiv g^{2j} \pmod{P}$, and we just take $s$ to be $s \equiv \pm g^j \pmod{p}$, $1 \le s < p$.    □

Frame also proved the next statement (while Bang had noted in 1935 that $p \mid W_n$):

**(2H)**    *If $2n + 1 = p$ is a prime then $p^{\lfloor (n-1)/2 \rfloor}$ divides $W_n$.*

PROOF. To begin, let $\xi = \cos 2\pi/n + \sqrt{-1}\sin 2\pi/n$. Given $u, v$, $1 \le u$, $v < n$, $\gcd(u,n) = \gcd(v,n) = 1$, for every $j$, $1 \le j < n$, such that $\gcd(j,n) = 1$ let $f_j = 1 - \xi^{ju} - \xi^{jv}$.

Let $d_{u,v} = \prod_{\gcd(j,n)=1} f_j$. We show that $d_{u,v} \in \mathbb{Z}$. Indeed, let $\sigma$ be any automorphism of $\mathbb{Q}(\xi)$, so $\sigma(\xi) = \xi^l$ with $\gcd(l,n) = 1$. Then $\sigma(1 - \xi^{ju} - \xi^{jv}) = 1 - \xi^{ku} - \xi^{kv}$ where $1 \le k < n$ and $k \equiv jl \pmod{n}$, i.e., $\sigma(f_j) = f_k$. Every factor $f_k$ of $d_{u,v}$ is so obtained, therefore $\sigma$ defines a permutation of the set of factors $f_j$ of $d_{u,v}$. Thus

$$\sigma(d_{u,v}) = \prod_{\gcd(j,n)=1} \sigma(f_j) = \prod_{\gcd(j,n)=1} f_j = d_{u,v}.$$

This proves that $d_{u,v} \in \mathbb{Q}$ and since $d_{u,v}$ is an algebraic integer, then $d_{u,v} \in \mathbb{Z}$. If $P$ is any prime ideal of $\mathbb{Q}(\xi)$ such that $P \cap \mathbb{Z} = \mathbb{Z}p$, if $s$ is a primitive root modulo $p$ such that $\xi \equiv s^2 \pmod{P}$ then

$$d_{u,v} \equiv \prod_{\gcd(j,n)=1} (1 - s^{2ju} - s^{2jv}) \pmod{P},$$

so

$$d_{u,v} \equiv \prod_{\gcd(j,n)=1} (1 - s^{2ju} - s^{2jv}) \pmod{p}.$$

For every $h$, $2 \leq h \leq p - 2$, and such that $h^2 \not\equiv -1 \pmod{p}$ in case $p \equiv 1 \pmod 4$, we define $a_h, b_h, 0 \leq a_h, b_h \leq p - 1$ by

$$\begin{cases} a_h \equiv 2h/(h^2 + 1) \pmod{p}, \\ b_h \equiv -(h^2 - 1)/(h^2 + 1) \pmod{p}. \end{cases}$$

Then $a_h \not\equiv 0, 1 \pmod{p}$ and $b_h \not\equiv 0, 1 \pmod{p}$. Hence there exist $u = u_h$, $v = v_h$ such that $1 \leq u, v \leq p - 1$ and

$$\begin{cases} a_h \equiv s^u \pmod{p}, \\ b_h \equiv s^v \pmod{p}. \end{cases}$$

We have $a_h^2 + b_h^2 \equiv 1 \pmod{p}$ so $1 - s^{2u} - s^{2v} \equiv 0 \pmod{p}$, therefore $p$ divides $d_{u,v}$.

If $h, h'$ with $2 \leq h, h' \leq p-2$ are such that $h^2 \not\equiv -1 \pmod{p}$, $h'^2 \not\equiv -1 \pmod{p}$ when $p \equiv 1 \pmod 4$, and if $h' \equiv \pm h$ or $\pm \overline{h} \pmod{p}$, where $h\overline{h} \equiv 1 \pmod{p}$ then it is easy to check that $a_{h'} \equiv \pm a_h \pmod{p}$ or $a_{h'} \equiv \pm b_h \pmod{p}$ while $b_{h'} \equiv \pm b_h \pmod{p}$ or $b_{h'} \equiv \pm a_h \pmod{p}$, respectively; hence, with obvious notations, $1 - s^{2u} - s^{2v} = 1 - s^{2u'} - s^{2v'}$. Conversely, if $h, h'$ are such that this equality holds then either $a_h \equiv \pm a_{h'} \pmod{p}$, $b_h \equiv \pm b_{h'} \pmod{p}$ or $a_h \equiv \pm b_{h'} \pmod{p}$, $b_h \equiv \pm a_{h'} \pmod{p}$. Examining all possible cases, this leads to $h' \equiv \pm h$ or $\pm \overline{h} \pmod{p}$.

If $p \not\equiv 1 \pmod 4$, the number of possible values for $h$ is $p - 3 = 2n - 2$; this yields $[(2n - 2)/4] = [(n - 1)/2]$ factors $1 - s^{2u} - s^{2v}$ which are multiples of $p$.

If $p \equiv 1 \pmod 4$, the number of possible values for $h$ is $p - 5 = 2n - 4$; this yields $[(2n - 4)/4] = [(n - 1)/2]$ (because $n$ is even) factors $1 - s^{2u} - s^{2v}$ which are multiples of $p$.

Thus, in all cases, $p^{[(n-1)/2]}$ divides $\prod_{u,v} d_{u,v}$. Let $W_n = \prod_{u,v} d_{u,v} t$ where $t \in \mathbb{Z}$ (since $t$ is an algebraic integer and $t \in \mathbb{Q}$); thus $p^{[(n-1)/2]}$ divides $W_n$. $\square$

For example, $47^{11}$ divides $W_{23}$ and $101^{24}$ divides $W_{50}$.

The next divisibility result concerns Lucas numbers. For the convenience of the reader we shall recall some relevant facts about Fibonacci and Lucas numbers (see also Chapter V, §3).

The *Fibonacci numbers* $F_n$ $(n \geq 0)$ are defined as follows:

$$F_0 = 0, \qquad F_1 = 1,$$

and for $n \geq 2$:

$$F_n = F_{n-1} + F_{n-2}.$$

Similarly, the *Lucas numbers* $L_n$ $(n \geq 0)$ are defined as follows:

$$L_0 = 2, \qquad L_1 = 1,$$

and for $n \geq 2$:

$$L_n = L_{n-1} + L_{n-2}.$$

Let $\alpha, \beta$ be the roots of the polynomial $X^2 - X - 1$, so

$$\alpha = \frac{1 + \sqrt{5}}{2} = 1.6180\ldots,$$
$$\beta = \frac{1 - \sqrt{5}}{2} = -0.6180\ldots,$$

and

$$\alpha + \beta = 1, \qquad \alpha - \beta = \sqrt{5}, \qquad \alpha\beta = -1.$$

As is known, $\alpha$ is called the *golden number* (or *golden ratio*).

The following lemma is attributed to Binet (1843):

LEMMA 2.3. *For every $n \geq 0$ :*

$$F_n = \frac{\alpha^n - \beta^n}{\alpha - \beta}, \qquad L_n = \alpha^n + \beta^n.$$

PROOF. We have $\alpha^2 = \alpha + 1$, $\beta^2 = \beta + 1$, so $\alpha^n = \alpha^{n-1} + \alpha^{n-2}$ and also $\beta^n = \beta^{n-1} + \beta^{n-2}$ (for $n \geq 2$).

Let $T_n = (\alpha^n - \beta^n)/(\alpha - \beta)$, so $T_0 = 0$, $T_1 = 1$ and

$$T_{n-1} + T_{n-2} = \frac{\alpha^{n-1} - \beta^{n-1}}{\alpha - \beta} + \frac{\alpha^{n-2} - \beta^{n-2}}{\alpha - \beta} = \frac{\alpha^n - \beta^n}{\alpha - \beta} = T_n$$

(for $n \geq 2$). Thus, the sequence $\{T_n \mid n \geq 0\}$ coincides with the Fibonacci sequence. Similarly, let $U_n = \alpha^n + \beta^n$, so $U_0 = 2$, $U_1 = 1$ and $U_{n-1} + U_{n-2} = \alpha^{n-1} + \beta^{n-1} + \alpha^{n-2} + \beta^{n-2} = \alpha^n + \beta^n = U_n$ (for $n \geq 2$). Thus, the sequence $\{U_n \mid n \geq 0\}$ coincides with the Lucas sequence. $\square$

For more results about Fibonacci and Lucas numbers, see, for example, the books by Vorob'ev (1961), Hoggatt (1969) or Ribenboim (1995).

Frame proved:

**(2I)**    *If $n$ is odd then $L_n^2$ divides $W_n$.*

PROOF. By (2C) we may assume that $3 \nmid n$. We have

$$\prod_{j=1}^{n-1}(1 - \xi^j - \xi^{2j}) = -\prod_{j=0}^{n-1}(1 - \xi^j \alpha)(1 - \xi^j \beta)$$
$$= -(1 - \alpha^n)(1 - \beta^n)$$
$$= -1 + (\alpha^n + \beta^n) - (-1)^n$$
$$= L_n,$$

since $n$ is odd, and using the preceding lemma. Similarly, $\prod_{k=1}^{n-1}(1 - \xi^{2k} - \xi^k) = L_n$. Now we note that if $1 \leq j, k \leq n-1$, then the pairs $(j \bmod n, 2j \bmod n)$ and $(2k \bmod n, k \bmod n)$ are distinct. Indeed, otherwise $j \equiv 2k \pmod{n}$ and $2j \equiv k \pmod{n}$, hence $3j \equiv 3k \pmod{n}$ and $j \equiv -k \pmod{n}$. Therefore $j = k$ and $j = n - k$, so $n = 2k$ is even, contrary to the hypothesis.

This shows that $W_n = L_n^2 v$ where $v \in \mathbb{Q}$, and $v$ is an algebraic integer, so $v \in \mathbb{Z}$. $\square$

For example, the squares of the Lucas numbers

$$L_{47} = 6\,643\,838\,879$$

and

$$L_{53} = 119\,218\,851\,371$$

(which are known to be primes), divide $W_{47}$, $W_{53}$, respectively.

   Improving on a result of Lubelski (1935) and independently of Bang (1935), E. Lehmer indicated in 1935 the following divisibility property of the Wendt determinant:

**(2J)**     *If $p$ is an odd prime then $p^{p-2}(2^{p-1} - 1)/p$ divides $W_{p-1}$.*

PROOF. Consider the matrix $C$ whose determinant is $W_{p-1}$:

$$C = \begin{bmatrix} 1 & \binom{p-1}{1} & \binom{p-1}{2} & \cdots & \binom{p-1}{p-2} \\ \binom{p-1}{p-2} & 1 & \binom{p-1}{1} & \cdots & \binom{p-1}{p-3} \\ \binom{p-1}{p-3} & \binom{p-1}{p-2} & 1 & \cdots & \binom{p-1}{p-4} \\ \vdots & \vdots & \vdots & \ddots & \vdots \\ \binom{p-1}{1} & \binom{p-1}{2} & \binom{p-1}{3} & \cdots & 1 \end{bmatrix}.$$

Adding every column of $C$ to its last column, we obtain a matrix $C'$ whose last column has all its elements equal to

$$1 + \binom{p-1}{1} + \binom{p-1}{2} + \cdots + \binom{p-1}{p-2} = 2^{p-1} - 1.$$

Adding to each column of $C'$ (up to the column $p - 3$) the next column, we obtain a matrix $C''$ such that the elements of the first $p - 3$ columns are of the form

$$\binom{p-1}{k} + \binom{p-1}{k+1} = \binom{p}{k+1},$$

for $k = 0, 1, \ldots, p - 2$. These elements are all multiples of $p$. Thus $W_{p-1} = \det C''$ is a multiple of $p^{p-3}(2^{p-1} - 1) = p^{p-2}(2^{p-1} - 1)/p$.   □

   For a recent related result, see Helou (1997).

Frame observed experimentally that if $n \leq 50$ and $6 \nmid n$ then

$$\left|\log_{10}|W_n| - n^2 \log_{10} c\right| < 0.33,$$

where

$$\log c = \frac{2}{\pi} \int_0^{\pi/3} \log(2\cos\theta)\,d\theta,$$

so $\log_{10} c = 0.140305$. Thus $W_n$ has about $0.1403\, n^2$ digits. The knowledge of the size of $W_n$ and of some prime factors of $W_n$ may guarantee that the factorization of $W_n$ is already complete.

In 1982, Boyd showed that the sequence $\{\triangle_n\}$, with

$$\triangle_n = \log_{10}|W_n| - n^2 \log_{10} c,$$

is bounded and has exactly three limit points $0$, $\frac{1}{2}\log_{10} 3$, $\frac{2}{3}\log_{10} 2$, corresponding, respectively, to $n \equiv \pm 1$, $\pm 2$ or $3 \pmod 6$.

The values of $W_n$ (for $n = 2k \geq 20$) in the table below have been kindly provided to me by J.S. Frame.

Wendt determinant:

$$
\begin{aligned}
W_1 &= 1, \\
W_2 &= -3, \\
W_3 &= 28 = 2^2 \times 7, \\
W_4 &= -375 = -3 \times 5^3, \\
W_5 &= 3751 = 11^2 \times 31, \\
W_6 &= 0, \\
W_7 &= 2^6 \times 29^2 \times 127, \\
W_8 &= -3^7 \times 5^3 \times 17^3, \\
W_9 &= 2^2 \times 7 \times 19^4 \times 37^2 \times 73, \\
W_{10} &= -3 \times 11^9 \times 31^3, \\
W_{11} &= 23^5 \times 67^2 \times 89 \times 199^2, \\
W_{12} &= 0, \\
W_{13} &= 3^6 \times 53^2 \times 79^2 \times 131^2 \times 521^2 \times 8191, \\
W_{14} &= -2^{24} \times 3 \times 29^6 \times 43^3 \times 127^3, \\
W_{15} &= 2^{14} \times 7 \times 11^2 \times 31^7 \times 61^4 \times 151 \times 271^2, \\
W_{16} &= -3^7 \times 5^3 \times 7^6 \times 17^{15} \times 257^3, \\
W_{20} &= -3 \times 5^{24} \times 11^9 \times 31^3 \times 41^9 \times 61^6, \\
W_{22} &= -3 \times 23^{21} \times 67^6 \times 89^3 \times 199^6 \times 683^3, \\
W_{26} &= -3^{25} \times 53^6 \times 79^6 \times 131^6 \times 521^6 \times 2731^3 \times 8191^3, \\
W_{28} &= -2^{60} \times 3 \times 5^3 \times 13^6 \times 29^{21} \times 43^3 \times 113^6 \times 127^3 \times 197^6, \\
W_{32} &= -3^7 \times 5^6 \times 7^6 \times 17^{15} \times 47^6 \times 97^{12} \times 193^6 \times 257^6 \times 353^6 \\
&\quad \times 449^6 \times 65537^3,
\end{aligned}
$$

$$W_{34} = -3 \times 103^{12} \times 137^{12} \times 239^6 \times 307^6 \times 409^6 \times 613^6 \times 3571^6$$
$$\times 43691^3 \times 131071^3,$$
$$W_{38} = -3 \times 7^6 \times 191^{12} \times 229^{12} \times 419^6 \times 457^6 \times 647^6 \times 761^6$$
$$\times 1483^6 \times 9349^6 \times 174763^3 \times 524287^3,$$
$$W_{40} = -3^{19} \times 5^{24} \times 11^{15} \times 31^9 \times 41^{39} \times 241^6 \times 281^{12} \times 641^6 \times 881^6$$
$$\times 1048577^3,$$
$$W_{44} = -3 \times 5^3 \times 23^{18} \times 67^6 \times 89^{21} \times 199^6 \times 397^9 \times 617^6 \times 1013^6$$
$$\times 2113^9 \times 2333^6 \times 3257^6 \times 3257^6 \times 4357^6 \times 15709^3,$$
$$W_{46} = -3 \times 47^{45} \times 139^{12} \times 461^6 \times 599^{12} \times 691^{12} \times 829^6 \times 1151^6$$
$$\times 2347^6 \times 4357^6 \times 178481^3 \times 2796203^3,$$
$$W_{50} = -3 \times 11^6 \times 101^{24} \times 151^{18} \times 251^9 \times 401^6 \times 601^3 \times 1151^{12}$$
$$\times 1301^6 \times 1601^6 \times 1951^6 \times 3851^6 \times 4651^6 \times 5801^6 \times 6101^6$$
$$\times 44561^3 \times 55831^3,$$
$$W_{52} = -3^{25} \times 5^3 \times 53^{51} \times 79^6 \times 131^6 \times 157^{15} \times 233^6 \times 313^{12}$$
$$\times 521^6 \times 677^6 \times 1301^6 \times 1613^9 \times 2731^3 \times 6709^6 \times 8191^3$$
$$\times 13417^6 \times 20593^6.$$

In 1991, Fee and Granville computed the factors of (the modified) $W_n$ for every even $n \leq 200$ (including when 6 divides $n$). From these calculations, it followed:

**(2K)**    *If $k \leq 100$, if $p$ and $2kp + 1$ are primes, then the first case of Fermat's last theorem is true for the exponent $p$.*

PROOF. The result follows from the explicit determination of the factors of $W_{2k}$ (for $2k \leq 200$) and the verification that the conditions of (2B) are satisfied. $\square$

From the values of $W_{2k}$ (for $2k = 2, 4, 8, 16$) it follows at once that Fermat's congruence $X^p + Y^p + Z^p \equiv 0 \pmod{q}$, where $p$ and $q = 2kp + 1$ are odd primes, has only the trivial solution. However $X^3 + Y^3 + Z^3 \equiv 0 \pmod{31}$ and $X^3 + Y^3 + Z^3 = 0 \pmod{43}$ have nontrivial solution, because $31 \mid W_{10}$ and $43 \mid W_{14}$.

According to a result of Dickson (1909) (see Chapter X, (2C)), if $p, q$ are primes and $q \geq (p-1)^2(p-2)^2 + 6p - 2$ then the congruence (2.1) has a nontrivial solution. Therefore, if $q = 2kp + 1$ and $6 \nmid 2k$ then by (2B) $q \mid W_{2k}$.

We conclude this section by referring to statements made by Gandhi (1975, 1976) if the first case of Fermat's theorem is false for *p*. These conditions are expressed in terms of appropriate circulant determinants, however the proofs were never published, due to the untimely death of Gandhi.

## Bibliography

1843 Binet, J., *Mémoire sur l'intégration des équations linéaires aux différences finies d'un ordre quelconque à coefficients variables*, C. R. Acad. Sci. Paris, **17** (1843), 559–567.

1853 Spottiswoode, W., *Elementary theorems relating to determinants* (rewritten and much enlarged by the author), J. Reine Angew. Math., **51** (1853), 209–271 and 328–381.

1871 Stern, M.A., *Einige Bemerkungen über eine Determinante*, J. Reine Angew. Math., **73** (1871), 374–380.

1894 Wendt, E., *Arithmetische Studien über den letzten Fermatschen Satz, welcher aussagt, daß die Gleichung $a^n = b^n + c^n$ für $n > 2$ in ganzen Zahlen nicht auflösbar ist*, J. Reine Angew. Math., **113** (1894), 335–346.

1895 Matthews, G.B., *A note in connexion with Fermat's last theorem*, Messenger Math., **24** (1895), 97–99.

1909 Dickson, L.E., *On the congruence $x^n + y^n + z^n \equiv 0 \pmod{p}$*, J. Reine Angew. Math., **135** (1909), 134–141.

1909 Dickson, L.E., *Lower limit for the number of sets of solutions of $x^e + y^e + z^e \equiv 0 \pmod{p}$*, J. Reine Angew. Math., **135** (1909), 181–188.

1910 Bachmann, P., *Niedere Zahlentheorie*, Teubner, Leipzig, 1910; reprinted by Chelsea, New York, 1966.

1920 Muir, T., *The Theory of Determinants in the Historical Order of Development*, Vol. III, Macmillan, London, 1920.

1935 Bang, A.S., *Om tal af Formen $a^m + b^m - c^m$*, Mat. Tidsskrift, **B** (1935), 49–59.

1935 Lehmer, E., *On a resultant connected with Fermat's last theorem*, Bull. Amer. Math. Soc., **41** (1935), 864–867.

1935 Lubelski, S., *Studien über den großen Fermat'schen Satz*, Prace Matematyczne Fyz., **42** (1935), 11–44.

1961 Vorob'ev, N.N., *Fibonacci Numbers*, Blaisdell, New York, 1961.

1968 Rivoire, P., *Dernier Théorème de Fermat et Groupe de Classes dans Certains Corps Quadratiques Imaginaires*, Thèse,

Université Clermont-Ferrand, 1968, 59 pp.; reprinted in Ann. Sci. Univ. Clermont-Ferrand II, **68** (1979), 1–35.

1969 Hoggatt, V.E., *Fibonacci and Lucas Numbers*, Houghton-Mifflin, Boston, 1969.

1975 Gandhi, J.M., *Fermat's last theorem, II: A new circulant condition for the first case*, Notices Amer. Math. Soc., **22** (1975), A-450.

1975 Gandhi, J.M., *Fermat's last theorem, III: A new circulant condition for the first case, for primes of the form 6m − 1*, Notices Amer. Math. Soc., **22** (1975), A-502.

1975 Gandhi, J.M., *Fermat's last theorem, IV: A new circulant condition for the first case, for primes of the form 6m + 1*, Notices Amer. Math. Soc., **22** (1975), A-541.

1975 Gandhi, J.M., *Fermat's last theorem, V*, Notices Amer. Math. Soc., **23** (1976), A-56.

1980 Frame, J.S., *Factors of the binomial circulant determinant*, Fibonacci Quart., **18** (1980), 9–23.

1982 Boyd, D.W., *The asymptotic behaviour of the binomial circulant determinant*, J. Math. Anal. Appl., **86** (1982), 30–32.

1991 Fee, G. and Granville, A., *The prime factors of Wendt's binomial circulant determinant*, Math. Comp., **57** (1991), 839–848.

1995 Ribenboim, P., *The New Book of Prime Number Records*, Springer-Verlag, New York, 1995.

1997 Helou, C., *On Wendt's determinant*, Math. Comp., **66** (1997), 1341–1346.

## IV.3. Appendix: Sophie Germain's Primes

We have mentioned that it is a very difficult problem to know whether there exist infinitely many primes $p$ such that $2p + 1$ (or $4p + 1$, or $8p + 1$, etc ... ) is also a prime. A heuristic argument points to the validity of a much more general statement, as we shall explain.

If $x$ is any positive number, we denote by $\pi(x)$ the number of primes $p$ such that $p \leq x$. The famous prime number theorem of Hadamard and de la Vallée Poussin (1899) states that

$$\lim_{x \to \infty} \frac{\pi(x)}{x / \log x} = 1,$$

which we write also as $\pi(x) \sim x/\log x$.

Dirichlet's theorem states that if $a, m$ are relatively prime (positive) integers then there exist infinitely many primes in the arithmetic progression $\{a + km \mid k \geq 0\}$. If $\pi_{a,m}(x)$ denotes the number of primes $p$ in this arithmetic progression and such that $p \leq x$, then

$$\lim_{x \to \infty} \frac{\pi_{a,m}(x)}{\pi(x)} = \frac{1}{\varphi(m)},$$

that is,

$$\pi_{a,m}(x) \sim \frac{1}{\varphi(m)} \cdot \frac{x}{\log x}.$$

Considering the polynomial $f(X) = mX + a$, $\pi_{a,m}(x)$ represents the number of integers $n$, $1 \leq n \leq (x - a)/m$ such that $f(n)$ is a prime.

More generally, we may consider the following situation. Let $f_1(X), f_2(X), \ldots, f_k(X)$ be polynomials with integral coefficients, and positive leading coefficient. Let $d_j \geq 1$ be the degree of $f_j(X)$. Assume moreover that these polynomials are irreducible over $\mathbb{Q}$ and that none is a constant multiple of another. Let $N$ be any positive integer and let $Q(N) = Q_{f_1,\ldots,f_k}(N)$ denote the number of integers $n$, $1 \leq n \leq N$, such that $f_1(n), f_2(n), \ldots, f_k(n)$ are primes.

The probability that a large positive integer $m$ be a prime is $\pi(m)/m \sim 1/\log m$, by virtue of the prime number theorem.

Since we shall be interested in the values of the polynomials $f_1(X)$, $f_2(X), \ldots, f_k(X)$, we have to discount the fact that $k$-tuples of such values are not randomly distributed.

If $p$ is an arbitrary prime, let $s_p$ denote the chance that none of the integers of a random $k$-tuple be divisible by $p$. Then

$$s_p = \left(\frac{p-1}{p}\right)^k = \left(1 - \frac{1}{p}\right)^k.$$

Similarly, let $r_p$ denote the chance that for a random integer $n$, none of the integers $f_1(n), f_2(n), \ldots, f_k(n)$ be divisible by $p$. If $w(p)$ denotes the number of solutions of the congruence

$$f_1(X)f_2(X) \cdots f_k(X) \equiv 0 \pmod{p},$$

then $r_p = (p - w(p))/p = 1 - w(p)/p$. It may be shown that the product $\prod_p r_p/s_p$ is convergent, say to a limit $C = C(f_1, \ldots, f_k)$. If we agree that this number measures the extent to which the values of $f_1(X), \ldots, f_k(X)$ form a nonrandom $k$-tuple, then the probability that $f_1(n), f_2(n), \ldots, f_k(n)$ are all primes (for large $n$) is equal to

$$C(f_1, \ldots, f_n) \frac{1}{\log f_1(n) \cdots \log f_k(n)}$$

$$\sim C(f_1, \ldots, f_n) \frac{1}{d_1 \cdots d_k} \times \frac{1}{(\log n)^k},$$

since $\log f_i(n) \sim d_i \log n$. Then

$$Q(N) \sim C(f_1, \ldots, f_k) \frac{1}{d_1 \cdots d_k} \sum_{n=2}^{N} \frac{1}{(\log n)^k}.$$

In particular $\lim_{N \to \infty} Q(N) = \infty$. Thus, we see in a heuristic way that there should exist an infinite number of primes $p$ with the required properties.

Now we consider the following special cases:

(1) $f_1(X) = X$, $f_2(X) = 2X+$; and
(2) $f_1(X) = X$, $f_2(X) = X + 2$.

Case (1) deals with Sophie Germain's primes while case (2) refers to twin primes. In both cases $w(2) = 1$ and $w(p) = 2$ if $p > 2$. Hence the constant is

$$C = \frac{1 - \frac{1}{2}}{\left(1 - \frac{1}{2}\right)^2} \prod_{p>2} \frac{1 - \frac{2}{p}}{\left(1 - \frac{1}{p}\right)^2} = 1.3203236,$$

so

$$Q(N) \sim 1.3203236 \sum_{n=2}^{N} \frac{1}{(\log n)^2}.$$

This expression had been conjectured by Hardy and Littlewood in 1923, for the count of twin primes less than $N$. It agrees rather closely with the actual number of twin primes; see Sexton (1954), Wrench (1961), Shanks (1962), and Brent (1975).

## Bibliography

1923 Hardy, G.H. and Littlewood, J.E., *Some problems of "partitio numerorum", III. On the expression of a number as a sum of primes*, Acta. Math., **44** (1923), 1–70.

1954 Sexton, C.R., *Counts of twin primes less than* 100000, Math. Tables and Aids to Comp., **8** (1954), 47–49.

1961 Wrench, J.W., *Evaluation of Artin's constant and the twin prime constant*, Math. Comp., **15** (1961), 396–398.

1962 Bateman, P.T. and Horn, R.A., *A heuristic asymptotic formula concerning the distribution of prime numbers*, Math. Comp., **16** (1962), 363–367.

1962 Shanks, D., *Solved and Unsolved Problems in Number Theory*, Vol. I, Spartan, Washington, DC, 1962; reprinted by Chelsea, New York, 1979.

1963 Schinzel, A., *A remark on a paper by Bateman and Horn*, Math. Comp., **17** (1963), 445–447.

1965 Bateman, P.T. and Horn, R.A., *Primes represented by irreducible polynomials in one variable*, Proc. Symposia in Pure Mathematics, Theory of Numbers, Amer. Math. Soc, Vol. 8, 1965, pp. 119–132.

1975 Brent, R.P., *Irregularities in the distribution of primes and twin primes*, Math. Comp., **29** (1975), 43–56.

# V
# Interludes 5 and 6

In this chapter, we give more background material.

## V.1. $p$-Adic Numbers

**A. The Field of $p$-Adic Numbers.** In order to study divisibility properties by a prime $p$, it is often convenient to consider the development of integers in the base $p$:

$$a = a_0 + a_1 p + \cdots + a_m p^m,$$

with $0 \le a_i \le p - 1$, $p^m \le a < p^{m+1}$.

Hensel also introduced infinite $p$-adic developments. The new numbers so defined are the *$p$-adic integers*. He described the operations of addition and multiplication among the $p$-adic integers and proved a very important theorem concerning the existence of $p$-adic integers which are roots of certain polynomials.

The $p$-adic numbers may be considered as being limits of sequences of integers, relative to the $p$-adic distance. These considerations allowed the introduction of methods of analysis in the study of divisibility properties.

We shall describe here very briefly the concepts of $p$-adic numbers and the results which we shall require to discuss Fermat's equation.

In Chapter II, §1, we have defined, for every prime $p$, the $p$-adic valuation $v_p$ of $\mathbb{Q}$. We recall that the set of $p$-integral rational numbers is

$$\mathbb{Z}_p = \left\{ \frac{a}{b} \mid a, b \in \mathbb{Z}, b \neq 0, p \nmid b, \gcd(a,b) = 1 \right\} \cup \{0\}.$$

It is a subring of $\mathbb{Q}$ containing $\mathbb{Z}$. Moreover, if $r \in \mathbb{Q}$ then $v_p(r) \geq 0$ if and only if $r \in \mathbb{Z}_p$.

$\mathbb{Z}_p$ is called the *valuation ring* of $v_p$. The only maximal ideal of $\mathbb{Z}_p$ is equal to $\mathbb{Z}_p p = \{a/b \in \mathbb{Z}_p \mid p \text{ divides } a\}$. The field $\mathbb{Z}_p/\mathbb{Z}_p p$ is isomorphic with the field $\mathbb{F}_p$ (with $p$ elements); it is called the *residue field* of $v_p$.

The valuation $v_p$ defines, on the field $\mathbb{Q}$, the function $d_p$ given as follows. If $a, b \in \mathbb{Q}$, then

$$\begin{cases} d_p(a,b) = p^{-v_p(a-b)} & \text{when } a \neq b, \\ d_p(a,a) = 0. \end{cases}$$

It is easy to verify the following properties:

$$\begin{aligned}
d_p(a,b) &\geq 0, \\
d_p(a,b) &= d_p(b,a), \\
d_p(a,b) &= d_p(a-b,0), \\
d_p(a+c,b+c) &= d_p(a,b), \\
d_p(a,b) &\leq \max\{d_p(a,c), d_p(b,c)\}, \\
&\leq d_p(a,c) + d_p(b,c).
\end{aligned}$$

So $d_p$ is a distance, compatible with the operation of addition. It is called the *$p$-adic distance* of $\mathbb{Q}$.

The completion of $\mathbb{Q}$ relative to the $p$-adic distance is again a field, denoted $\hat{\mathbb{Q}}_p$ and called the *field of $p$-adic numbers*. The nonzero elements $\alpha$ of $\hat{\mathbb{Q}}_p$ are represented by $p$-adic developments

$$\alpha = \sum_{i=m}^{\infty} a_i p^i,$$

with $0 \leq a_i \leq p-1$, $m \in \mathbb{Z}$, and $a_m \neq 0$. If $\alpha_n = \sum_{i=m}^{n} a_i p^i$ (with $n \geq m$) then $\alpha = \lim_{n \to} \alpha_n$ (the limit is relative to the $p$-adic distance).

The *p*-adic valuation may be extended by continuity to a valuation $\hat{v}_p$ of the field $\hat{\mathbb{Q}}_p$, which is defined by

$$\hat{v}_p\left(\sum_{i=m}^{\infty} a_i p^i\right) = m \quad (\text{if } a_m \neq 0).$$

Thus, the values of $\hat{v}_p$ are also integers or infinity.

The topological closure of $\mathbb{Z}_p$ in the field $\hat{\mathbb{Q}}_p$ is a ring, denoted by $\hat{\mathbb{Z}}_p$. Its elements are called the *p-adic integers*. Thus $\alpha \in \hat{\mathbb{Q}}_p$ is a *p*-adic integer exactly when $\hat{v}_p(\alpha) \geq 0$. It is also clear that $\hat{\mathbb{Z}}_p \cup \mathbb{Q} = \mathbb{Z}_p$. The only nonzero prime ideal of $\hat{\mathbb{Z}}_p$ is $\hat{\mathbb{Z}}_p p$, consisting of the multiples of $p$. The residue field of $\hat{v}_p$ is $\hat{\mathbb{Z}}_p/\hat{\mathbb{Z}}_p p$, which is isomorphic to the field $\mathbb{F}_p$.

If $\alpha, \beta \in \hat{\mathbb{Q}}_p$, we say that $\alpha$ *divides* $\beta$ if there exists $\gamma \in \hat{\mathbb{Z}}_p$ such that $\alpha\gamma = \beta$; this means that $\hat{v}_p(\alpha) \leq \hat{v}_p(\beta)$. $\alpha \in \hat{\mathbb{Z}}_p$ is a *unit* in $\hat{\mathbb{Z}}_p$ when $\alpha$ divides 1, i.e., $\hat{v}_p(\alpha) = 0$. The set $U_p$ of units of $\hat{\mathbb{Z}}_p$ is a multiplicative group.

If $\alpha, \beta, \gamma \in \hat{\mathbb{Q}}_p$, $\gamma \neq 0$, we write $\alpha \equiv \beta \pmod{\gamma}$ if $\gamma$ divides $\alpha - \beta$. Similarly, if $\gamma \in \hat{\mathbb{Q}}_p$, $\gamma \neq 0$ and $F(X), G(X) \in \hat{\mathbb{Q}}_p[X]$ we write $F(X) \equiv G(X) \pmod{\gamma}$ when $\gamma$ divides each coefficient of $F(X) - G(X)$.

These congruence relations satisfy the usual properties of congruences of integers.

## B. Polynomials with *p*-Adic Coefficients.

We discuss briefly polynomials with coefficients in the field $\hat{\mathbb{Q}}_p$.

If $f(X) = a_0 X^n + a_1 X^{n-1} + \cdots + a_n \in \hat{\mathbb{Q}}_p[X]$, we make the definition $\tilde{v}_p(f) = \min_{0 \leq i \leq n}\{v_p(a_i)\}$. If $f, g \in \hat{\mathbb{Q}}_p[X]$, with $g \neq 0$, we define

$$\tilde{v}_p\left(\frac{f}{g}\right) = \tilde{v}_p(f) - \tilde{v}_p(g),$$

which is well defined. Then $\tilde{v}_p$ is a valuation of the field $\hat{\mathbb{Q}}_p(X)$, whose restriction to $\mathbb{Q}_p$ is the valuation $v_p$. For simplicity, we shall write $v_p$, instead of $\tilde{v}_p$.

If $f, g \in \hat{\mathbb{Q}}_p[X]$, we write $f \equiv g \pmod{p^n}$ when $v_p(f - g) \geq n$, or equivalently, $p^n$ divides each coefficient of $f - g$. For every $f = \sum_{i=0}^{m} a_i X^i \in \hat{\mathbb{Z}}_p[X]$, we denote by $\overline{f} = f \bmod p$ the polynomial $\sum_{i=0}^{m} \overline{a_i} X^i \in \mathbb{F}_p[X]$.

We recall now some well-known facts about polynomials in $\hat{\mathbb{Q}}_p[X]$. The polynomial $f \in \hat{\mathbb{Z}}_p[X]$ is said to be *primitive* when $v_p(f) = 0$. Every polynomial $f \in \mathbb{Z}_p[X]$ may be written as $f = af_1$, where $a \in \mathbb{Z}$, $f_1 \in \hat{\mathbb{Z}}_p[X]$ and $f_1$ is primitive.

**(1A)   Gauss' Lemma.** *If $f, g \in \hat{\mathbb{Z}}_p[X]$ are primitive polynomials, then $f \cdot g$ is also a primitive polynomial.*

**(1B)** *If $f \in \mathbb{Z}_p[X]$ is primitive and $f = g \cdot h$ with $g, h \in \hat{\mathbb{Q}}_p[X]$, then $f = f_1 \cdot g_1$ for some primitive polynomials $f_1, g_1 \in \mathbb{Z}_p[X]$, such that $\deg g_1 = \deg g$, $\deg h_1 = \deg h$.*

The nonconstant polynomial $f \in \mathbb{Z}_p[X]$ (respectively, $f \in \mathbb{Q}_p[X]$) is irreducible in $\mathbb{Z}_p[X]$ (respectively, in $\mathbb{Q}_p[X]$) if it is impossible to write $f = g \cdot h$, with $g, h$ nonconstant polynomials in $\mathbb{Z}_p[X]$ (respectively, $\mathbb{Q}_p[X]$).

**(1C)** *If $f \in \hat{\mathbb{Z}}_p[X]$, then $f$ is irreducible in $\hat{\mathbb{Z}}_p[X]$ if and only if it is irreducible in $\hat{\mathbb{Q}}_p[X]$.*

The nonconstant polynomials $f, g \in \hat{\mathbb{Z}}_p[X]$ are said to be *relatively prime* whenever, if $h \in \mathbb{Z}_p[X]$ and $h$ divides $f$ and $g$, then $\deg(h) = 0$.

**(1D)** *If $f \in \hat{\mathbb{Z}}_p[X]$ is nonconstant and primitive, and if $f$ does not divide the nonconstant polynomial $g \in \hat{\mathbb{Z}}_p[X]$, then $f$ and $g$ are relatively prime.*

**(1E)** *If $f, g \in \hat{\mathbb{Z}}_p[X]$ are nonconstant and relatively prime, then there exist polynomials $s, t \in \hat{\mathbb{Z}}_p[X]$ such that $s \cdot f + t \cdot g$ is a nonzero element of $\hat{\mathbb{Z}}_p$.*

**(1F)** *If $f, g, h \in \hat{\mathbb{Z}}_p[X]$, if $f$ is irreducible and if $f$ divides $g \cdot h$, then either $f$ divides $g$ or $f$ divides $h$.*

**(1G)** *If $g, h \in \hat{\mathbb{Z}}_p[X]$ are nonconstant and relatively prime, if $g$ or $h$ is primitive and both $g$ and $h$ divide $f$, then $g \cdot h$ divides $f$.*

**(1H)**     *Every nonzero polynomial $f \in \hat{\mathbb{Z}}_p[X]$, may be written as a product $f = a g_1 \cdots g_m$, where $a \in \hat{\mathbb{Z}}_p$, and $g_1, \ldots, g_n \in \hat{\mathbb{Z}}_p[X]$ are primitive irreducible polynomials with $n \geq 0$. Moreover, $a$ and $g_1, \ldots, g_n$ are uniquely defined up to a unit in $\hat{\mathbb{Z}}_p$.*

Now we shall consider the resultant and the discriminant of polynomials in $\hat{\mathbb{Z}}_p[X]$.

**(1I)**     *Let $f, g$ be nonconstant polynomials in $\hat{\mathbb{Z}}_p[X]$. Then the following conditions are equivalent:*

(a) *There exists a nonconstant polynomial $h \in \hat{\mathbb{Z}}_p[X]$ which divides both $f$ and $g$.*
(b) *There exist nonzero polynomials $f_1, g_1 \in \hat{\mathbb{Z}}_p[X]$ such that $\deg(f_1) < \deg(f)$, $\deg(g_1) < \deg(g)$, and*

$$g_1 \cdot f + f_1 \cdot g = 0.$$

PROOF. (a) $\Rightarrow$ (b)     We assume that $f = h \cdot f_1$ and $g = -h \cdot g_1$ with $f_1, g_1 \in \hat{\mathbb{Z}}_p[X]$; then $\deg(f_1) < \deg(f)$, $\deg(g_1) < \deg(g)$, and $g_1 \cdot f + f_1 \cdot g = 0$.

(b) $\Rightarrow$ (a)     Conversely, we assume that there exist polynomials $f_1$ and $g_1 \in \hat{\mathbb{Z}}_p[X]$, such that $\deg(f_1) < \deg(f)$, $\deg(g_1) < \deg(g)$, and $g_1 f + f_1 g = 0$. If $f, g$ are relatively prime, then by (1E) there would exist polynomials $s, t \in \hat{\mathbb{Z}}_p[X]$ such that $s \cdot f + t \cdot g = c \in \hat{\mathbb{Z}}_p$, with $c \neq 0$. Eliminating $g$ from the above relations, we obtain $f(s f_1 - t g_1) = c f_1$ where $\deg(f_1) < \deg(f)$, which is impossible.   $\square$

**(1J)**     *In order that $f = \sum_{i=0}^{m} a_i X^{m-i}$ and $g = \sum_{j=0}^{n} b_j X^{n-j}$ (where $m, n > 0$, and $f, g \in \hat{\mathbb{Z}}_p[X]$) have a common nonconstant factor, it is necessary and sufficient that $R(f, g) = 0$.*

PROOF. It was seen in Chapter II, (4B), that if $f, g$ have a common nonconstant factor, then $R(f, g) = 0$.

Conversely, by (1I) it is equivalent to show the existence of nonzero polynomials $f_1 \in \hat{\mathbb{Z}}_p[X], g_1 \in \hat{\mathbb{Z}}_p[X]$, $f_1 = \sum_{i=0}^{m-1} c_i X^{m-1-i}$, $g_1 = \sum_{i=0}^{n-1} d_i X^{n-1-i}$, such that $g_1 \cdot f + f_1 \cdot g = 0$ (it is not excluded that $c_0 = d_0 = 0$). This relation is equivalent to the following system of

$m+n$ equations in the unknown quantities $c_0, \ldots, c_{m-1}, d_0, \ldots, d_{m-1}$ (obtained by equating to zero the coefficients of the powers of $X$):

$$
\begin{cases}
d_0 a_0 + c_0 b_0 & = 0, \\
d_0 a_1 + d_1 a_0 + c_0 b_1 + c_1 b_0 & = 0, \\
d_0 a_2 + d_1 a_1 + d_2 a_0 + c_0 b_2 + c_1 b_1 + c_2 b_0 & = 0, \\
\qquad\qquad\qquad\qquad\qquad\qquad\vdots
\end{cases}
$$

This homogeneous linear system has a nontrivial solution in $\hat{\mathbb{Q}}_p$ if and only if the determinant of its matrix vanishes, or equivalently, the determinant of the matrix obtained after exchanging rows and columns vanishes; in other words, $R(f,g) = 0$. Now we finish the proof by noticing that if there exists a nontrivial solution in $\hat{\mathbb{Q}}_p$, by multiplying by the common denominator of these elements, we obtain a nontrivial solution in $\hat{\mathbb{Z}}_p$. □

**(1K)**   *Let $f, g \in \mathbb{Z}_p[X]$ be relatively prime nonconstant polynomials, such that $v_p(R(f,g)) = \rho$. Then every nonzero polynomial $h \in \hat{\mathbb{Z}}_p[X]$ such that $v_p(h) \geq \rho$ and $\deg(h) < \deg(f) + \deg(g)$ may be written in a unique way as $h = g_1 \cdot f + f_1 \cdot g$, where $f_1, g_1 \in \hat{\mathbb{Z}}_p[X]$, $v_p(f_1) \geq v_p(h) - \rho$, $v_p(g_1) \geq v_p(h) - \rho$, $\deg(f_1) < \deg(f)$, and $\deg(g_1) < \deg(g)$.*

PROOF. Let

$$
f = \sum_{i=0}^{m} a_i X^{m-i},
$$

$$
g = \sum_{i=0}^{n} b_i X^{n-i},
$$

$$
h = \sum_{i=0}^{m+n-1} e_1 X^{m+n-i}.
$$

We want to determine $f_1 = \sum_{i=0}^{m-1} c_i X^{m-1-i}, g_1 = \sum_{i=0}^{n-1} d_i X^{n-1-i}$ in $\mathbb{Z}_p[X]$ such that $h = g_1 \cdot f + f_1 \cdot g$. Comparing the coefficients of $X$ in both sides of the above relation, we obtain a linear system of $m + n$ equations in the $m + n$ unknown quantities $c_i, d_j$, whose determinant is exactly $R(f,g)$.

Since $f, g$ are relatively prime, by (1J) we have $R(f,g) \neq 0$; hence the above system has a unique solution.

The coefficients $c_i, d_j$ may be computed by Cramer's rule; their numerators are linear forms in the $e_i$ with coefficients in $\hat{\mathbb{Z}}_p$ (because $f, g \in \hat{\mathbb{Z}}_p[X]$), and their denominators are equal to $R(f, g)$. From $v_p(e_i) \geq v_p(h) \geq 0 = v_p(R(f,g))$, it follows that $v_p(c_i) \geq 0$, $v_p(d_j) \geq 0$, so that $f_1, g_1 \in \hat{\mathbb{Z}}_p[X]$ and $v_p(f_1) \geq v_p(h) - \rho$, $v_p(g_1) \geq v_p(h) - \rho$.  $\square$

**(1L)**    Let $g \in \hat{\mathbb{Z}}_p[X]$ be a nonconstant polynomial. In order that there exist a nonconstant polynomial $g \in \hat{\mathbb{Z}}_p[X]$ such that $g^2$ divides $f$, it is necessary and sufficient that $\mathrm{Discr}(f) = 0$.

PROOF. In fact, if $g^2$ divides $f$, then $g$ divides $f$ and $f'$, hence $\mathrm{Discr}(f) = R(f, f') = 0$. Conversely, if $\mathrm{Discr}(f) = 0$, by (1J) there exists a nonconstant polynomial $f \in \hat{\mathbb{Z}}_p[X]$ dividing $f$ and $f'$; by (1H), we may assume that $g$ is irreducible. We have $f = g \cdot h$, hence $f' = g' \cdot h + g \cdot h'$; since $g$ divides $f'$, it follows that $g$ divides $g' \cdot h$; from $\deg(g') < \deg(g)$ we see that $g$ does not divide $g'$, hence $g$ divides $h$ (by (1F)), and so $g^2$ divides $f$.  $\square$

We shall now investigate the behavior of the resultant $R(f, g)$ when $f, g$ are replaced by sufficiently close polynomials, relative to the metric defined by the valuation $v_p$ on $\hat{\mathbb{Q}}_p(X)$.

**(1M)**    If $f, g, f_1, g_1 \in \hat{\mathbb{Z}}_p[X]$ are nonconstant polynomials and $v_p(f_1 - f) \geq \alpha$, $v_p(g_1 - g) \geq \beta$, then $v_p(R(f_1, g_1) - R(f, g)) \geq \min\{\alpha, \beta\}$.

PROOF. Let $a \in \hat{\mathbb{Z}}_p$ be such that $v_p(a) = \min\{v_p(f_1 - f), v_p(g_1 - g)\}$; then $f_1 = f + ah$ and $g_1 = g + ak$, where $h, g \in \mathbb{Z}_p[X]$. Thus $R(f_1, g_1) = R(f + ah, g + ak)$. Writing the eliminating matrix between $f + ah$, $g + ak$, and computing the determinant, we obtain $R(f, g) + as$, where $s \in \hat{\mathbb{Z}}_p$ is a certain sum of products of elements equal to $a$ or to coefficients of $f, h, g, k$. Thus

$$v_p(R(f_1, g_1) - R(f, g)) \geq v_p(a) \geq \min\{\alpha, \beta\}.  \quad \square$$

**(1N)**    With the above notations, if $f, f_1 \in \hat{\mathbb{Z}}_p[X]$ are nonconstant polynomials and $v_p(f - f_1) \geq \alpha$, then $v_p(\mathrm{Discr}(f) - \mathrm{Discr}(f_1)) \geq \alpha$.

PROOF. In fact, since $\text{Discr}(f) = R(f, f')$, $\text{Discr}(f_1) = R(f_1, f_1')$, we have only to remark that $v_p(f_1 - f) \geq \alpha$ implies $v_p(f_1' - f') \geq \alpha$. Indeed, if $m = \max\{\deg(f), \deg(f_1)\}$ and $f = \sum_{i=0}^{m} a_i X^{m-i}$, $f_1 = \sum_{i=0}^{m} b_i X^{m-i}$, then

$$f' = \sum_{i=0}^{m-1} a_i X^{m-1-i}, \qquad f_1' = \sum_{i=0}^{m-1} (m-1) b_i X^{m-1-i};$$

thus $v_p((m-1)(b_i - a_i)) \geq v_p(b_i - a_i)$ for every $i = 0, \ldots, m-1$, and so $v_p(f_1' - f') \geq \min\{v_p(b_i - a_i) : i = 0, \ldots, m-1\} \geq v_p(f_1 - f) \geq \alpha$. $\square$

We also have at once:

**(1O)**    If $f, g \in \hat{\mathbb{Z}}_p[X]$ and $\overline{f} = f \bmod p$, $\overline{g} \bmod p$, *then the resultant of* $\overline{f}$ *and* $\overline{g}$ (*computed in* $\mathbb{F}_p[X]$) *is* $R(\overline{f}, \overline{g}) = \overline{R(f, g)}$ *and the discriminant of* $\overline{f}$ (*computed in* $\mathbb{F}_p[X]$) *is* $\text{Discr}(\overline{f}) = \overline{\text{Discr}(f)}$.

We say that the monic nonconstant polynomials $f, g \in \hat{\mathbb{Z}}_p[X]$ are *relatively prime modulo* $p$ when $\overline{f}, \overline{g}$ are relatively prime polynomials in $\mathbb{F}_p[X]$. Similarly, $f$ is said to be *irreducible modulo* $p$ whenever $\overline{f}$ is an irreducible polynomial of $\mathbb{F}_p[X]$. Every polynomial $f \in \hat{\mathbb{Z}}_p[X]$ is congruent modulo $p$ to a product of polynomials in $\hat{\mathbb{Z}}_p[X]$, which are irreducible modulo $p$, and are uniquely defined modulo $p$.
With these definitions, we have:

**(1P)**    $f, g \in \hat{\mathbb{Z}}_p[X]$ *are relatively prime modulo* $p$ *if and only if the resultant is a unit in* $\hat{\mathbb{Z}}_p$, *i.e., if and only if* $v_p(R(f, g)) = 0$.

PROOF. By definition, $f, g$ are relatively prime modulo $p$ when $\overline{f}, \overline{g}$ are relatively prime polynomials in $\mathbb{F}_p[X]$; by Chapter II, (4B), this means that $R(\overline{f}, \overline{g}) \neq 0$; by (1O), this is equivalent to $\overline{R(f, g)} \neq 0$, that is, $p$ does not divide $R(f, g)$, or equivalently, $v_p(R(f, g)) = 0$. $\square$

**(1Q)**    *Let* $f, g \in \mathbb{Z}_p[X]$ *be irreducible polynomials modulo* $p$. *Then* $p$ *divides* $R(f, g)$ *if and only if* $f \equiv g \pmod{p}$.

PROOF. In fact, $p$ divides $R(f, g)$ exactly when $f, g$ are not relatively prime modulo $p$; hence there exists a nonconstant polynomial $h \in \mathbb{Z}_p[X]$ such that $f \equiv h \cdot f_1 \pmod{p}$, $g \equiv h \cdot g_1 \pmod{p}$; by hypothesis, we must have $f \equiv h \pmod{p}$, $g \equiv h \pmod{p}$, hence $f \equiv g \pmod{p}$. The converse is trivial. $\square$

We observe that if $f = \sum_{i=0}^{m} \in \mathbb{Z}_p[X]$ and $f'$ is a multiple of $p$, then $f \equiv \sum_{i \geq 0} a_{pi} X^{pi} \pmod{p}$. Indeed, the coefficients of $f'$ are $ja_j$; thus if $p$ divides $f'$, then $p$ divides $a_j$ when $p$ does not divide $j$. Therefore,

$$f \equiv \left( \sum_{i \geq 0} a_{pi} X^{pi} \right)^p \pmod{p}.$$

In particular, if $f$ is irreducible modulo $p$, then $\overline{f'} \neq 0$ and hence we may consider the discriminant of $f$ modulo $p$.

**(1R)**    If $f \in \hat{\mathbb{Z}}_p[X]$ is irreducible modulo $p$, then $p$ does not divide $\mathrm{Discr}(f)$.

PROOF. We write $f = f_1 + pf_2$ where all the coefficients of $f_1$ are not multiples of $p$. Then $f' = f_1' + pf_2'$ and by (1N), $\mathrm{Discr}(f) \equiv \mathrm{Discr}(f_1)$ mod $p$. If $p$ divides $\mathrm{Discr}(f)$, then $p$ divides $\mathrm{Discr}(f_1) = R(f_1, f_1')$. By (1P), there exists $h \in \hat{\mathbb{Z}}_p[X]$ such that $\overline{h}$ is nonconstant and $\overline{h}$ is a common factor of $\overline{f}_1, \overline{f}_1'$. Thus $\overline{f}_1 = \overline{h} \cdot \overline{g}$, $\overline{f}_1' = \overline{h} \cdot \overline{k}$ with $g, k \in \hat{\mathbb{Z}}_p[X]$. Since $\overline{f}_1 = \overline{f}$ is irreducible, then $\overline{g} = \overline{c}$ with $c \in \hat{\mathbb{Z}}_p$. So $\overline{c} \cdot \overline{f}_1' = \overline{f}_1 \cdot k$. Therefore

$$\deg(\overline{f}_1') \leq \deg(f_1') = \deg(\overline{f}_1) \leq \deg(\overline{f}_1 \cdot \overline{k}) = \deg(\overline{f}_1'),$$

which is absurd. $\square$

**(1S)**    Let $f \in \hat{\mathbb{Z}}_p[X]$ be such that $\overline{f}$ is not constant. Then $f$ has a multiple irreducible factor modulo $p$ if and only if $p$ divides $\mathrm{Discr}(f)$.

PROOF. We have $f \equiv g_1 g_2 \cdots g_n \pmod{p}$, where $g_1, g_2, \ldots, g_n$ are irreducible modulo $p$. Hence, by (1M) and Chapter II, (4D),

$$\mathrm{Discr}(f) \equiv \mathrm{Discr}(g_1 g_2 \cdots g_n)$$
$$= \pm \prod_{i=1}^{n} \mathrm{Discr}(g_i) \cdot \prod_{i<j} [R(g_i, g_j)]^2 \pmod{p}.$$

By (1R), $p$ does not divide $\mathrm{Discr}(g_i)$ for $1 \leq i \leq n$. Then $p$ divides $\mathrm{Discr}(f)$ if and only if there exist indices $i < j$ such that $p$ divides

$R(g_i, g_j)$. By (1Q) this means that $g_i \equiv g_j \pmod{p}$, concluding the proof. $\square$

**C. Hensel's Lemma.** This very important result, proved by Hensel in 1908, is certainly the *raison d'être* of the $p$-adic numbers. It asserts the existence, under appropriate conditions, of $p$-adic roots of polynomials. We shall prove here Hensel's lemma in its strong form:

**(1T)**     *Let* $F, g, h \in \hat{\mathbb{Z}}_p[X]$ *be such that:*

    (i) $\deg(g) = m > 0$, $\deg(h) = n > 0$, $\deg(F) = m + n$, $g$ *is monic and* $\deg(F - gh) < \deg(F)$;

    (ii) $v_p(R(g, h)) = \rho \geq 0$; *and*

    (iii) $v_p(F - gh) = \alpha > 2\rho$.

*Then there exist* $G, H \in \hat{\mathbb{Z}}_p[X]$ *such that* $v_p(G - g) \geq \alpha - \rho$, $v_p(H - h) \geq \alpha - \rho$, $\deg(G) = \deg(g)$, $\deg(H) = \deg(h)$, $G$ *is monic*, $H, h$ *have the same leading coefficient, and finally,* $F = G \cdot H$.

PROOF. We shall prove the following assertion, for $j \geq 0$:

    (*) If $g, h \in \hat{\mathbb{Z}}_p[X]$, $\deg(g) = m$, $\deg(h) = n$, $g$ is monic, $\deg(F - gh) < \deg(F)$, $v_p(F - gh) \geq \alpha + j$, and $v_p(R(g, h)) = \rho$, then there exist polynomials $g^*, h^* \in \hat{\mathbb{Z}}_p[X]$, such that $\deg(g^*) < m$, $\deg(h^*) < n$, $v_p(g^*) \geq \alpha + j - \rho$, $v_p(h^*) \geq \alpha + j - \rho$, and $v_p(F - (g + g^*)(h - h^*)) \geq \alpha + j - 1$.

Indeed, since $v_p(R(g, h)) = \rho$ then $R(g, h) \neq 0$. By (1J), $g$ and $h$ are relatively prime.

We note that $v_p(F - gh) \geq \alpha + j\alpha\rho$ and $\deg(F - gh) < \deg(F) = \deg(g) + \deg(h)$, and it follows from (1K) that there exist uniquely defined polynomials $g^*, h^* \in \hat{\mathbb{Z}}_p[X]$, such that the following hold:

$$F - gh = h^* g + g^* h,$$
$$v_p(g^*) \geq v_p(F - gh) - \rho \geq \alpha + j - \rho,$$
$$v_p(h^*) \geq v_p(F - gh) - \rho \geq \alpha + j - \rho,$$
$$\deg(g^*) < \deg(g),$$
$$\deg(h^*) < \deg(h).$$

Therefore

$$
\begin{aligned}
v_p(F - (g + g^*)(h + h^*)) &= v_p((F - gh) - (h^*g + g^*h) - g^*h^*) \\
&= v_p(g^*h^*) \\
&\geq 2(\alpha + j - \rho) \\
&= (\alpha - 2\rho) + (\alpha + 2j) \\
&\geq \alpha + j + 1.
\end{aligned}
$$

We apply this result, beginning with $g = g_0$, $h = h_0$, $j \geq 0$, and letting $g_1 = g_0 + g_0^*$, $h_1 = h_0 + h_0^*$. Then we apply the result for $g_1, h_1, j = 1$, obtaining $g_2 = g_1 + g_1^*$, $h_2 = h_1 + h_1^*$ and so on.

We have to note that

$$
v_p(R(g_{j+1}, h_{j+1})) = v_p(R(g_j, h_j)) = \rho
$$

for every $j \geq 0$, because

$$
\begin{aligned}
v_p\left[R(g_{j+1}, h_{j+1}) - R(g_j, h_j)\right] &\geq \min\{v_p(g^*),\, v_p(h^*)\} \\
&\geq \alpha + j - \rho > \rho,
\end{aligned}
$$

as follows from (1M).

Thus, we have the sequences of polynomials $(g_j)_{j \geq 0}$ and $(h_j)_{j \geq 0}$ such that $\deg(g_j) = m$, $\deg(h_j) = n$, each $g_j$ is monic, $h_j$ and $h$ have the same leading coefficient, and finally

$$
\begin{aligned}
v_p(g_{j+1} - g_j) &\geq \alpha + j - \rho, \\
v_p(h_{j+1} - h_j) &\geq \alpha + j - \rho.
\end{aligned}
$$

Thus $(g_j)_{j \geq 0}$ and $(h_j)_{j \geq 0}$ are Cauchy sequences of polynomials of degree $m, n$, respectively. This means that if

$$
g_j = \sum_{i=0}^{m} b_{ij} X^i,
$$

$$
h_j = \sum_{i=0}^{n} c_{ij} X^i,
$$

the sequences $(b_{ij})_{j \geq 0}$, $(c_{ij})_{j \geq 0}$ (for every $i$), are Cauchy sequences in $\mathbb{Q}_p$. Since $\hat{\mathbb{Q}}_p$ is complete, let $b_i = \lim b_{ij}$, $c_i = \lim c_{ij}$, and $G = \sum_{i=0}^{m} b_i X^i$, $H = \sum_{i=0}^{n} c_i X^i$.

Then $v_p(G - g_j) \geq \alpha + j - \rho$, because

$$G - g_j = \lim_{s \to \infty} \left( \sum_{i=0}^{s} g_{j+1}^* \right)$$

and

$$v_p\left( \sum_{i=0}^{s} g_{j+1}^* \right) \geq \alpha + j - \rho,$$

for every $s \geq 1$. Similarly, $v_p(H - h_j) \geq \alpha + j - \rho$ for every $j \geq 0$. Finally,

$$\begin{aligned}
v_p(F - GH) &= v_p[(F - g_j h_j) + (g_j - G)H + g_j(h_j - H)] \\
&\geq \min\{v_p(F - g_j h_j), v_p(g_j - G) + v_p(H), \\
&\qquad v_p(g_j) + v_p(h_j - H)\} \\
&\geq \alpha + j - \rho,
\end{aligned}$$

for every $j \geq 0$. It follows that $F = GH$ with $G$ monic, and $H, h$ having the same leading coefficient. $\square$

Now we give Hensel's lemma in its more customary form:

**(1U)**     *Let $F, g, h \in \hat{\mathbb{Z}}_p[X]$ be such that:*
  (i) $\deg(g) = m > 0$, $\deg(h) = n > 0$, $\deg(F) = m + n$, $g$ is *monic and* $\deg(F - gh) < \deg(F)$;
  (ii) *$g, h$ are relatively prime modulo $p$; and*
  (iii) *$F \equiv g \cdot h \pmod{p}$.*
*Then there exist polynomials $G, H \in \mathbb{Z}_p[X]$ such that $G \equiv g \pmod{p}$, $H \equiv h \pmod{p}$, $\deg(G) = \deg(g)$, $\deg(H) = \deg(h)$, $G$ is monic, $H, h$ have the same leading coefficient, and $F = G \cdot H$.*

PROOF. This is an immediate corollary of the preceding result. Indeed, by (1P), $v_p(R(g, h)) = 0$. Since $v_p(F - gh) \geq 1$, the above result may be applied. $\square$

Another commonly encountered form of Hensel's lemma concerns the lifting of roots modulo $p$.

**(1V)**     *Let $F \in \hat{\mathbb{Z}}_p[X]$ with $\deg(F) \geq 1$, let $a \in \hat{\mathbb{Z}}_p$ be a simple root of the congruence $F(X) \equiv 0 \pmod{p}$. Then there exists $b \in \hat{\mathbb{Z}}_p$ such that $\bar{b} = \bar{a}$ and $F(b) = 0$.*

PROOF. By hypothesis, $F \equiv (X - a)h \pmod{p}$ where $h(A) \not\equiv 0$ $\pmod{p}$. So $X - a, h$ are relatively prime modulo $p$. By (1U), $F = GH$, with $G$ monic, $\deg(G) = 1$, $\overline{G} = X - a$, thus $G = X - b$ with $b \in \hat{\mathbb{Z}}_p$, $\overline{b} = \overline{a}$, and therefore $F(b) = 0$. □

We shall apply Hensel's lemma to the polynomial $X^{p-1} - 1$:

**(1W)**    *If $p$ is a prime then $\hat{\mathbb{Z}}_p$ contains $p - 1$    $(p - 1)$th roots of 1. More precisely, for every $j = 1, 2, \ldots, p - 1$ there exists a unique element $\omega_j \in \hat{\mathbb{Z}}_p$ such that $\omega_j^{p-1} = 1$ and $\omega_j \equiv j \pmod{p}$.*

PROOF. For every $j = 1, \ldots, p-1$, $j^{p-1} \equiv 1 \pmod{p}$, so $X^{p-1} - 1 \equiv \prod_{j=1}^{p-1}(X - j) \pmod{p}$. Thus $1, 2, \ldots, p - 1$ are all the roots of the congruence, and they are simple. By (1V), for every $j$ there exists $\omega_j \in \hat{\mathbb{Z}}_p$ such that $\omega_j^{p-1} = 1$ and $\omega_j \equiv j \pmod{p}$.

For the uniqueness, we observe that if $\omega \in \hat{\mathbb{Z}}_p$, $\omega^{p-1} = 1$, and $\omega \equiv k \pmod{p}$, then $\omega$ must coincide with one of the roots of $X^{p-1} - 1$, say $\omega = \omega_j$; then $j \equiv \omega_j = \omega \equiv k \pmod{p}$, so $j = k$, i.e., $\omega = \omega_k$. □

Let $p$ be a prime and let $(\hat{\mathbb{Z}}/p)^\times$ denote the multiplicative group of nonzero residue classes modulo $p$. Let $\Omega$ denote the multiplicative group of $(p - 1)$th roots of unity in $\hat{\mathbb{Z}}_p$.

As a corollary, we have:

**(1X)**    *The mapping which associates to each nonzero residue class $j$ modulo $p$ the $(p - 1)$th root of unity $\omega_j$ in $\hat{\mathbb{Z}}_p$ such that $\omega_j \equiv j \pmod{p}$, establishes an isomorphism between the multiplicative groups $(\mathbb{Z}/p)^\times$ and $\Omega$. Moreover, $\omega_g$ is a generator of $\Omega$ if and only if $g$ is a primitive root modulo $p$.*

PROOF. Indeed, if $1 \le j, k, h \le p - 1$ and $jk \equiv h \pmod{p}$, by (1W) it follows that $\omega_j \omega_k \equiv \omega_h \pmod{p}$. Since $\omega_j \equiv j \pmod{p}$, the mapping $j$ modulo $p \mapsto \omega_j$ is an isomorphism. The last assertion is trivial. □

**(1Y)**    *With the above notations:*
   (1) *iF $P - 1 \nmid r$ then $\sum_{\omega \in \Omega} \omega^r = 0$; and*
   (2) *if $p - 1 \mid r$ then $\sum_{\omega \in \Omega} \omega^r = p - 1$.*

PROOF. (1)　Let $g$ be a primitive root modulo $p$, so $\omega_g$ is a generator of the multiplicative group $\Omega$. Then

$$\sum_{\omega \in \Omega} \omega^r = \sum_{j=0}^{p-2} \omega_g^{jr} = \frac{1 - \omega_g^{(p-1)r}}{1 - \omega_g^r} = 0,$$

when $p - 1 \nmid r$.

(2)　If $p - 1 \mid r$ then $\omega^r = 1$ for every $\omega \in \Omega$, hence $\sum_{\omega \in \Omega} \omega^r = p - 1$. $\square$

## Bibliography

1908　Hensel, K., *Theorie der algebraischen Zahlen*, Teubner, Leipzig, 1908.

1999　Ribenboim, P., *The Theory of Classical Valuations*, Springer-Verlag, New York, 1999.

## V.2. Linear Recurring Sequences of Second Order

Let $A, B$ be nonzero integers such that $D = A^2 - 4B \neq 0$. Let $R_0, R_1$ be given integers and for every $k \geq 2$ let $R_k = AR_{k-1} - BR_{k-2}$.

If $A = 1$, $B = -1$ and $R_0 = 0$, $R_1 = 1$, then $R_k$ is the $k$th *Fibonacci number*. If $A = 1$, $B = -1$ and $R_0 = 2$, $R_1 = 1$, then $R_k$ is the $k$th *Lucas number*. These sequences of numbers have been briefly considered in Chapter IV, §2.

We shall now indicate several elementary properties of the sequence of numbers $(R_k)_{k \geq 0}$. For our purpose, we shall assume that $R_0 = 0$ and $R_1 = 1$. There are analogous results when $R_0 = 2$, $R_1 = A$.

Part (1) of the following lemma was given by Siebeck (1846); it is a generalization of Binet's result for Fibonacci numbers (Chapter IV, Lemma 2.3). In 1878, Lucas published a classical paper on this subject; see also Lehmer (1930). For further references on recurring sequences, see Dickson (1920, Vol. I, pp. 393–411).

LEMMA 2.1. *With above notations and hypotheses:*

(1) *If $\alpha, \beta$ are the roots of the equation $X^2 - AX + B = 0$, that is,*

$$\alpha = \frac{A + \sqrt{A^2 - 4B}}{2}, \qquad \beta = \frac{A - \sqrt{A^2 - 4B}}{2},$$

*then for every $k \geq 0$:*

$$R_k = \frac{\alpha^k - \beta^k}{\alpha - \beta}.$$

(2) *If $k, h \geq 1$ then $R_{k+h} = R_k R_{h+1} - B R_{k-1} R_h$.*

(3) $(\alpha^h + \beta^h)^2 = D R_h^2 + 4 B^h$.

(4) *If $k \mid h$ then $R_k \mid R_h$.*

(5) $A R_n + (\alpha^h + \beta^h) = 2 R_{h+1}$ *for $h \geq 0$,*
   $A R_h - (\alpha^h + \beta^h) = 2 B R_{h-1}$ *for $h \geq 1$.*

(6) *If $k \geq 1$ is odd then*

$$2^{k-1} R_k = \sum_{h=0}^{(k-1)/2} \binom{k}{2h+1} A^{k-2h-1} D^h,$$

$$2^{k-1}(\alpha^k + \beta^k) = \sum_{h=0}^{(k-1)/2} \binom{k}{2h} A^{k-2h} D^h.$$

(7) *If $k \geq 1$ is odd and $n \geq 1$ then*

$$R_{nk} = D^{(k-1)/2} R_n^k + \sum_{h=1}^{(k-1)/2} \frac{k}{h}\binom{k-h-1}{h-1} B^{nh} D^{(k-2h-1)/2} R_n^{k-2h}.$$

PROOF. (1)    From $\alpha^2 - A\alpha + \beta = 0$ it follows that $\alpha^{k+2} = A\alpha^{k+1} - B\alpha^k$ and similarly $\beta^{k+2} = A\beta^{k+1} - B\beta^k$. Noting that $\alpha \neq \beta$ since $A^2 - 4B \neq 0$, by subtraction and division by $\alpha - \beta$, we obtain

$$\frac{\alpha^{k+2} - \beta^{k+2}}{\alpha - \beta} = A \frac{\alpha^{k+1} - \beta^{k+1}}{\alpha - \beta} - B \frac{\alpha^k - \beta^k}{\alpha - \beta}.$$

Let $R'_k = (\alpha^k - \beta^k)/(\alpha - \beta)$ for every $k \geq 0$, so $R'_{k+2} = A R'_{k+1} - B R'_k$ for $k \geq 0$. Since the sequences $(R_k)_{k \geq 0}$, $(R'_k)_{k \geq 0}$ satisfy the same recurrence relation and $R'_0 = R_0 = 0$, $R'_1 = R_1 = 1$, then $R_k = R'_k = (\alpha^k - \beta^k)/(\alpha - k)$ for every $k \geq 0$.

(2)

$$R_k R_{h+1} - B R_{k-1} R_h$$
$$= \frac{(\alpha^k - \beta^k)(\alpha^{h+1} - \beta^{h+1}) - \alpha\beta(\alpha^{k-1} - \beta^{k-1})(\alpha^h - \beta^h)}{(\alpha - \beta)^2}$$
$$= \frac{\alpha^{k+h} - \beta^{k+h}}{\alpha - \beta} = R_{k+h}.$$

(3)

$$(\alpha^h + \beta^h)^2 - D R_h^2 = \alpha^{2h} + 2\alpha^h\beta^h + \beta^{2h} - D\left(\frac{\alpha^h - \beta^h}{\alpha - \beta}\right)^2$$
$$= \alpha^{2h} + 2\alpha^h\beta^h + \beta^{2h} - (\alpha^{2h} - 2\alpha^h\beta^h + \beta^{2h})$$
$$= 4\alpha^h\beta^h = 4B^h,$$

since $(\alpha - \beta)^2 = D$, $\alpha\beta = B$.

(4)  Let $h = nk$. If $n = 1$ it is trivial and we proceed by induction:

$$R_{(n+1)k} = R_{nk} R_{k+1} - B R_{nk-1} R_k$$

is a multiple of $R_k$.

(5)   We have $A = \alpha + \beta$, so by (1),

$$AR_h + (\alpha^h + \beta^h)$$
$$= (\alpha + \beta)\frac{\alpha^h - \beta^h}{\alpha - \beta} + (\alpha^h + \beta^h)$$
$$= \frac{\alpha^{h+1} - \alpha\beta^h + \alpha^h\beta - \beta^{h+1} + \alpha^{h+1} + \alpha\beta^h - \alpha^h\beta - \beta^{h+1}}{\alpha - \beta}$$
$$= 2\frac{\alpha^{h+1} - \beta^{h+1}}{\alpha - \beta} = 2R_{h+1}.$$

Similarly,

$$AR_h - (\alpha^h + \beta^h)$$
$$= (\alpha + \beta)\frac{\alpha^h - \beta^h}{\alpha - \beta} - (\alpha^h + \beta^h)$$
$$= \frac{\alpha^{h+1} - \alpha\beta^h + \alpha^h\beta - \beta^{h+1} - \alpha^{h+1} - \alpha\beta^h - \alpha^h\beta + \beta^{h+1}}{\alpha - \beta}$$
$$= 2\alpha\beta\frac{\alpha^{h-1} - \beta^{h-1}}{\alpha - \beta} = 2BR_{h-1}.$$

(6)   Let $k \geq 1$ be odd. From

$$\begin{cases} 2\alpha = A + \sqrt{D}, \\ 2\beta = A - \sqrt{D}, \end{cases}$$

by raising to the power $k$, and subtracting,

$$2^k(\alpha^k - \beta^k) = 2\left[ \sum_{h=0}^{(k-1)/2} \binom{k}{2h+1} A^{k-2h-1} D^{(2h+1)/2} \right].$$

Dividing by $2(\alpha - \beta) = 2\sqrt{D}$, we have

$$2^{k-1} R_k = \sum_{h=0}^{(k-1)/2} \binom{k}{2h+1} A^{k-2h-1} D^h.$$

Similarly, by raising the expressions of $2\alpha$ and $2\beta$ to the power $k$, by subtracting and dividing by 2, we obtain

$$2^{k-1}(\alpha^k + \beta^k) = \sum_{h=0}^{(k-1)/2} \binom{k}{2h} A^{k-2h} D^h.$$

(7)   By Chapter VII, (1D), we have the identity

$$X^k + Y^k = (X+Y)^k + \sum_{h=1}^{(k-1)/2} (-1)^h \frac{k}{h} \binom{k-h-1}{h-1} X^h Y^h (X+Y)^{k-2h}.$$

We take $X = \alpha^n$, $Y = -\beta^n$, and since $k$ is odd,

$$\alpha^{nk} - \beta^{nk} = (\alpha^n - \beta^n)^k + \sum_{h=1}^{(k-1)/2} \frac{k}{h} \binom{k-h-1}{h-1} (\alpha\beta)^{nh} (\alpha^n - \beta^n)^{k-2h}.$$

Dividing by $\alpha - \beta$, noting that $\alpha\beta = B$ and $\alpha - \beta = \sqrt{D}$ then by (1) we have

$$R_{nk} = D^{(k-1)/2} R_n^k + \sum_{h=1}^{(k-1)/2} \frac{k}{h} \binom{k-h-1}{n-1} B^{nh} D^{(k-2h-1)/2} R_n^{k-2h}. \quad \square$$

Now we investigate the divisibility properties of the terms of the recurring sequence $(R_k)_{k \geq 0}$.

If $m \geq 1$ and if there exists an index $k$ such that $m$ divides $R_k$, we denote by $r(m)$ the smallest such index. It is called the *rank of appearance* of $m$ in the sequence $(R_k)_{k \geq 0}$.

LEMMA 2.2.     (1) *If $m \geq 1$ and $\gcd(m, B) = 1$ then there exists the rank of appearance of $m$.*
(2) *If $m \geq 1$ and $\gcd(m, B) = 1$ then $m$ divides $R_k$ if and only if $r(m)$ divides $k$.*
(3) *If $m, n \geq 1$, $\gcd(m, B) = \gcd(n, B) = \gcd(m, n) = 1$ then $r(mn) = \mathrm{lcm}\{r(m), r(n)\}$.*
(4) *If $p$ is an odd prime, $p \nmid B$, then $R_p \equiv (D/r) \pmod{p}$ (where $(D/p)$ denotes the Jacobi symbol).*
(5) *If $p$ is an odd prime, $p \nmid B$, then $r(p)$ divides $p - (D/p)$.*
(6) *If $p$ is an odd prime, $p \nmid B$, if $k = v_p(R_{r(p)})$ and $e \geq k$, then $r(p^e) = p^{e-k} r(p)$.*
(7) *If $k \geq 1$, $h \geq 1$ then $R_{kh+1} \equiv R_{h+1}^k \pmod{R_h^2}$.*
(8) *If $k \geq 1$, $h \geq 1$ then $R_{kh} \equiv k R_h R_{h+1}^{k-1} \pmod{R_h^2}$.*

PROOF. (1) Consider the set of pairs $\{(R_k \mod m, R_{k-1} \mod m \mid k = 1, 2, \ldots\}$. Since there exist only finitely many couples of residue classes modulo $m$, there exist integers $k, l$, with $k < l$ such that

$$\begin{cases} R_k \equiv R_l \pmod{m}, \\ R_{k-1} \equiv R_{l-1} \pmod{m}. \end{cases}$$

Since $BR_{k-2} = AR_{k-1} - R_k$ then $BR_{k-2} \equiv BR_{l-2} \pmod{m}$ hence $R_{k-2} \equiv R_{l-2} \pmod{m}$, because $\gcd(m, B) = 1$. Repeating this argument, it follows that $0 = R_0 \equiv R_{l-k} \pmod{m}$, so there exists an integer $l - k \geq 1$ such that $m \mid R_{l-k}$. Hence there exists a smallest integer $r(m)$ such that $m \mid R_{r(m)}$.

(2)    First we note that for every $k \geq 2$, if $m$ divides $R_k$ then $\gcd(m, R_{k-1}) = 1$. Otherwise there exists a prime $p$ dividing $m, R_k$ and $R_{k-1}$. From $R_k = AR_{k-1} - BR_{k-2}$ and $\gcd(m, B) = 1$, it follows that $p \mid R_{k-2}$. Repeating the argument, we would conclude that $p \mid R_1 = 1$, which is a contradiction.

Now let $S$ be the set of all indices $k$ such that $m \mid R_k$; by (1), $S \neq 0$. We show that if $k, h \in S$ then $k + h \in S$. Indeed, by Lemma 2.1(2), $m \mid k + h$.

Similarly, if $k, h \in S$, $k < h$, then $h - k \in S$. In fact, $R_h = R_{k+(h-k)} = R_k R_{h-k+1} - BR_{k-1}R_{h-k}$, hence $m \mid BR_{k-1}R_{h-k}$. But $\gcd(m, B) = 1$ and, as shown above, $\gcd(m, R_{k-1}) = 1$, hence $m \mid R_{h-k}$.

This suffices to show that $S$ is the set of multiples of its smallest element, namely $r(m)$.

(3)    We have $m \mid mn$ and $mn \mid R_{r(mn)}$ so $m \mid R_{r(mn)}$ hence $r(m) \mid r(mn)$; similarly $r(n) \mid r(mn)$ hence $l = \text{lcm}\{r(m), r(n)\}$ divides $r(mn)$.

Conversely, $m \mid R_{r(m)}$ and $r(m) \mid l$ hence $m \mid R_l$; similarly $n \mid R_l$, hence from $\gcd(m, n) = 1$ then $mn \mid R_l$, that is, $r(mn) \mid l$.

(4)    By Chapter II, (3.3), we have

$$R_p = \frac{\alpha^p - \beta^p}{\alpha - \beta} = (\alpha - \beta)^{p-1} + pf = D^{(p-1)/2} + pf.$$

Since $D^{(p-1)/2} \equiv (D/p) \pmod{p}$ then $R_p \equiv \left(\frac{D}{p}\right) \pmod{p}$.

(5)    If $p \mid D$ then $(D/p) = 0$ and $R_p \equiv 0 \pmod{p}$. Let $(D/p) = 1$; we show that $R_{p-1} \equiv 0 \pmod{p}$ hence by (2), $r(p) \mid p-1$. Indeed, by Lemma 2.1(5), and by (4) above,

$$2BR_{p-1} = AR_p - (\alpha^p + \beta^p) \equiv A - (\alpha^p + \alpha^p) \pmod{p}.$$

By Lemma 2.1(6), $2^{p-1}(\alpha^p + \beta^p) \equiv A^p \pmod{p}$. Hence

$$2BR_{p-1} \equiv 2^p BR_{p-1} \equiv 2^{p-1}A - A^p \equiv A - A^p \equiv 0 \pmod{p}.$$

Since $p \nmid B$, $p \neq 2$, then $R_{p-1} \equiv 0 \pmod{p}$.

Now let $(D/p) = -1$; we show that $R_{p+1} \equiv 0 \pmod{p}$, hence by (2), $r(p) \mid p+1$. Indeed, by Lemma 2.1(5), and by (4) above,

$$2R_{p+1} = AR_p + (\alpha^p + \beta^p) \equiv -A + (\alpha^p + \beta^p) \pmod{p}.$$

By Lemma 2.1(6), $2^{p-1}(\alpha^p + \beta^p) \equiv A^p \pmod{p}$, hence

$$2R_{p+1} \equiv 2^p R_{p+1} \equiv -2^{p-1}A + A^p \equiv -A + A^p \equiv 0 \pmod{p}.$$

Since $p \neq 2$ then $R_{p+1} \equiv 0 \pmod{p}$.

(6)    Let $R_{r(p)} = p^k t$, with $p \nmid t$, $k \geq 1$. By Lemma 2.1(7), we have for $m \geq 1$,

$$R_{r(p)p^m} = D^{(p^m - 1)/2} R_{r(p)}^{p^m}$$

$$+ \sum_{h=1}^{(p^m-1)/2} \frac{p^m}{h} \binom{p^m - h - 1}{h - 1} B^{r(p)h} D^{(p^m - 2h - 1)/2} R_{r(p)}^{p^m - 2h}.$$

For $h = (p^m - 1)/2$ the summand is equal to $p^m B^{r(p)(p^m - 1)/2} R_{r(p)}$ and its $p$-adic value is $m + k$, since $p \nmid B$.

If $0 < h < (p^m - 1)/2$ then

$$v_p\left(\frac{p^m}{h}\binom{p^m - h - 1}{h - 1} B^{r(p)h} D^{(p^m - 2h - 1)/2} R_{r(p)}^{p^m - 2h}\right)$$

$$\geq v_p\left(\frac{p^m}{h} R_{r(p)}^{p^m - 2h}\right) = m + (p^m - 2h)k - v_p(h).$$

However, if $m \geq 1$ and $0 < h < (p^m - 1)/2$ then $p^m > 2h + v_p(h) + 1$. Indeed, $2h = p^m - p^s r$ with $p \nmid r$, $0 \leq s < m$ and if $s = 0$ then $r > 1$; so $2h + v_p(h) + 1 = p^m - p^s r + s + 1 < p^m$ because $s + 1 < p^s r$ (since $p \neq 2$). Hence,

$$m + (p^m - 2h)k - v_p(h) > m + v_p(h)(k - 1) + k \geq m + k.$$

We have also

$$v_p\left(D^{(p^m - 1)/2} R_{r(p)}^{p^m}\right) \geq p^m k > m + k,$$

when $m \geq 1$ (because $p \neq 2$). Thus, for $m \geq 1$ we have $v_p\left(R_{r(p)p^m}\right) = m + k$. This is also true when $m = 0$.

Taking $m = e - k \geq 0$ then $v_p\left(R_{r(p)p^{e-k}}\right) = e$ so $p^e \mid R_{r(p)p^{e-k}}$ and therefore, by (2), $r(p^e) \mid r(p)p^{e-k}$. Since $p^e \mid R_{r(p^e)}$ then $r(p) \mid r(p^e)$. Hence $r(p^e) = r(p)p^m$ with $0 \leq m \leq e - k$. If $m < e - k$ then $v_p(R_{r(p)p^m}) = m + k < e$, so $p^e \nmid R_{r(p)p^m}$ hence $r(p^e) \neq r(p)p^m$ and this shows that $r(p^e) = r(p)p^{e-k}$.

(7)  The proof is by induction, being trivial when $k = 1$. We have by Lemma 2.1(2):

$$R_{(k+1)h+1} = R_{(kh+1)+h}$$
$$= R_{kh+1}R_{h+1} - BR_{kh}R_h$$
$$\equiv R_{kh+1}R_{h+1}$$
$$\equiv R_{h+1}^{k+1} \pmod{R_h^2}$$

by induction and since $R_h$ divides $R_{kh}$ (by Lemma 2.1(4)).

(8)  The proof is by induction, being trivial when $k = 1$. We have, by Part (7), by Lemma 2.1(2) and (4), and by induction:

$$R_{(k+1)h} = R_{kh+h} = R_h R_{kh+1} - BR_{h-1}R_{kh}$$
$$\equiv R_h R_{h+1}^k - BR_{h-1}kR_h R_{h+1}^{k-1}$$
$$\equiv R_h R_{h+1}^{k-1}[R_{h+1} - B_k R_{h-1}]$$
$$\equiv R_h R_{h+1}^{k-1}(k + 1)R_{h+1} \equiv (k + 1)R_h R_{h+1}^k \pmod{R_h^2},$$

noting that

$$R_{h+1} = R_h R_2 - BR_{h-1} \equiv -BR_{h-1} \pmod{R_h}. \quad \square$$

## Bibliography

1843 Binet, J., *Mémoire sur l'intégration des équations linéaires aux différences finies d'un ordre quelconque à coefficients variables*, C. R. Acad. Sci. Paris, **17** (1843), 559–567.

1846 Siebeck, H., *Die recurrenten Reihen vom Standpunkt der Zahlentheorie aus betrachtet*, J. Reine Angew. Math., **33** (1846), 71–77.

1878 Lucas, E., *Théorie des fonctions numériques simplement périodiques*, Amer. J. Math., **1** (1878), 184–239 and 289–321.

1920 Dickson, L.E., *History of the Theory of Numbers*, Vol. I. Carnegie Institution, Washington, DC, 1920; reprinted by Chelsea, New York, 1952.

1930 Lehmer, D.H., *An extended theory of Lucas' functions*, Ann. of Math., **31** (1930), 419–448.

1995 Ribenboim, P., *The Fibonacci numbers and the Arctic Ocean*, Sympos. Gaussiana, Conf. A (editors, M. Behara, R. Fritsch, and R.G. Lintz), W. de Gruyter, Berlin, 1995, pp. 41–83.

# VI
# Arithmetic Restrictions on Hypothetical Solutions and on the Exponent

Let $p$ be an odd prime and assume that $x, y, z$ are nonzero pairwise relatively prime integers such that $x^p + y^p + z^p = 0$.

In this chapter we indicate congruences and divisibility properties satisfied by expressions involving the numbers $x, y, z, p$. In some instances, we will be able to reach a contradiction, proving that Fermat's last theorem (or the first case) holds for certain exponents $p$. In Section 3 we focus on a conjecture of Abel, which has not yet been completely established by a direct proof.

## VI.1. Congruences

Let $p$ be an odd prime and assume that $x, y, z$ are nonzero relatively prime integers such that $x^p + y^p + z^p = 0$. For easy reference, we recall results from Chapter III, §1.

If $p \nmid xyz$ then there exist nonzero integers $r, s, t, r_1, s_1, t_1$ such that

$$(1.1) \quad \begin{cases} x + y = t^p, & (x^p + y^p)/(x+y) = t_1^p, & z = -tt_1, \\ y + z = r^p, & (y^p + z^p)/(y+z) = r_1^p, & x = -rr_1, \\ z + x = s^p, & (z^p + x^p)/(z+x) = s_1^p, & y = -ss_1, \end{cases}$$

$p \nmid rstr_1s_1t_1$ and $r, s, t, r_1, s_1, t_1$ are pairwise relatively prime. Moreover $r^p + s^p + t^p \neq 0$.

If $p \nmid xy$ but $p \mid z$, then there exist nonzero integers $r, s, t, r_1, s_1, t_1$, and $n \geq 2$ such that

(1.2)
$$\begin{cases} x + y = p^{pn-1}t^p, & (x^p + y^p)/(x + y) = pt_1^p, & z = -p^n tt_1, \\ y + z = r^p, & (y^p + z^p)/(x + y) = r_1^p, & x = -rr_1, \\ x + y = s^p, & (z^p + x^p)/(x + y) = s_1^p, & y = -ss_1, \end{cases}$$

$p \nmid rstr_1s_1t_1$ and $r, s, t, r_1, s_1, t_1$ are pairwise relatively prime. We have $r^p + s^p + p^{pn-1}t^p \neq 0$.

Moreover: If $p \nmid xyz$ then $r_1 \equiv 1 \pmod{2p^2}$, $s_1 \equiv 1 \pmod{2p^2}$, and $t_1 \equiv 1 \pmod{2p^2}$. If $p \mid z$, $p \nmid xy$ then $r_1 \equiv 1 \pmod{2p}$, $s_1 \equiv 1 \pmod{2p}$, and $t_1 \equiv 1 \pmod{2p^2}$. If $p \nmid xyz$ then

(1.3)
$$\begin{cases} x = -r^p + k, \\ y = -s^p + k, \\ z = -t^p + k, \end{cases}$$

where $k = (r^p + s^p + t^p)/2$. If $p \mid z$, $p \nmid xy$ then

(1.4)
$$\begin{cases} x = -r^p + k, \\ y = -s^p + k, \\ z = -p^{pn-1}t^p + k, \end{cases}$$

where $k = (r^p + s^p + p^{pn-1}t^p)/2$.

We begin with an easy congruence, soon to be reinforced. From $x^p \equiv x \pmod{p}$, $y^p \equiv y \pmod{p}$, $z^p \equiv z \pmod{p}$ then $-z \equiv -z^p = x^p + y^p \equiv x + y \pmod{p}$, so $x^p + y^p = -z^p \equiv (x + y)^p \pmod{p^2}$.

The first result is due to Fleck (1909). It was given by Lind in a weaker form, in 1910. It was rediscovered by Frobenius in 1914 and again by Vandiver (1914), Pomey (1923) and Pérez-Cacho (1958).

**(1A)**    Let $p$ be an odd prime, and let $x, y, z$ be nonzero relatively prime integers, such that $x^p + y^p + z^p = 0$.

  (1) If $p$ does not divide $x$ then $x^{p-1} \equiv 1 \pmod{p^3}$.
  (2) If $p \nmid xyz$ then $(x + y)^p \equiv x^p + y^p \pmod{p^4}$.

PROOF. (1)  1st Case:  $p$ does not divide $yz$. As recalled above, $r_1 \equiv 1 \pmod{p^2}$. Hence $x = -rr_1 \equiv -r \pmod{p^2}$ and $x^p \equiv -r^p$ $\pmod{p^3}$.

By symmetry, we have also $y^p \equiv -s^p \pmod{p^3}$ and $z^p \equiv -t^p$ $\pmod{p^3}$. Since $x^p + y^p + z^p = 0$ then $r^p + s^p + t^p \equiv 0 \pmod{p^3}$. It follows from the above relations (1.3) that $x \equiv -r^p \pmod{p^3}$. Hence $x^p \equiv x \pmod{p^3}$ and $x^{p-1} \equiv 1 \pmod{p^3}$.

2nd Case:  Assume that $p$ divides $z$, hence $p$ does not divide $y$. As recalled above, $t_1 \equiv 1 \pmod{p^2}$, hence $t_1^p \equiv 1 \pmod{p^3}$.

On the other hand, $pn - 1 \geq 4$, so $x \equiv -y \pmod{p^4}$ and $pt_1^p = (x^p + y^p)/(x+y) = x^{p-1} - x^{p-2}y + \cdots - xy^{p-2} + y^{p-1} \equiv px^{p-1} \pmod{p^4}$ hence $x^{p-1} \equiv t_1^p \equiv 1 \pmod{p^3}$.

(2)   We have $0 = x^p + y^p + z^p \equiv x + y + z \pmod{p^3}$, hence $x + y \equiv -z \pmod{p^3}$. Then $(x+y)^p \equiv -z^p \equiv x^p + y^p \pmod{p^4}$.  □

With methods from Class Field Theory, as a consequence of theorems of Furtwängler, Vandiver proved (1914, 1919):

**(1B)**    *With above notations: $x^p \equiv x \pmod{p^3}$, $y^p \equiv y \pmod{p^3}$, $z^p \equiv z \pmod{p^3}$, and $x + y + z \equiv 0 \pmod{p^3}$.*

The result which follows is again due to Fleck (1909). Partial assertions were rediscovered by Pomey (1923), Vandiver (1925), James (1934), Niewiadomski (1938), and Inkeri (1946).

**(1C)**    *Let $x, y, z$ be nonzero relatively prime integers such that $x^p + y^p + z^p = 0$.*

(1) *If $p \nmid xyz$ then $x + y + z$ is a multiple of 6 and of $rstp^3$ and $r + s + t$ is a multiple of $p^2$.*

(2) *If $p \mid z$ then $x + y + z$ is a multiple of 6 and of $rstp^2$ and $r + s$ is a multiple of $p$, while $r + s + t$ is not a multiple of $p$.*

PROOF. It follows from the relations (1.1) (respectively, (1.2)), that in both cases $r, s, t$ divide $x + y + z$. Also, $x + y + z$ is even and $x^p \equiv x \pmod{3}$, $y^p \equiv y \pmod{3}$, $z^p \equiv z \pmod{3}$, hence $x + y + z \equiv x^p + y^p + z^p = 0 \pmod{3}$.

In the first case, it follows from (1A) that $x^p \equiv x \pmod{p^3}$, $y^p \equiv y$ $\pmod{p^3}$, $z^p \equiv z \pmod{p^3}$, hence $x + y + z \equiv 0 \pmod{p^3}$. Since $p \nmid rst$ then $rstp^3$ divides $x + y + z$.

As it was recalled above, $r_1 \equiv 1 \pmod{p^2}$, $s_1 \equiv 1 \pmod{p^2}$, $t_1 \equiv 1 \pmod{p^2}$. It follows from relations (1.1) that $x \equiv -r \pmod{p^2}$, $y \equiv -s \pmod{p^2}$, $z \equiv -t \pmod{p^2}$, hence $r + s + t \equiv 0 \pmod{p^2}$.

Assuming that $p \mid z$, then $p \nmid xy$, hence it follows from relations (1.2) that $p^n$ divides $z$ and $x + y$ (with $n \geq 2$), hence $p^2$ divides $x + y + z$; since $p \nmid rst$ then $p^2 rst$ divides $x + y + z$.

As recalled above, $r_1 \equiv 1 \pmod{p}$, $s_1 \equiv 1 \pmod{p}$ so by relations (1.2), $x \equiv -r \pmod{p}$, $y \equiv -s \pmod{p}$. Hence $r + s \equiv -(x + y) \equiv z \equiv 0 \pmod{p}$. Thus $r + s + t \equiv t \not\equiv 0 \pmod{p}$. $\square$

The next proposition was proved, in the first case, by Spunar (1929) and James (1934); a simpler proof was given by Segal (1938).

**(1D)**    *Let $x, y, z$ be nonzero relatively prime integers such that $x^p + y^p + z^p = 0$.*

(1) *If $p \nmid xyz$ then $r + s + t \neq 0$.*
(2) *If $p \mid z$ then $r + s + p^n t \neq 0$ and also $r + s + t \neq 0$ (where $n$ was defined in (1.2)).*

PROOF. (1) Assume that $p \nmid xyz$ and $r + s + t = 0$. Since $\gcd(r, s, t) = 1$ we may suppose, for example, that $r, s$ are odd while $t$ is even. Then $r^p + s^p = (y + z) + (z + x) = x + y + 2z = t^p - 2tt_1$ (by relations (1.1)). Hence

$$\frac{r^p + s^p}{r + s} = -\frac{r^p + s^p}{t} = 2t_1 - t^{p-1}.$$

The left-hand side is equal to $r^{p-1} - r^{p-2}s + \cdots - rs^{p-2} + s^{p-1}$, hence it is the sum of $p$ odd numbers, so it is odd. On the other hand, $2t_1 - t^{p-1}$ is even, which is impossible. Hence $r + s + t \neq 0$ in the first case.

(2) If $p \mid z$ then $p \nmid xy$ and by relations (1.2), $r^p + s^p = (y + z) + (z + x) = x + y + 2z = p^{pn-1}t^p - 2p^n tt_1$. Hence, if $r + s + p^n t = 0$ then

$$\frac{r^p + s^p}{r + s} = -\frac{r^p + s^p}{p^n t} = 2t_1 - p^{pn-n-1}t^{p-1}.$$

Since $p \nmid t_1$ and $n \geq 2$ then $p$ does not divide $(r^p + s^p)/(r + s)$. On the other hand, by Chapter II, (3R)(4), $\gcd(p, r+s) = \gcd(r+s, (r^p + s^p)/(r+s))$. By (1C), $p \mid r+s$, hence $p$ divides $(r^p + s^p)/(r+s)$, which is a contradiction. From (1C), since $p \nmid r+s+t$, then $r+s+t \neq 0$. $\square$

In connection with this result, Racliş considered in 1944 the following conjecture, where $p$ is an odd prime:

$(R'_p)$   If $a, b, c$ are nonzero integers and $pabc$ divides $a^p + b^p + c^p$ then $a + b + c = 0$ or $a^p + b^p + c^p = 0$.

Racliş showed:

**(1E)**    Let $p$ be an odd prime. If $(R'_p)$ is true then the first case of Fermat's theorem is true for the exponent $p$.

PROOF. Assume that $x, y, z$ are nonzero pairwise relatively prime integers, not multiples of $p$, such that $x^p + y^p + z^p = 0$. Let $r, s, t$ be defined as in relations (1.1), so $r, s, t$ are nonzero integers. We remarked after (1.1) that $r^p + s^p + t^p \neq 0$, and by (1D), $r + s + t \neq 0$. By (1C), $p$ divides $r + s + t$. By (1.1) and (1.3), $r, s, t$ divide $r^p + s^p + t^p$, and since $p, r, s, t$ are pairwise relatively prime then $prst$ divides $r^p + s^p + t^p$, showing that $(R'_p)$ is not true.   □

The validity of $(R'_p)$ is very questionable. $(R'_3)$ and $(R'_5)$ are false: $p = 3$, $a = b = c = 1$ and $p = 5$, $a = 33$, $b = -2$, $c = -1$ provide counterexamples.

Similarly, consider the following statement:

$(R''_p)$   If $a, b, c$ are nonzero integers and $pabc$ divides $a^p + b^p + p^{pn-1}c^p$ (for some $n \geq 2$) then $a + b + p^n c = 0$ or $a^p + b^p + p^{pn-1}c^p = 0$.

Then:

**(1F)**    Let $p$ be an odd prime. If $(R''_p)$ is true then the second case of Fermat's theorem is true for the exponent $p$.

PROOF. Assume that $x, y, z$ are pairwise relatively prime integers, such that $p \mid z$ and $x^p + y^p + z^p = 0$. Let $r, s, t$, and $n$ be defined as in relations (1.2), so $r, s, t$ are nonzero integers, $n \geq 2$. Then by (1D), $r + s + p^n t \neq 0$, and by the remark after (1.2), $r^p + s^p + p^{pn-1}t^p \neq 0$.

However, as already shown in (1C), $r + s \equiv 0 \pmod{p}$, so $p$ divides $r^p + s^p + p^{pn-1}t^p$. By (1.4), $r, s, t$ divide $r^p + s^p + p^{pn-1}t^p$ and since $p, r, s, t$ are pairwise relatively prime then $prst$ divides $r^p + s^p + p^{pn-1}t^p$. This contradicts the assumption $(R''_p)$.   □

In 1946, Inkeri generalized (1A). He also proved the corresponding result for the second case, with a more powerful method from Class Field Theory.

**(1G)**    *If $p$ is an odd prime, $n \geq 1$, if $x, y, z$ are nonzero relatively prime integers such that $x^{p^n} + y^{p^n} + z^{p^n} = 0$, and if $p$ does not divide $x$ then $x^{p-1} \equiv 1 \pmod{p^{2n+1}}$.*

PROOF. Since $x, y, z$ are pairwise relatively prime, we may assume, for example, that $p \nmid y$. We have

$$-x^{p^n} = y^{p^n} + z^{p^n} = (y+z) \prod_{m=1}^{n} Q_p\left(y^{p^{m-1}}, -z^{p^{m-1}}\right).$$

By Chapter II, (3C)(2), the factors on the right are pairwise relatively prime, so $y + z = a^{p^n}$ where $a \mid x$, $p \nmid a$. Similarly, $x + z = b^{p^n}$ where $b \mid y$, $p \nmid b$.

If $p \nmid z$, we have similarly $x + y = c^{p^n}$ where $c \mid z$, $p \nmid c$ and $(x^p + y^p)/(x+y) = d^{p^n}$ where $d \mid z$, $p \nmid d$ and $\gcd(c, d) = 1$. However, if $p \mid z$, by Chapter II, (3C)(3), $x + y = p^h c^{p^n}$, $(x^p + y^p)/(x+y) = pd^{p^n}$ where $c \mid z$, $d \mid z$, $p \nmid c$, $p \nmid d$, $\gcd(c, d) = 1$, $h \geq 1$.

We show that if $q$ is any prime dividing $d$ (whether $p \mid z$ or $p \nmid z$) then $q \equiv 1 \pmod{p^{n+1}}$. Indeed, since $q \mid d$ then $q \mid x^p + y^p$, $q \mid z$, $q \neq p$, and $q \nmid c$, hence $q \nmid x + y$. So

$$\begin{cases} y \equiv a^{p^n} \pmod{q}, \\ x \equiv b^{p^n} \pmod{q}, \end{cases}$$

hence $q \nmid a^{p^n} + b^{p^n}$. But $a^{p^{n+1}} + b^{p^{n+1}} \equiv x^p + y^p \equiv 0 \pmod{q}$. Therefore $q$ is a primitive factor of the binomial $a^{p^{n+1}} + b^{p^{n+1}}$. By Chapter II, (3G), $q \equiv 1 \pmod{p^{n+1}}$. It follows that $d \equiv 1 \pmod{p^{n+1}}$, hence $d^{p^n} \equiv 1 \pmod{p^{2n+1}}$.

To conclude the proof, we examine separately the two cases.

If $p \nmid x$ then $(x^p + y^p)/(x+y) \equiv 1 \pmod{p^{2n+1}}$ hence $x^p + y^p \equiv x + y \pmod{p^{2n+1}}$. By symmetry, $y^p + z^p \equiv y + z \pmod{p^{2n+1}}$ and $x^p + z^p \equiv x + z \pmod{p^{2n+1}}$. Adding up these congruences, dividing by 2, and subtracting the second one, we obtain $x^p \equiv x \pmod{p^{2n+1}}$, hence $x^{p-1} \equiv 1 \pmod{p^{2n+1}}$.

If $p \mid z$, let $v_p(z) = k \geq 1$, so $v_p(z^{p^n}) = p^n k \geq n + 1$. By Chapter II, (3C)(4), $v_p(x + y) = p^n k - n$, so $x \equiv -y \pmod{p^{kp^n - n}}$.

Hence

$$pd^{p^n} = \frac{x^p + y^p}{x + y} = x^{p-1} - x^{p-2}y + \cdots - xy^{p-2} + y^{p-1}$$

$$\equiv px^{p-1} \pmod{p^{kp^n - n}}.$$

But $kp^n - n \geq 2n + 2$, because $p > 3$ (otherwise the hypothesis is not satisfied); hence $x^{p-1} \equiv d^{p^n} \equiv 1 \pmod{p^{2n+1}}$.  $\square$

The next group of results will involve congruences of the type

$$(1 + x)^{p^k} \equiv 1 + x^{p^k} \pmod{p^{k+1}}.$$

It is convenient to precede the discussion with the following easy result, which is given explicitly by Ferentinou-Nicolacopoulou (1965):

LEMMA 1.1. *Let $p$ be an odd prime not dividing the integer $a$.*

(1) *If $n > m \geq 0$ then $a^{p^n} \equiv a^{p^{n-m}} \pmod{p^{n-m+1}}$.*

(2) *If $k \geq 1$ then $a^{p^k} - a^{p^{k-1}} \equiv ap^{k-1}(a^{p-1} - 1) \pmod{p^{k+1}}$.*

*If $p$ also does not divide $a + 1$:*

(3) *If $k \geq 1$ then $[(a + 1)^{p^k} - a^{p^k} - 1] - [(a + 1)^{p^{k-1}} - a^{p^{k-1}} - 1]$*
   *$\equiv p^{k-1}[(a + 1)^p - a^p - 1] \pmod{p^{k+1}}$.*

(4) *If $k \geq 2$ then $(a + 1)^{p^k} \equiv a^{p^k} + 1 \pmod{p^{k+1}}$ if and only if*
   *$(a + 1)^{p^{k-1}} \equiv a^{p^{k-1}} + 1 \pmod{p^{k+1}}$.*

(5) *If $k \geq 1$ and $(a+1)^{p^k} \equiv a^{p^k} + 1 \pmod{p^{k+2}}$ then $(a+1)^{p^{k-1}} \equiv a^{p^{k-1}} + 1 \pmod{p^{k+1}}$.*

PROOF. (1)    $a^{p^m} \equiv \cdots \equiv a^p \equiv a \pmod{p}$, hence raising to the power $p^{n-m}$:

$$a^{p^n} \equiv a^{p^{n-m}} \pmod{p^{n-m+1}}.$$

(2)    $a^{p^k} - a^{p^{k-1}} = a^{p^{k-1}}[a^{p^{k-1}(p-1)} - 1]$. If $a^{p-1} = 1 + bp$ then

$$\left(a^{p-1}\right)^{p^{k-1}} \equiv 1 + bp^k \pmod{p^{k+1}},$$

hence

$$\left(a^{p-1}\right)^{p^{k-1}} - 1 \equiv bp^k \equiv (a^{p-1} - 1)\, p^{k-1} \pmod{p^{k+1}}.$$

Since $a^{p^{k-1}} \equiv a \pmod{p}$ then

$$a^{p^k} - a^{p^{k-1}} \equiv a(a^{p-1} - 1)p^{k-1} \pmod{p^{k+1}}.$$

(3)   By (2) and the hypothesis, $p$ does not divide $a + 1$:

$$(a + 1)^{p^k} - (a + 1)^{p^{k-1}} \equiv (a + 1)^{p^{k-1}}((a + 1)^{p-1} - 1) \pmod{p^{k+1}}$$

and similarly

$$a^{p^k} - a^{p^{k-1}} \equiv a p^{k-1}(a^{p-1} - 1) \pmod{p^{k+1}}.$$

Therefore, by subtraction,

$$[(a + 1)^{p^k} - a^{p^k} - 1] - [(a + 1)^{p^{k-1}} - a^{p^{k-1}} - 1]$$
$$\equiv p^{k-1}[(a + 1)^p - a^p - 1] \pmod{p^{k+1}}.$$

(4)   First note that if $k \geq 2$ then $a^{p^k} \equiv a^{p^{k-1}} \equiv a^p \pmod{p^2}$ and also

$$(a + 1)^{p^k} \equiv (a + 1)^{p^{k-1}} \equiv (a + 1)^p \pmod{p^2}.$$

Thus

$$(a + 1)^{p^k} - a^{p^k} - 1 \equiv (a + 1)^{p^{k-1}} - a^{p^{k-1}} - 1$$
$$\equiv (a + 1)^p - a^p - 1 \pmod{p^2}.$$

If

$$(a + 1)^{p^k} - a^{p^k} - 1 \equiv 0 \pmod{p^{k+1}}$$

then also

$$(a + 1)^{p^k} - a^{p^k} - 1 \equiv 0 \pmod{p^2},$$

so $(a + 1)^p - a^p - 1 \equiv 0 \pmod{p^2}$. By (3),

$$(a + 1)^{p^{k-1}} - a^{p^{k-1}} - 1 \equiv 0 \pmod{p^{k+1}}.$$

The converse is proved in the same way.

(5)   By hypothesis, $(a + 1)^{p^k} \equiv a^{p^k} + 1 \pmod{p^{k+1}}$ hence also $(a + 1)^{p^{k-1}} \equiv a^{p^{k-1}} + 1 \pmod{p^{k+1}}$.  □

In particular, as was noted by Birkhoff and was published by Carmichael in his second note of 1913, if $p$ does not divide $a$, nor $a + 1$, then $(1 + a)^p \equiv 1 + a^p \pmod{p^3}$ if and only if $(1 + a)^{p^2} \equiv 1 + a^{p^2} \pmod{p^3}$.

The following result was given by Klösgen in 1970:

(1H)   *Let $p$ be an odd prime, and $m \geq 1$. The following conditions are equivalent:*

(1) *There exist integers $x, y, z$, not multiples of $p$, satisfying the congruence*

$$x^{p^m} + y^{p^m} + z^{p^m} \equiv 0 \pmod{p^{m+1}}.$$

(2) *There exists an integer $a$, $1 \le a \le (p-3)/2$, such that*

$$1 + a^{p^m} \equiv (1+a)^{p^m} \pmod{p^{m+1}}.$$

*Moreover, if any two of the numbers $x, y, z$ are congruent modulo $p$ then $2^{p^m - 1} \equiv 1 \pmod{p^{m+1}}$.*

PROOF. We need only to show that (1) implies (2), since the other implication is trivial. Let $z'$ be an integer such that $z'z \equiv 1 \pmod{p}$. Let $a \equiv z'x \pmod{p}$, $-b \equiv z'y \pmod{p}$, where $1 \le a \le p - 1$, $1 \le b \le p - 1$. Then

$$a^{p^m} \equiv z'^{p^m} x^{p^m} \pmod{p^{m+1}}, \qquad -b^{p^m} \equiv z'^{p^m} y^{p^m} \pmod{p^{m+1}},$$

so $a^{p^m} + 1 \equiv b^{p^m} \pmod{p^{m+1}}$. If $b \equiv a + t \pmod{p}$ then $a^{p^m} + 1 \equiv b^{p^m} \equiv (a + t)^{p^m} \equiv a + t \pmod{p}$ so $t \equiv 1 \pmod{p}$, and $a^{p^m} + 1 \equiv (a+1)^{p^m} \pmod{p^{m+1}}$, with $1 \le a < a + 1 \le p - 1$.
   If $a = (p-1)/2$ then $a + 1 = (p+1)/2 \equiv -(p-1)/2 \pmod{p}$ so

$$(p-1)^{p^m} + 2^{p^m} \equiv -(p-1)^{p^m} \pmod{p^{m+1}}$$

hence $2^{p^m} \equiv 2 \pmod{p^{m+1}}$ and we take $a = 1$, since $1^{p^m} + 1 = (1 + 1)^{p^m} \pmod{p^{m+1}}$. If $(p-1)/2 < a \le p-2$ then taking $a_1 = p - 1 - a$ we have $1 \le a_1 \le (p-3)/2$ and

$$1 + a_1^{p^m} \equiv (1 + a_1)^{p^m} \pmod{p^{m+1}}.$$

   For the last assertion, we may assume, for example, that $x \equiv y \pmod{p}$, the other cases being similar. In the course of the proof, we had $a \equiv z'x \equiv z'y \equiv -b \pmod{p}$, $b \equiv a + 1 \pmod{p}$ hence $2a \equiv -1 \pmod{p}$, so $a = (p-1)/2$. Therefore $2^{p^m - 1} \equiv 1 \pmod{p^{m+1}}$. $\square$

   An immediate corollary of this result is the following.
   If the first case of Fermat's theorem fails for the exponent $p$, that is, if there exist integers $x, y, z$, not multiples of $p$, such that $x^p + y^p + z^p = 0$, then we also have $x^p + y^p + z^p \equiv 0 \pmod{p^2}$. There exists $a$, $1 \le a \le (p-3)/2$, such that $1 + a^p \equiv (1+a)^p$

(mod $p^2$). Carmichael (1913) and Meissner (1914) obtained the following more precise results (the statement for the first case is also given by Gandhi, in 1975).

**(1I)** *Let $x, y, z$ be nonzero relatively prime integers such that $x^p + y^p + z^p = 0$.*

(1) *If $p \nmid xyz$ there exists an integer $a$, $1 \leq a \leq (p-3)/2$ such that*

$$(1+a)^p \equiv 1 + a^{p^2} \pmod{p^3},$$

*or equivalently, $(1+a)^p \equiv 1 + a^p \pmod{p^3}$.*

(2) *If $p \mid xyz$ there exists an integer $a$, $1 \leq a \leq (p-3)/2$, such that $(1+a)^p \equiv 1 + a^p \pmod{p^2}$.*

PROOF. (1)   By (1A), $x^p \equiv x \pmod{p^3}$ so $x^{p^2} \equiv x^p \equiv x \pmod{p^3}$. Similarly $y^{p^2} \equiv y \pmod{p^3}$ and $z^{p^2} \equiv z \pmod{p^3}$. Hence, by (1C), $x^{p^2} + y^{p^2} + z^{p^2} \equiv x + y + z \equiv 0 \pmod{p^3}$.

By (1H) there exists $a$, $1 \leq a \leq (p-3)/2$, such that $(1+a)^{p^2} \equiv 1 + a^{p^2} \pmod{p^3}$ and by Birkhoff's remark following Lemma 1.1, $(1+a)^p \equiv 1 + a^p \pmod{p^3}$.

(2)   If $p \mid xyz$, we may assume, for example, that $p \nmid xy$, $p \mid z$. According to (1C), $x + y + z \equiv 0 \pmod{p^2}$. Then $(x+y)^p \equiv -z^p = x^p + y^p \pmod{p^3}$. Since $x \not\equiv 0 \pmod{p}$, $y \not\equiv 0 \pmod{p}$, there exists an integer $b$, $1 \leq b \leq p-1$ such that $y \equiv bx \pmod{p}$. Then $y^p \equiv b^p x^p \pmod{p^2}$ and $(x+y)^p \equiv x^p(1+b)^p \pmod{p^2}$. Thus $x^p(1+b)^p \equiv x^p(1+b^p) \pmod{p^2}$. Since $p \nmid x$ then $(1+b)^p \equiv 1 + b^p \pmod{p^2}$.

As in (1H), if $1 \leq b \leq (p-3)/2$ we take $a = b$, if $(p-1)/2 < b \leq p-1$ we take $a = p-1-b$ and if $b = (p-1)/2$ then we take $a = 1$.   □

We remark that the above criterion for the first case is useless when $p \equiv 1 \pmod 6$, as pointed out by Birkhoff (in Carmichael's second note, 1913). For this purpose, we establish the following easy lemma:

LEMMA 1.2. *Let $p$ be an odd prime. Then $p \equiv 1 \pmod 6$ if and only if $p \neq 3$ and there exists an integer $t$, $1 \leq t \leq p-1$, such that $t^2 + t + 1 \equiv 0 \pmod p$.*

PROOF. Let $p = 6k + 1$, let $g$ be a primitive root modulo $p$. Let $t$ be such that $1 \leq t \leq p - 1$ and $t \equiv g^{2k}$ (mod $p$). So $t \not\equiv 1$ (mod $p$), but $t^3 \equiv 1$ (mod $p$). Hence $t^2 + t + 1 = (t^3 - 1)/(t - 1) \equiv 0$ (mod $p$).

Conversely, if $t^2 + t + 1 \equiv 0$ (mod $p$), then $(2t+1)^2 = 4t^2 + 4t + 1 = 4(t^2 + t + 1) - 3 \equiv -3$ (mod $p$). So $-3$ is a square modulo $p \neq 3$. Hence $1 = (-3/p) = (p/3)$ so $p \equiv 1$ (mod 3), hence $p \equiv 1$ (mod 6).  □

Now we use the following fact established by Cauchy in 1841: if $p \equiv 1$ (mod 6) then the polynomial $pX(X+1)(X^2+X+1)^2$ divides $(X+1)^p - X^p - 1$ (see Chapter VII, (2A)).

Thus, if $p \equiv 1$ (mod 6), by the lemma there exists $t$, $1 \leq t \leq p-1$, such that $t^2 + t + 1 \equiv 0$ (mod $p$). It follows that $(1 + t)^p \equiv 1 + t^p$ (mod $p^3$) and by a remark of Birkhoff, $(1 + t)^{p^2} \equiv 1 + t^{p^2}$ (mod $p^3$). Proceeding as in the proof of (1H), there exists $a$, $1 \leq a \leq (p - 3)/2$, such that $(1 + a)^{p^2} \equiv 1 + a^{p^2}$ (mod $p^3$).

This establishes the assertion that if $p \equiv 1$ (mod 6) then the criterion of (1I) is useless in the first case.

Wagstaff verified in 1975 that for every prime $p < 100\,000$, $p \equiv -1$ mod 6, the congruence $(1 + x)^p \equiv 1 + x^p$ (mod $p^3$) has no solution in integers $a$, $1 \leq a \leq (p - 3)/2$. In this way, it was proved that the first case is true for such exponents.

In the same year of 1975, Gandhi had independently suggested that such computations be performed. (1I) may be rephrased as follows:

**(1J)**  *Let $g$ be a primitive root modulo $p$.*

(1) *If $1 + g^{jp^2} + g^{kp^2} \not\equiv 0$ (mod $p^3$) for all indices $j, k = 1, \ldots, p-1$ then the first case of Fermat's theorem holds for the exponent $p$.*

(2) *If $1 + g^{jp} + g^{kp} \not\equiv 0$ (mod $p^2$) for all indices $j, k = 1, \ldots, p-1$ then Fermat's theorem holds for the exponent $p$.*

PROOF. (1) If the first case fails for the exponent $p$, there exists $a$, $1 \leq a \leq (p - 3)/2$, such that $(1 + a)^{p^2} \equiv 1 + a^{p^2}$ (mod $p^3$). Note that $a \not\equiv 0, -1$ (mod $p$). Let $j, k$ be indices such that $a \equiv g^j$ (mod $p$), $-(1 + a) \equiv g^k$ (mod $p$). Then $a^{p^2} \equiv g^{jp^2}$ (mod $p^3$), $(1 + a)^{p^2} \equiv -g^{kp^2}$ (mod $p^3$) and therefore $1 + g^{jp^2} + g^{kp^2} \equiv 0$ (mod $p^3$).

(2) If Fermat's theorem fails for the exponent $p$, there exists $a$, $1 \leq a \leq (p - 3)/2$ such that $(1 + a)^p \equiv 1 + a^p$ (mod $p^2$). We conclude

similarly that there exist indices $j, k$ such that $1 + g^{jp} + g^{kp} \equiv 0$ (mod $p^2$).   $\square$

In 1950, Trypanis announced without proof the following strengthening of Carmichael's result. It was rediscovered by Ferentinou-Nicolacopoulou in 1965 (we present her proof) and generalized by Klösgen in 1970 (see (1J)).

**(1K)**    *If the first case of Fermat's theorem fails for the exponent $p > 5$, there exists an integer $a$, $1 \le a \le (p-5)/2$, such that*

$$(1+a)^{p^2} \equiv 1 + a^{p^2} \pmod{p^4},$$

*or equivalently*

$$(1+a)^{p^3} \equiv 1 + a^{p^3} \pmod{p^4}.$$

PROOF. Assume that $x, y, z$ are positive integers, not multiples of $p$, such that

$$x^p + y^p = z^p.$$

Let $t$ be the order of $p$ modulo $z$, so $t$ is the smallest positive integer such that

$$p^t \equiv 1 \pmod{z}.$$

We write $p^t - 1 = dz$, for some integer $d > 0$. Clearly $p$ does not divide $d$. Let $m = dx$, $n = dy$, hence $p$ does not divide $m, n$ and

$$m^p + n^p = (p^t - 1)^p.$$

So $m, n, -(p^t-1)$ satisfy Fermat's equation. By (1C), $m+n \equiv p^t - 1$ (mod $p^3$).

Let us note that $m, n$ are less than $p^t - 1$ and since $(m+n)^p > m^p + n^p$ then $m + n > p^t - 1$, so $p^t < m+n+1 < 2p^t$. This implies that $t \ge 4$, because if $t \le 3$ then

$$m + n \equiv p^t - 1 \pmod{p^t},$$

that is, $m + n + 1 \equiv 0$ (mod $p^t$), which is not possible.

Since $m+n \equiv -1$ (mod $p^3$) then $(n+1)^p \equiv -m^p$ (mod $p^4$). Also from $m^p + n^p = (p^t - 1)^p$ we deduce that

$$n^p + 1 \equiv -m^p \pmod{p^{t+1}},$$

therefore $n^p + 1 \equiv -m^p \pmod{p^5}$, and combining with the previous congruence,
$$(n+1)^p \equiv n^p + 1 \pmod{p^4}.$$
By (1A), $m^{p-1} \equiv n^{p-1} \equiv (p^t - 1)^{p-1} \equiv 1 \pmod{p^3}$ and from $n+1 \equiv -m \pmod{p^3}$ we have
$$(n+1)^{p-1} \equiv m^{p-1} \equiv n^{p-1} \equiv 1 \pmod{p^3}.$$

As is known, there are $\varphi(p^3) = p^2(p-1)$ invertible residue classes modulo $p^3$, and they form a multiplicative cyclic group. If the residue class of $w$ modulo $p^3$ is any generator of this group, from
$$n^{p-1} \equiv 1 \pmod{p^3} \qquad \text{and} \qquad (n+1)^{p-1} \equiv 1 \pmod{p^3}$$
it follows that the orders of the residue classes $n \bmod p^3$ and $(n+1) \bmod p^3$ divide $p-1$; so there exist positive integers $h, k$, $0 \le h, k \le p-2$, such that
$$n \equiv w^{hp^2} \pmod{p^3}$$
and
$$n+1 \equiv w^{kp^2} \pmod{p^3}.$$
Let $b \equiv w^h \pmod{p}$. Then
$$n \equiv b^{p^2} \equiv b \pmod{p}$$
and
$$n+1 \equiv w^{kp^2} \equiv w^k \pmod{p},$$
so $w^k \equiv b+1 \pmod{p}$ and we deduce that
$$n^p \equiv b^{p^3} \pmod{p^4}$$
and
$$(n+1)^p \equiv (b+1)^{p^3} \pmod{p^4}.$$
Since $(n+1)^p \equiv n^p + 1 \pmod{p^4}$ it follows that
$$(b+1)^{p^3} \equiv b^{p^3} + 1 \pmod{p^4}.$$
Let us note that $p$ does not divide $b$ nor $b+1$. If $1 \le b \le (p-5)/2$ we take $a = b$. If $b = (p-3)/2$, from
$$\left(\frac{p-1}{2}\right)^{p^3} \equiv \left(\frac{p-3}{2}\right)^{p^3} + 1 \pmod{p^4}$$
we deduce that
$$-1 \equiv -3^{p^3} + 2^{p^3} \pmod{p^4},$$

hence we may take $a = 2$. If $b = (p-1)/2$ then

$$\left(\frac{p+1}{2}\right)^{p^3} \equiv \left(\frac{p-1}{2}\right)^{p^3} + 1 \pmod{p^4},$$

hence

$$1 \equiv -1 + 2^{p^3} \pmod{p^4},$$

so we take $a = 1$. If $(p+1)/2 \le b < p-1$, let $a = p-1-b$, so $1 \le a \le (p-3)/2$ and

$$(a+1)^{p^3} \equiv -b^{p^3} \equiv 1 - (b+1)^{p^3} \equiv 1 + a^{p^3} \pmod{p^4}.$$

By Lemma 1.1(4), it follows that

$$(a+1)^{p^2} \equiv a^{p^2} + 1 \pmod{p^4}. \quad \square$$

Let us note that according to Lemma 1.1(5), (1K) is in fact a strengthening of (1I).

From these results, we obtain as an immediate corollary the one indicated by Gandhi (1976):

**(1L)**    *If the first case of Fermat's theorem fails for the exponent $p$ then there exists an integer $b$, not a multiple of $p$, such that $(1+b)^p \equiv 1 + b^p \pmod{p^4}$.*

PROOF. By (1J) there exists $a$, not a multiple of $p$, such that

$$(1+a)^{p^2} \equiv 1 + a^{p^2} \pmod{p^4}.$$

By Lemma 1.1,

$$(1+a)^p \equiv 1 + a^p \pmod{p^3}.$$

Raising to the $p$th power:

$$(1+a)^{p^2} \equiv (1+a^p)^p \pmod{p^4},$$

so

$$(1+a^p)^p \equiv 1 + a^{p^2} \pmod{p^4}.$$

Letting $b = a^p$ then $(1+b)^p \equiv 1 + b^p \pmod{p^4}$. $\quad \square$

Brčić-Kostić proved in 1952 a result of the same kind, under some special conditions on the prime exponent $p$. Klösgen proved in 1970 the following generalization of (1K):

**(1M)**    If $n \geq 1$, if $x, y, z$ are integers, not multiples of the odd prime $p$, such that $x^{p^n} + y^{p^n} + z^{p^n} = 0$, then

$$x^{p^{3n}} + y^{p^{3n}} + z^{p^{3n}} \equiv 0 \pmod{p^{3n+1}},$$

or equivalently, there exists an integer $a$, $1 \leq a \leq (p-3)/2$, such that

$$(1+a)^{p^{3n}} \equiv 1 + a^{p^{3n}} \pmod{p^{3n+1}}.$$

PROOF. By (1G), $x^p \equiv x \pmod{p^{2n+1}}$. Raising to the power $p^n$ : $x^{p^{n+1}} \equiv x^{p^n} \pmod{p^{3n+1}}$. Again, $x^{p^{n+2}} \equiv x^{p^{n+1}} \equiv x^{p^n} \pmod{p^{3n+1}}$ and repeating this procedure, $x^{p^{3n}} \equiv x^{p^n} \pmod{p^{3n+1}}$. Similarly, $y^{p^{3n}} \equiv y^{p^n} \pmod{p^{3n+1}}$ and $z^{p^{3n}} \equiv z^{p^n} \pmod{p^{3n+1}}$. Hence

$$x^{p^{3n}} + y^{p^{3n}} + z^{p^{3n}} \equiv x^{p^n} + y^{p^n} + z^{p^n} = 0 \pmod{p^{3n+1}}.$$

The last assertion was proved in (1H).    □

Taking $n = 1$, we obtain (1J).

Johnson investigated (in 1977) whether the congruences of Carmichael and Trypanis may be further strengthened, modulo every power $p^{n+2}$ ($n \geq 1$). This is a typical situation to be handled by $p$-adic methods.

Let $\alpha_j \in \hat{\mathbb{Z}}_p$ denote the unique $p$-adic integer which is a $(p-1)$th root of unity and is such that $\alpha_j \equiv j \pmod{p}$, for every $j = 1, 2, \ldots, p-1$ (see Chapter V, (1W)).

**(1N)**    Let $p$ be a prime, $p > 3$, and let $a$ be an integer such that $p$ does not divide $a$ nor $a + 1$. Then the following conditions are equivalent:

    (1) for every $n \geq 1$: $(1+a)^{p^n} \equiv 1 + a^{p^n} \pmod{p^{n+2}}$;
    (2) $a + \alpha_a = \alpha_{1+a}$; and
    (3) $a^2 + a + 1 \equiv 0 \pmod{p}$.

PROOF. The $p$-adic development of $\alpha_a$ is of the form $\alpha_a = a + \rho_a p$, with $\rho_a \in \hat{\mathbb{Z}}_p$ and $\rho_a$ is uniquely defined by $\alpha_a$. We show that for every $n \geq 0$ we have the congruence

(1.5)                    $\alpha_a \equiv a^{p^n} + \rho_a p^{n+1} \pmod{p^{n+2}}$.

For $n = 0$ it is trivial. We assume that (1.5) is true for $n$. Then from $\alpha_a^{p-1} = 1$ it follows that

$$\begin{aligned} \alpha_a = (\alpha_a)^p &\equiv (a^{p^n} + \rho_a p^{n+1})^p \\ &\equiv a^{p^{n+1}} + \rho_a p^{n+2} a^{p^n(p-1)} \\ &\equiv a^{p^{n+1}} + \rho_a p^{n+2} \pmod{p^{n+3}}, \end{aligned}$$

since $a^{p^n(p-1)} \equiv 1 \pmod{p}$.

Now we show the equivalence of the statements (1), (2), (3).

(1) $\rightarrow$ (2)    For every $n \geq 1$, $1 + a^{p^n} \equiv (1 + a)^{p^n} \pmod{p^{n+2}}$. Hence by the above congruence (1.5):

$$1 + \alpha_a \equiv 1 + a^{p^n} + \rho_a p^{n+1} \equiv (1 + a)^{p^n} + \rho_a p^{n+1} \pmod{p^{n+2}}.$$

On the other hand, $\alpha_{1+a} \equiv (1 + a)^{p^n} + \rho_{1+a} p^{n+1} \pmod{p^{n+2}}$. Hence $1 + \alpha_a \equiv \alpha_{1+a} \pmod{p^{n+1}}$ for every $n \geq 1$. From the uniqueness of the $p$-adic development it follows that $1 + \alpha_a = \alpha_{1+a}$.

(2) $\rightarrow$ (1)    We have

$$\alpha_{1+a} \equiv (1 + a)^{p^n} + \rho_{1+a} p^{n+1} \pmod{p^{n+2}}$$

and

$$1 + \alpha_a \equiv 1 + a^{p^n} + \rho_a p^{n+1} \pmod{p^{n+2}}.$$

By hypothesis, $\alpha_{1+a} = 1 + \alpha_a$, so $(1 + a) + \rho_{1+a} p = 1 + (a + \rho_a p)$. By the uniqueness of the $p$-adic developments, $\rho_{1+a} = \rho_a$, hence $(1 + a)^{p^n} \equiv 1 + a^{p^n} \pmod{p^{n+2}}$.

(2) $\rightarrow$ (3)    The $p$-adic $(p-1)$th roots of unity constitute a multiplicative cyclic group. Let $\alpha_j$ be a generator. We consider the subfield $\mathbb{Q}(\alpha_j)$ of $\hat{\mathbb{Q}}_p$. It is a Galois extension of $\mathbb{Q}$. Let $\sigma$ be the automorphism such that $\sigma(\alpha_j) = \alpha_j^{-1}$. Since $\alpha_a = (\alpha_j)^k$ (for some exponent $k$) then $\sigma(\alpha_a) = (\alpha_j)^{-k} = \alpha_a^{-1}$. Similarly $\sigma(\alpha_{1+a}) = \alpha_{1+a}^{-1}$.

From the hypothesis $1 + \alpha_a = \alpha_{1+a}$, we deduce by applying $\sigma$ that $1 + \alpha_a^{-1} = \alpha_{1+a}^{-1}$. Hence

$$\alpha_a \alpha_{1+a} + \alpha_{1+a} = \alpha_a,$$

so

$$\alpha_a + \alpha_a^2 + 1 + \alpha_a = \alpha_a$$

and

$$\alpha_a^2 + \alpha_a + 1 = 0.$$

Since $\alpha_a \equiv a \pmod{p}$, it follows that $a^2 + a + 1 \equiv 0 \pmod{p}$.

$(3) \rightarrow (2)$    Since $p \neq 3$ and $a^2 + a + 1 \equiv 0 \pmod{p}$ then $a \not\equiv 1 \pmod{p}$. Multiplying with $a - 1$ we have $a^3 \equiv 1 \pmod{p}$. Then $(\alpha_a)^3 \equiv 1 \pmod{p}$. But $(\alpha_a^3)^{p-1} = 1$, so $\alpha_a^3 = 1$, which is the unique $(p-1)$th root of unity in $\hat{\mathbb{Z}}_p$ congruent to 1 modulo $p$. Since $\alpha_a \neq 1$, it is a primitive cubic root of 1, and therefore

$$\alpha_a^2 + \alpha_a + 1 = 0.$$

So $\alpha_a + 1 = -\alpha_a^2 \equiv -a^2 \equiv a + 1 \pmod{p}$, and therefore $\alpha_a + 1 \equiv \alpha_{a+1}$, the unique $(p-1)$th root of 1 in $\hat{\mathbb{Z}}_p$ which is congruent to $a+1$ modulo $p$.  $\square$

Let us note that in view of Lemma 1.1, condition (1) above is equivalent to:

(1′)   *For every $n \geq 1$*: $(1 + a)^{p^{n+1}} \equiv 1 + a^{p^{n+1}} \pmod{p^{n+2}}$.

According to Lemma 1.2, if $p > 3$ there exists an integer $a$, $1 \leq a \leq p - 1$, satisfying the equivalent conditions of (1N) if and only if $p \equiv 1 \pmod{6}$.

As a corollary, we have:

**(1O)**    *If $p$ is a prime, $p \equiv 5 \pmod{6}$, then there exists an integer $n_0 > 0$ such that if $n \geq n_0$ and $a = 1, 2, \ldots, p - 2$ then*

$$(1 + a)^{p^n} \not\equiv 1 + a^{p^n} \pmod{p^{n+2}}.$$

PROOF. By Lemma 1.2, if $a = 1, 2, \ldots, p - 2$ then $a^2 + a + 1 \not\equiv 0 \pmod{p}$. By (1N), for every $a$ there exists an index $n(a)$ such that

$$(1 + a)^{p^{n(a)}} \not\equiv 1 + a^{p^{n(a)}} \pmod{p^{n(a)+2}}.$$

By Lemma 1.1(5), if $n \geq n_0 = \max\{n(a) \mid a = 1, 2, \ldots, p - 2\}$ then

$$(1 + a)^{p^n} \not\equiv 1 + a^{p^n} \pmod{p^{n+2}}$$

for every $a = 1, 2, \ldots, p - 1$.  $\square$

It is interesting to find out whether the following implication is true:

**(I)**  *If $a$ is an integer, $a = 1, 2, \ldots, p - 2$, and $(1 + a)^p \equiv 1 + a^p$ (mod $p^3$) then $a^2 + a + 1 \equiv 0$ (mod $p$).*

Indeed, if (I) is true then the first case of Fermat's theorem holds for every exponent $p \equiv 5$ (mod 6). Because, otherwise by (1I) there exists $a$, $1 \le a \le p - 2$, such that $(1 + a)^p \equiv 1 + a^p$ (mod $p^3$). Hence by (I), $a^2 + a + 1 \equiv 0$ (mod $p$) and therefore $p \equiv 1$ (mod 6), contrary to the hypothesis.

Arwin showed in 1920 that there exist integers $a$ and primes $p$ such that $(1 + a)^p \equiv 1 + a^p$ (mod $p^2$) but $a^2 + a + 1 \not\equiv 0$ (mod $p$). So such a strengthening of the implication (I) is false.

### Bibliography

1841 Cauchy, A., *Exercices d'analyse et de physique mathémati-que*, **2** (1841), 137–144 (Notes sur quelques théorèmes d'algè-bre); *Oeuvres Complètes*, (2), Vol. XII, pp. 157–166, Gaut-hier-Villars, Paris, 1916.

1909 Fleck, A., *Miszellen zum großen Fermatschen Problem*, Sit-zungsber. Berliner Math. Ges., **8** (1909), 133–148.

1910 Lind, B., *Über das letzte Fermatsche Theorem*, Abh. Ge-schichte Math. Wiss், **26** (1910), 23–65.

1913 Carmichael, R.D., *Note on Fermat's last theorem*, Bull. Amer. Math. Soc., **19** (1913), 233–236.

1913 Carmichael, R.D., *Second note on Fermat's last theorem*, Bull. Amer. Math. Soc., **19** (1913), 402–403.

1914 Frobenius, G., *Über den Fermatschen Satz, III*, Sitzungsber. Königl. Preussischen Akad. Wiss., Berlin, **22** (1914), 653–681; reprinted in *Collected Works*, Vol. 3, pp. 648–676, Springer-Verlag, Berlin, 1968.

1914 Vandiver, H.S., *A note on Fermat's last theorem*, Trans. Amer. Math. Soc., **15** (1914), 202–204.

1914 Meissner, W., *Über die Lösungen der Kongruenz $x^{p-1} \equiv 1$ (mod $p^m$) und ihre Verwertung zur Periodenbestimmung mod $p^k$*, Sitzungsber. Berliner Math. Ges., **13** (1914), 96–107.

1917 Pollaczek, F., *Über den großen Fermat'schen Satz*, Sitzungs-ber. Akad. Wiss., Wien, Abt. IIa, **126** (1917), 45–59.

1919 Vandiver, H.S., *A property of cyclotomic numbers and its relation to Fermat's last theorem*, Ann. of Math., **21** (1919), 73–80.

1920 Arwin, A., *Über die Lösung der Kongruenz $(\lambda+1)^p - \lambda^p - 1 \equiv 0 \pmod{p^2}$*, Acta Math., **42** (1920), 173–190.

1923 Pomey, L., *Sur le dernier théorème de Fermat*, C. R. Acad. Sci. Paris, **177** (1923), 1187–1190.

1925 Vandiver, H.S., *A property of cyclotomic integers and its relation to Fermat's last theorem*, Ann. of Math., **26** (1925), 217–232.

1929 Spunar, V.A., *On Fermat's last theorem, II*, J. Washington Acad. Sci., **19** (1929), 395–401.

1934 James, G., *On Fermat's last theorem*, Amer. Math. Monthly, **41** (1934), 419–424.

1938 Segal, D., *A note on Fermat's last theorem*, Amer. Math. Monthly, **45** (1938), 438–439.

1938 Niewiadomski, R., *Sur la grandeur absolue et relation mutuelle des nombres entiers qui peuvent résoudre l'équation $x^p + y^p = z^p$* (in Polish), Wiadom. Mat., **44** (1938), 113–127.

1944 Racliş, N., *Démonstration du grand théorème de Fermat pour des grandes valeurs de l'exposant*, Bull. École Polytechnique Bucarest, **15** (1944), 45–61.

1946 Inkeri, K., *Untersuchungen über die Fermatschen Vermutung*, Ann. Acad. Sci. Fenn., Ser. AI, **33** (1946), 1–60.

1950 Trypanis, A.A., *On Fermat's last theorem*, Proc. Intern. Congress Math., Cambridge, 1950, Vol. 1, 301–302.

1952 Brčić-Kostić, M., *L'extension de la congruence $(a+b)^n - a^n - b^n \equiv 0 \pmod{n}$ (n un nombre premier)* (in Croatian, French summary), Hrvatsko Prirodoslovno Društvo, Glas. Mat. Ser. II, **7** (1952), 7–11.

1958 Pérez-Cacho, L., *Sobre algunas cuestiones de la teoria de números*, Rev. Mat. Hisp.-Amer., (4), **18** (1958), 10–27 and 113–124.

1965 Ferentinou-Nicolacopoulou, J., *A new necessary condition for the existence of a solution to the equation $x^p + y^p = z^p$ of Fermat* (Greek, French summary), Bull. Soc. Math. Grèce (N.S.), **6** (1965), 222–236.

1965 Ferentinou-Nicolacopoulou, J., *Remarks on the preceding article* (Greek, French summary), Bull. Soc. Math. Grèce

(N.S.), **6** (1965), 356–357.

1970 Klösgen, W., *Untersuchungen über Fermatsche Kongruen-zen*, Gesellschaft Math. Datenverarbeitung, Bonn, No. 37, 1970, 124 pp.

1975 Everett, C.J. and Metropolis, N., *On the roots of $x^m \pm 1$ in the p-adic field $\mathbb{Q}_p$*, Notices Amer. Math. Soc., **22** (1975), A-619; preprint, Los Alamos Sci. Lab., LA-UR-74-1835.

1975 Gandhi, J.M., *Fermat's last theorem, I*, Notices Amer. Math. Soc., **22** (1975), A-486.

1975 Wagstaff, S., *Fermat's last theorem is true for any exponent less than* 100 000, Notices Amer. Math. Soc., **23** (1975), A-53, Abstract 731-10-35.

1976 Gandhi, J.M., *On the first case of Fermat's last theorem*, preprint.

1977 Johnson, W., *On the congruences related to the first case of Fermat's last theorem*, Math. Comp., **31** (1977), 519–526.

## VI.2. Divisibility Conditions

Let $p$ be an odd prime and assume that $x, y, z$ are nonzero relatively prime integers such that $x^p + y^p + z^p = 0$. In this section we shall indicate some divisibility conditions which the integers $x, y, z$ or some of their combinations must satisfy.

The following proposition was given by Pérez-Cacho in 1958. However, statement (2) had already been proved by Massoutié in 1931; a simpler proof was also given by Pomey in 1931.

**(2A)**    *Let $p$ be an odd prime number and assume that there exist nonzero pairwise relatively prime integers $x, y, z$ such that $x^p + y^p + z^p = 0$.*

  (1) *If 3 does not divide $xyz$ then $x \equiv y \equiv z \not\equiv 0 \pmod 3$, the integers $x^2 - yz, y^2 - xz, z^2 - xy$ are divisible by 3, but not by 9, and if $q$ is a prime, $q \neq 3$, dividing one of the numbers $x^2 - yz, y^2 - xz, z^2 - xy$, then $q \equiv 1 \pmod 6$.*
  (2) *If $p \equiv -1 \pmod 6$ then 3 divides $xyz$.*

PROOF. (1)    Since $3 \nmid xyz$ then $x, y, z$ are congruent to 1 or to $-1$ modulo 3. From $(\pm 1)^p + (\pm 1)^p + (\pm 1)^p \equiv 0 \pmod 3$ the only possibility is that $x \equiv y \equiv z \not\equiv 0 \pmod 3$. Therefore $x^2 \equiv yz$

(mod 3), thus 3 divides $x^2 - yz$, and similarly 3 divides $y^2 - xz$, $z^2 - xy$.

We note that one, and only one, of the integers $x, y, z$ is even. Thus $x^2 \neq yz$, $y^2 \neq xz$, $z^2 \neq xy$.

We show that $9 \nmid z^2 - xy$. We have $x^{2p} + x^p y^p + x^p z^p = 0$ and $xy = (xy - z^2) + z^2$ so $-(x^{2p} + x^p z^p) = x^p y^p = [(xy - z^2) + z^2]^p \equiv p(xy - z^2)z^{2(p-1)} + z^{2p}$ (mod $(z^2 - xy)^2$). Noting that $x, y, z$ are distinct (since 2 is not a $p$th power), we have

$$Q_3(x^p, z^p) = x^{2p} + x^p z^p + z^{2p} \equiv p(z^2 - xy)z^{2(p-1)} \quad (\text{mod } (z^2 - xy)^2).$$

Since $3 \mid x^p - z^p$, by Chapter II, (3B)(6), we have $v_3(Q_3(x^p, z^p)) = v_3(3) = 1$. Thus $9 \nmid z^2 - xy$.

Now let $q$ be a prime dividing $z^2 - xy$, $q \neq 3$ (the argument is similar if $q$ divides $x^2 - yz$ or $y^2 - xz$). Then $q \nmid z$ (otherwise $q \mid z$ and $q \mid x$ or $q \mid y$, contrary to the hypothesis). Let $z'$ be such that $zz' \equiv 1$ (mod $q$). Multiplying with $z'^{2p}$ we have $(x^p z'^p)^2 + (x^p z'^p) + 1 \equiv 0$ (mod $q$). By Lemma 1.2, $q \equiv 1$ (mod 6).

(2) If $3 \nmid xyz$ and $p \nmid z$ (the argument is similar when $p \nmid x$ or $p \nmid y$) then

$$\begin{aligned}
Q_p(z^2, xy) &= \frac{z^{2p} - x^p y^p}{z^2 - xy} \\
&= \frac{x^{2p} + x^p z^p + z^{2p}}{z^2 - xy} \equiv pz^{2(p-1)} \equiv p \ (\text{mod } z^2 - xy).
\end{aligned}$$

Also $z^2 - xy \equiv z^{2p} - x^p y^p$ (mod $p$). If $p \nmid z^2 - xy$ then from the above congruence, $p \equiv 1$ (mod $z^2 - xy$) so $p \equiv 1$ (mod 3), by (1). If $p \mid z^2 - xy$, by (1), $p \equiv 1$ (mod 3). □

Pomey claimed in 1931 to have shown that for any exponent $p$, if $x, y, z$ are nonzero and $x^p + y^p + z^p = 0$ then 3 divides $xyz$. He also claimed in 1934 that 5 divides $xyz$. However, his proofs were erroneous (see Brauer, 1934).

Inkeri has proved in 1946 the following statement:

**(2B)**  *If $p$ is an odd prime, $p \not\equiv 1, 9$ (mod 20), if $x, y, z$ are nonzero integers such that $x^p + y^p + z^p = 0$ then 5 divides $xyz$.*

PROOF. By hypothesis, $p \neq 5$ and we may assume that $x, y, z$ are relatively prime.

If $5 \nmid xyz$ then $x^p, y^p, z^p$ are congruent modulo 5 to $\pm 1, \pm 2$. Changing notation, if necessary, we may assume that $x^p \equiv y^p$ or $-y^p$ (mod 5). If we had $x^p \equiv -y^p$ (mod 5) then $z^p \equiv -x^p - y^p \equiv 0$ (mod 5), contrary to the hypothesis. Therefore, $x^p \equiv y^p$ (mod 5).

Let $h$ be an integer such that $ph \equiv 1$ (mod 4). If $p \equiv 3$ (mod 4) we may take $h = 3$. Raising the above congruence to the power $h$, we deduce that $x \equiv y$ (mod 5). Since $z^p = -x^p - y^p \equiv -2y^p$ (mod 5) then again $z^{2p} \equiv 4y^{2p} \equiv -x^p y^p \equiv (-xy)$ (mod 5) hence raising to the power $h$, $z^2 \equiv -xy$ (mod 5). In particular, since $x \equiv y \equiv \pm 1$ or $\pm 2$ (mod 5) then $z^2 \equiv \pm 1$ (mod 5).

We examine now the integer $z^{2p} + x^p y^p$, which is necessarily odd, since exactly one of the integers $x, y, z$ is even.

We show that if $q$ is a prime, $q \neq 5$ and $q$ divides $z^{2p} + x^p y^p$ then $q \equiv \pm 1$ (mod 5). Indeed, we have the relation $(2x^p + 3y^p)^2 - 5y^{2p} = 4(x^{2p} + 3x^p y^p + y^{2p}) = 4(z^{2p} + x^p y^p)$. So for such a prime $q$, we have the congruences $(2x^p + 3y^p)^2 \equiv 5y^{2p}$ (mod $q$). So 5 is a square modulo $q$. By the quadratic reciprocity law,

$$1 = \left(\frac{5}{q}\right) = \left(\frac{q}{5}\right) \qquad \text{thus} \quad q \equiv \pm 1 \quad \text{(mod 5)}.$$

It follows that any factor $k$ of $z^{2p} + x^p y^p$ which is not a multiple of 5 must be congruent to $\pm 1$ (mod 5). In particular, we take

$$k = \frac{z^{2p} + x^p y^p}{z^p + xy} = z^{2(p-1)} - z^{2(p-2)} xy + z^{2(p-3)}(xy)^2 - \cdots + (xy)^{p-1}.$$

Since $z^2 \equiv -xy$ (mod 5) then $k \equiv pz^{2(p-1)} \equiv p$ (mod 5) because $z^2 \equiv \pm 1$ (mod 5). Thus $k \not\equiv 0$ (mod 5) and therefore $k \equiv \pm 1$ (mod 5).

We conclude that $p \equiv \pm 1$ (mod 5).

We have still to show that $p \equiv 1$ (mod 4). If we assume that $p \equiv 3$ (mod 4) then $3p \equiv 1$ (mod 4) hence $z \equiv z^{3p} \equiv (-2)^3 y^{3p} \equiv 2y$ (mod 5). Considering the relation $(2z^p + 3y^p)^2 - 5y^{2p} = 4(z^{2p} + 3y^p z^p + y^{2p}) = 4(x^{2p} + y^p z^p)$ we deduce as before that every prime factor $q \neq 5$ of $x^{2p} + y^p z^p$ must be congruent to $\pm 1$ (mod 5). Hence every factor $k$ of $x^{2p} + y^p z^p$, $k \not\equiv 0$ (mod 5), must be congruent to $\pm 1$ (mod 5).

In particular, taking $k = x^2 + yz$, if $p \equiv 3$ (mod 4) we have $y \equiv -2z$ (mod 5) hence $x^2 \equiv y^2 \equiv -2yz$ (mod 5) thus $k = x^2 + yz \equiv -yz \not\equiv 0$ (mod 5). Therefore $k \equiv \pm 1$ (mod 5). However, on the other hand $k \equiv x^2 + yz \equiv y^2 + 2y^2 = 3y^2 \equiv \pm 2 \pmod 5$, which is a contradiction. This shows that $p \equiv 1 \pmod 4$.

Therefore, $p \equiv 1$ or $9 \pmod{20}$, contrary to the hypothesis, showing that $5 \mid xyz$.  □

Pérez-Cacho proved in 1958:

**(2C)**  *If $p$ is an odd prime number, if $x, y, z$ are nonzero relatively prime integers such that $x^p + y^p + z^p = 0$ then:*

(1) *if $q \neq 5$ is a prime factor of $(x^2 - yz)(y^2 - zx)(z^2 - xy)$ then $q \equiv \pm 1 \pmod{10}$; and*

(2) *none of the numbers $x^2 - yz$, $y^2 - zx$, $z^2 - xy$ is a multiple of 25.*

Inkeri also proved in 1946 the following result:

**(2D)**  *Let $p$ be an odd prime number, and assume that $x, y, z$ are relatively prime nonzero integers such that $x^{p^n} + y^{p^n} + z^{p^n} = 0$ (where $n \geq 1$). Then:*

(1) *$5$ divides $xyz(x - y)(x - z)(y - z)$.*

(2) *$7$ divides $xyz(x - y)(x - z)(y - z)(x^2 - yz)(y^2 - xz)(z^2 - xy)$ (if $p > 3$).*

(3) *$11$ divides $xyz(x - y)(x - z)(y - z)(x^2 + yz)(y^2 + xz)(z^2 + xy)$ (if $p > 5$).*

PROOF. First we note that if $l$ is a prime and $p \nmid l - 1$, if $a, b$ are nonzero relatively prime integers then $l \mid a + b$ if and only if $l \mid a^p + b^p$. Indeed, if $a$ or $b$ is a multiple of $l$, it is obvious.

Let $l \nmid ab$.

It is clear that if $l \mid a + b$ then $l \mid a^p + b^p$. Conversely, if $l \mid a^p + b^p$ but $l \nmid a + b$ then $l$ is a primitive factor of $a^p + b^p$ and by Chapter II, (3G), $l \equiv 1 \pmod p$, which is a contradiction.

Let $u = x^{p^n}$, $v = y^{p^n}$, $w = z^{p^n}$, so we have $u + v + w = 0$ and we need to show (in view of the above remark, because $p > 11$):

(1) $5$ divides $uvw(u - v)(u - w)(v - w)$.

(2) $7$ divides $uvw(u-v)(u-w)(v-w)(u^2 - vw)(v^2 - uw)(w^2 - uv)$.

(3) $11$ divides
$$uvw(u - v)(u - w)(v - w)(u^2 + vw)(v^2 + uw)(w^2 + uv).$$

Let $l = 5, 7$, or $11$.

Since $\gcd(u, v, w) = 1$, we may assume, for example, that $l \nmid v$. Let $v'$ be such that $v'v \equiv 1 \pmod{l}$. Hence multiplying with $v'$ we have $t + t' \equiv 1 \pmod{l}$, where $t \equiv -v'u \pmod{l}$, $t' \equiv -v'w \pmod{l}$. Let

$$
\begin{aligned}
T_1 &= t, \\
T_2 &= t - 1, \\
T_3 &= t + 1, \\
T_4 &= 2t - 1, \\
T_5 &= t - 2, \\
T_6 &= t^2 - t + 1, \\
T_7 &= t^2 + t - 1, \\
T_8 &= t^2 - t - 1, \\
T_9 &= t^2 - 3t + 1.
\end{aligned}
$$

Then

$$
\begin{aligned}
vT_1 &\equiv -u \pmod{l}, \\
vT_2 &\equiv -u - v \equiv w \pmod{l}, \\
vT_3 &\equiv -u + v \pmod{l}, \\
vT_4 &\equiv -2u - v \equiv -u + w \pmod{l}, \\
vT_5 &\equiv -u - 2v \equiv w - v \pmod{l}, \\
v^2 T_6 &\equiv u^2 + vu + v^2 \equiv u^2 - vw \\
        &\equiv v^2 - uw \equiv w^2 - vu \pmod{l}, \\
v^2 T_7 &\equiv u^2 - vu - v^2 \equiv u^2 + vw \pmod{l}, \\
v^2 T_8 &\equiv u^2 + vu - v^2 \equiv -v^2 - uw \pmod{l}, \\
v^2 T_9 &\equiv u^2 + 3vu + v^2 \equiv w^2 + uv \pmod{l}.
\end{aligned}
$$

(1)   Let $l = 5$, then it is easy to verify: if

$$
\begin{aligned}
t &\equiv 0 \pmod{5} & \text{then} \quad & 5 \mid T_1, \\
t &\equiv 1 \pmod{5} & \text{then} \quad & 5 \mid T_2, \\
t &\equiv 2 \pmod{5} & \text{then} \quad & 5 \mid T_5, \\
t &\equiv -1 \pmod{5} & \text{then} \quad & 5 \mid T_3, \\
t &\equiv -2 \pmod{5} & \text{then} \quad & 5 \mid T_4,
\end{aligned}
$$

hence $5$ divides $uvw(u - v)(u - w)(v - w)$.

(2)  Let $l = 7$; then it is easy to verify: if $t \equiv -3, -2, -1, 0, 1, 2, 3$ (mod 7) then 7 divides $T_4, T_6, T_3, T_1, T_2, T_5, T_6$, respectively. Hence 7 divides $uvw(u-v)(u-w)(v-w)(u^2-vw)(v^2-uw)(w^2-vu)$.

(3)  Let $l = 11$; then if $t \equiv -5, -4, -3, -2, -1, 0, 1, 2, 3, 4, 5$ (mod 11) then 11 divides $T_4, T_7, T_8, T_9, T_3, T_1, T_2, T_5, T_7, T_8, T_9$, respectively. So 11 divides $uvw(u-v)(u-w)(v-w)(u^2+vw)(v^2+uw)(w^2+uv)$.  □

Let us note that the argument breaks down for $l > 11$. Indeed, in this case if $t \equiv -2$ (mod $l$) then the values modulo $l$ of $T_i$ $(1 \le i \le 9)$ are distinct from 0 and have absolute value at most equal to 11.

Concerning divisibility by 4, we indicate a result of Pierre (1943), preceded by a lemma on Jacobi symbols.

LEMMA 2.1. *Let $a, b, c$ be pairwise relatively prime odd integers such that*

$$\left(\frac{a}{bc}\right) = \left(\frac{b}{ac}\right) = \left(\frac{c}{ab}\right) = +1.$$

*Then at most one of the numbers $a, b, c$ is congruent to 3 modulo 4.*

PROOF. Assume that $a \equiv b \equiv 3$ (mod 4). Then by the reciprocity law for Jacobi's symbol:

$$1 = \left(\frac{c}{ab}\right) = \left(\frac{ab}{c}\right)(-1)^{(c-1)/2 \times (ab-1)/2} = \left(\frac{ab}{c}\right) = \left(\frac{a}{c}\right)\left(\frac{b}{c}\right),$$

$$1 = \left(\frac{b}{ac}\right) = \left(\frac{b}{a}\right)\left(\frac{b}{c}\right),$$

$$1 = \left(\frac{a}{bc}\right) = \left(\frac{a}{b}\right)\left(\frac{a}{c}\right).$$

Hence $(a/b) = (a/c) = (b/c) = (b/a)$. However

$$\left(\frac{b}{a}\right) = \left(\frac{a}{b}\right)(-1)^{(b-1)/2 \times (a-1)/2} = -\left(\frac{a}{b}\right),$$

which is a contradiction.  □

(2E)    *Let $p$ be an odd prime number, and let $x, y, z$ be nonzero relatively prime integers such that $x^p + y^p + z^p = 0$. Then 4 divides one of the numbers $x, y, z$.*

PROOF. We may assume $z$ even, while $x, y$ are odd; we also assume $z \equiv 2 \pmod 4$ to derive a contradiction.

*First Case: $p \nmid xyz$.*
  By (2.1),

$$s_1^p = \frac{z^p + x^p}{z + x}$$
$$= z^{p-1} - z^{p-2}x + \cdots + z^2 x^{p-3} - z x^{p-2} + z^{p-1}$$
$$\equiv -2x^{p-2} + x^{p-1} \pmod 4.$$

Since $x$ is odd then $x^{p-1} \equiv 1 \pmod 4$ hence $s_1^p \equiv -2x^{p-2} + 1 \equiv -2(4k \pm 1)^{p-2} + 1 \equiv \mp 2 + 1 \equiv 3 \pmod 4$, and therefore $s_1 \equiv 3 \pmod 4$. Similarly, $r_1^p = (z^p + y^p)/(z + y) \equiv 3 \pmod 4$, so $r_1 \equiv 3 \pmod 4$.

By Chapter II, (3A), $(r_1/s_1 t_1) = (s_1/r_1 t_1) = (t_1/s_1 r_1) = 1$, hence by Lemma 2.1 at most one of the numbers $r_1, s_1, t_1$ may be congruent to 3 modulo 4, which is a contradiction.

*Second Case: $p \mid xyz$.*
  First we consider the case when $p \mid z$. As in the first case, $s_1 \equiv 3 \pmod 4$, $r_1 \equiv 3 \pmod 4$.

By Chapter III, (2A), the odd integers $r_1, s_1, pt_1$ satisfy $(r_1/pt_1 s_1) = (s_1/pt_1 r_1) = (pt_1/r_1 s_1) = 1$. This contradicts Lemma 2.2.

Now we assume that $p \nmid z$ and, for example, $p \mid x$. Proceeding as before, by (1.2) we have $pr_1^p \equiv 3 \pmod 4$ and since $r_1$ is odd then $r_1^{p-1} \equiv 1 \pmod 4$, so $pr_1 \equiv 3 \pmod 4$. Moreover, $s_1 \equiv 3 \pmod 4$ and by Chapter III, (2A), the odd integers $pr_1, s_1, t_1$ satisfy $(pr_1/s_1 t_1) = (s_1/pr_1 t_1) = (t_1/pr_1 s_1) = +1$, and this contradicts Lemma 2.1. $\square$

In 1910, Lind claimed that 9 divides $x + y + z$, but his proof was insufficient. As a consequence there are several inequalities and equations in his paper which are questionable (see Dickson, 1920, p. 769).

After these divisibility results by small numbers, we turn our attention to divisibility results by expressions built from the numbers $x, y, z$ which are hypothetical solutions of Fermat's equation.

In 1913, Niewiadomski showed:

**(2F)**    *Let $p$ be an odd prime number, and let $x, y, z$ be nonzero relatively prime integers, such that $x^p + y^p + z^p = 0$. Then $x^{2p+1} + y^{2p+1} + z^{2p+1}$ is divisible by $(x + y)(y + z)(z + x)$.*

PROOF. We have

$$x^{2p+1} + y^{2p+1} + z^{2p+1} = \left( \frac{x+y}{x^p + y^p} + \frac{y+z}{y^p + z^p} + \frac{z+x}{z^p + x^p} \right) x^p y^p z^p,$$

as may be easily verified. Hence

$$\frac{x^{2p+1} + y^{2p+1} + z^{2p+1}}{(x+y)(y+z)(z+x)} = -\frac{(y^p + z^p)(z^p + x^p)}{(y+z)(z+x)} - \frac{(x^p + y^p)(z^p + x^p)}{(x+y)(z+x)}$$
$$- \frac{(x^p + y^p)(y^p + z^p)}{(x+y)(y+z)}$$

and this number is an integer.    □

The following result of Rameswar Rao (1969) is also very simple:

**(2G)**    *If $p$ is an odd prime number, if $x, y, z$ are nonzero pairwise relatively prime integers such that $x^p + y^p + z^p = 0$, then $x+y$ divides $d^p$ where $d = \gcd(x+y, z)$ (similar statements hold by symmetry for $x + z$, $y + z$).*

PROOF. From $x + y + z \equiv 0 \pmod{p}$, there exists $k \neq 0$ such that $x + y + z = kp$. Since $p$ is odd then $x + y = kp - z$ divides both $x^p + y^p = -z^p$ and $(kp)^p - z^p$ so $kp - z$ divides $(kp)^p$. Since $d = \gcd(kp, z)$, we may write $kp = ud$, $z = vd$ with $\gcd(u, v) = 1$. But $u - v$ divides both $u^p d^{p-1}$ and $v^p d^{p-1}$ hence $(u - v)d = x + y$ divides $d^p$.    □

The following result, which appeared in a paper by Simmons (1966) was attributed to G. Reis; the assertion (1) was proved again by Rollero (1981):

**(2H)**    *Let $p$ be an odd prime number, and let $x, y, z$ be pairwise relatively prime positive integers such that $x^p + y^p = z^p$. Then there exist uniquely defined positive integers $k, a, b$ such that $x = k + a$, $y = k + b$, $z = k + a + b$.*
    *Moreover:*

(1) *pab divides $k^p$.*
(2) $\gcd(a, b) = 1$.
(3) *If $a \neq 1$ then $\gcd(k, a) \neq 1$; if $p \nmid a$ then $\gcd(k, a) \neq a$.*
(4) *If $b \neq 1$ then $\gcd(k, b) \neq 1$; if $p \nmid b$ then $\gcd(k, b) \neq b$.*

PROOF. We may assume $0 < x < y < z$. Since $z < x + y$, let $k$ be defined by $x + y = z + k$. From $y < z$ we have $k < x < y$. Let $a, b$ be defined by $x = k + a$, $y = k + b$. So $z = k + a + b$.

It is clear that $k, a, b$ are uniquely defined: if $x = k' + a'$, $y = k' + b'$, and $z = k' + a' + b'$ then $0 = (k - k') + (a - a') = (k - k') + (b - b') = (k - k') + (a - a') + (b - b')$ so $k = k'$, $a = a'$, $b = b'$. We have

$$(2.1) \qquad (k + a)^p + (k + b)^p = (k + a + b)^p.$$

It is easily seen that $\gcd(k + a, k + b, k + a + b) = 1$ because $\gcd(x, y, z) = 1$. Therefore $k+a$, $k+b$, $k+a+b$ are pairwise relatively prime.

Since $(k+a+b)^p = (k+a)^p + p(k+a)^{p-1}b + \cdots + b^p$ then $(k+b)^p = p(k + a)^{p-1}b + \cdots + b^p$ hence $k^p + pk^{p-1}b + \binom{p}{2}k^{p-2}b^2 + \cdots + b^p = p(k+a)^{p-1}b + \cdots + b^p$ so

$$k^p = pb[(k + a)^{p-1} - k^{p-1}] + \binom{p}{2}[(k + a)^{p-2} - k^{p-2}]b^2$$

$$+ \cdots + \binom{p}{p-1}ab^{p-1}.$$

Each bracketed expression is a multiple of $a$, thus $k^p$ is a multiple of $pab$. If $a \neq 1$ then $\gcd(k, a) \neq 1$, if $b \neq 1$ then $\gcd(k, b) \neq 1$. We have $\gcd(a, b) = 1$ for if a prime $q$ divides $a, b$ then it would also divide $k$, hence $k + a$, and $k + b$.

If $p \nmid a$ then $\gcd(k, a) \neq a$ otherwise $k = al$, $l$ an integer, and from (2.2) we have, after dividing by $a$,

$$a^{p-1}l^p = pba^{p-2}[(l + 1)^{p-1} - l^{p-1}] + \binom{p}{2}a^{p-3}\left[(l + 1)^{p-2} - l^{p-2}\right]b^2$$

$$+ \cdots + pb^{p-1}.$$

Hence $a$ divides $pb^{p-1}$. But $p \nmid a$ hence $a$ divides $b^{p-1}$, a contradiction. Similarly, if $p \nmid b$ then $\gcd(k, b) \neq b$.  □

Let $p$ be an odd prime and assume that $x, y, z$ are nonzero pairwise relatively prime integers such that $x^p + y^p + z^p = 0$.

Fleck (1909, 1910) began a more systematic study of divisibility properties of the following numbers (built from the hypothetical solution $(x, y, z)$):

$$A = y^2 + yz + z^2,$$
$$B = z^2 + zx + x^2,$$
$$C = x^2 + xy + y^2,$$

$$
\begin{array}{ll}
A_1 = x^2 - yz, & A_2 = x^2 + yz, \\
B_1 = y^2 - zx, & B_2 = y^2 + zx, \\
C_1 = z^2 - xy, & C_2 = z^2 + xy, \\
S = x + y + z, & T = -(xy + yz + zx), \\
U = xyz, & 2V = x^2 + y^2 + z^2.
\end{array}
$$

Let $r, s, t$ be defined as Barlow relations (Chapter III, §1).

**(2I)**  *With the above hypotheses and notations, there exist nonzero integers $G, M, J, K, L, J_1, K_1, L_1$ such that*

(1)

$$
S = \begin{cases} -rstp^3 GM & \text{(in the first case)}, \\ -rstp^2 GM & \text{(in the second case)}. \end{cases}
$$

(2) $A = GJ, B = GK, C = GL, A_1 = GJ_1, B_1 = GK_1, C_1 = GL_1$.

(3) $G$ *is the greatest common divisor of $S$ and the six expressions above.*

(4) $J, K, L, J_1, K_1, L_1$ *are pairwise relatively prime.*

(5) *The prime factors of $J, K, L$ are of the form $6hp + 1$.*

(6) *The prime factors of $J_1, K_1, L_1$ are of the form $6hp^2 + 1$.*

(7) $x^{3p} \equiv y^{3p} \equiv z^{3p} \pmod{GJKLJ_1K_1L_1}$.

In 1979, Inkeri gave corrections to some proofs of Fleck and obtained further results along the same lines.

We conclude the section with a result of Pollaczek, obtained in 1917. No elementary proof is known for it. Pollaczek's proof was based on congruences obtained by Kummer in 1857, which should hold if the first case of Fermat's theorem is assumed false for the exponent $p$. Kummer's proof of these congruences involves a detailed consideration of arithmetical properties of the cyclotomic field $\mathbb{Q}(\zeta_p)$, where $\zeta_p$ is a primitive $p$th root of 1. It is therefore not included in this book.

Here is Pollaczek's result:

**(2J)**    *Let $p$ be an odd prime number and assume that there exist pairwise relatively prime nonzero integers $x, y, z$, $x^p + y^p + z^p = 0$. Then $A = y^2 + yz + z^2$, $B = z^2 + zx + x^2$, and $C = x^2 + xy + y^2$ are not divisible by $p$.*

## Bibliography

1857  Kummer, E.E., *Einige Sätze über die aus den Wurzeln der Gleichung $\alpha^\lambda = 1$ gebildeten complexen Zahlen, für den Fall daß die Klassenzahl durch $\lambda$ theilbar ist, nebst Anwendungen derselben auf einen weiteren Beweis des letztes Fermat'schen Lehrsatzes*, Math. Abh. Königl. Akad. Wiss. zu Berlin, 1857, pp. 41–74.

1909  Fleck, A., *Miszellen zum großen Fermatschen Problem*, Sitzungsber. Berliner Math. Ges., **8** (1909), 133–148.

1910  Fleck, A., *Bemerkung zum großen Fermatschen Problem*, Sitzungsber. Berliner Math. Ges., **9** (1910), 50–53.

1910  Lind, B., *Über das letzte Fermatsche Theorem*, Abh. Geschichte Math. Wiss., no. 26, 1910, pp. 23–65.

1913  Niewiadomski, R., *Question 4194*, L'Interm. Math., **20** (1913), 76.

1917  Pollaczek, F., *Über den großen Fermat'schen Satz*, Sitzungsber. Akad. Wiss., Wien, Abt. IIa, **126** (1917), 45–59.

1920  Dickson, L.E., *History of the Theory of Numbers*, Vol. II, Carnegie Institution, Washington, DC, 1920; reprinted by Chelsea, New York, 1971.

1931  Massoutié, L., *Sur le dernier théorème de Fermat*, C. R. Acad. Sci. Paris, **193** (1931), 502–504.

1931  Pomey, L., *Nouvelles remarques relatives au dernier théorème de Fermat*, C. R. Acad. Sci. Paris, **193** (1931), 563–564.

1934  Pomey, L., *Sur le dernier théorème de Fermat (divisibilité par 3 et par 5)*, C. R. Acad. Sci. Paris, **199** (1934), 1562–1564.

1934  Brauer, A., *Review of the above paper by L. Pomey*, Jahrbuch Fortschritte Math., **60**, II (1934), 928.

1943  Pierre C., *Sur le théorème de Fermat $a^n + b^n = c^n$*, C. R.

Acad. Sci. Paris, **217** (1943), 37–39.

1946 Inkeri, K., *Untersuchungen über die Fermatsche Vermutung*, Ann. Acad. Sci. Fenn., Ser. AI, **33** (1946), 60 pp.

1958 Pérez-Cacho, L., *Sobre alcunas cuestiones de la teoría de números*, Rev. Mat. Hisp.-Amer., (4), **18** (1958), 10–27 and 113–124.

1966 Simmons, G.J., *Some results pertaining to Fermat's conjecture*, Math. Mag., **39** (1966), 18–21.

1969 Rameswar Rao, D., *Some theorems on Fermat's last theorem*, Math. Student, **37** (1969), 208–210.

1981 Rollero, A., *Un'osservazione sull'ultimo teorema di Fermat*, Atti Accad. Ligure Sci. Lett., **38** (1981), 3–9.

1979 Inkeri, K., *On some expressions associated with Fermat's last theorem*, J. Reine Angew. Math., **311/312** (1979), 178–190.

## VI.3.  Abel's Conjecture

Abel stated in 1823 that if $x, y, z$ are nonzero relatively prime integers such that $0 < x < y < z$ and $x^n + y^n = z^n$ ($n > 2$), then none of $x, y, z$ are prime-powers. No direct proof of this statement has ever been discovered. However, we shall see that it is correct when $n$ is not a prime number or when $n = p$ is a prime not dividing $xyz$. This last assertion will not be proved in this book, since it requires analytical methods.

Many partial results, obtained by various authors, are summarized in Table 7.

In 1887, Mansion claimed to have shown that if the exponent is an odd prime that $x$ is not a prime. His proof was erroneous. In 1891, Lucas published a proof that if $n$ is arbitrary then $x$ cannot be a prime-power; but his proof was incomplete, as pointed out by Markoff in 1895. In 1955, Möller established a theorem containing all the above results. However, the proof on page 27 of his paper was insufficient. We give below a simpler and correct proof:

**(3A)**    *Let $n \geq 3$ be an odd integer with $r$ distinct prime factors. If $1 \leq x < y$ are relatively prime integers and $a = y^n + x^n$, $b = y^n - x^n$, then:*

   (1) *$a$ and $b$ have at least $r$ distinct prime factors;*

(2) *If a has exactly r distinct prime factors then* $a = 2^3 + 1^3$ *(so* $n = 3$, $r = 1$*); and*

(3) *If b has exactly r distinct prime factors then* $y = x + 1$, $b = (x+1)^n - x^n$.

TABLE 7

| Year | Author | Exponent $n > 2$ | Result |
|------|--------|------------------|--------|
| 1857 | Talbot | arbitrary | (I) $y, z$ are not primes |
|      |        |           | (II) if $x$ is a prime then |
|      |        |           | $\quad z - y = 1$ |
| 1884 | Jonquières | arbitrary | (I) and (II) |
| 1887 | Borletti | odd prime | $z$ is not a prime |
|      |        | even | $x, y, z$ are not primes |
| 1901 | Gambioli | arbitrary | (II) $z$ not a prime-power |
|      |        | not a power of 2 | |
| 1905 | Sauer | arbitrary | $y, z$ are not prime-power |
| 1932 | Mileikowsky | arbitrary | $y, z$ are not prime-power |
|      |        | not a prime | (III) $x$ not a prime-power |
| 1949 | Izvekoff | arbitrary | (I) |
| 1952 | Bini | odd | $z$ not a prime-power nor |
|      |        |     | equal to $nq$, $q$ prime, $\neq n$ |
|      |        | odd prime | $z$ is not a multiple |
|      |        |           | of $n$, $z$ is not equal to |
|      |        |           | $q_1 q_2 \cdots q_r$ where $q_1$ |
|      |        |           | $< q_2 < \cdots < q_r$ are prime |
|      |        |           | and $q_1^n > 2 q_1 q_2 \cdots q_r$. |

PROOF. (1) Let $p_1, \ldots, p_r$ be the distinct prime factors of $n$. Since $y \pm x \neq 0$, by Chapter II, (3B)(3), $y^n \pm x^n = (y \pm x) \cdot Q_n(y, \mp x)$ is a multiple of $(y \pm x) \prod_{i=1}^{r} Q_{p_i}(y, \mp x)$. By the same result, Part (2), the integers $Q_{p_i}(y, \mp x)$ (for $i = 1, \ldots, r$) are pairwise relatively prime.

We observe that $x^{p_i} + y^{p_i} > x + y$ and $(x^{p_i} - y^{p_i})/(x - y) = x^{p_i - 1} + x^{p_i - 2} y + \cdots + y^{p_i - 1} > 1$. So each $Q_{p_i}(y, \mp x)$ has at least one prime factor. Hence, $a, b$ have at least $r$ distinct prime factors.

(2)    Let $m$ be the product of the $r$ distinct prime factors of $n$, so $m \mid n$. First we show that $n = m$. We have $a = (y^m + x^m)(y^n + x^n)/(y^m + x^m)$. By (1) and the hypothesis, $y^m + x^m$ has exactly $r$ distinct prime factors, which are the same as those of $a$. If

$n > m$ then $a$ has no primitive factors. By Chapter II, (3J), we have $a = 2^3 + 1^3$, so $n = 3 = m$, which is a contradiction. Thus $n = m$, i.e., $n$ is the product of $r$ distinct primes.

Let $r = 1$, $n = p$, so $a = (y + x)Q_p(y, -x)$. By hypothesis, $a$ is the power of some prime $q$. Since $q$ divides $y + x$ and $Q_p(y, -x) > 1$ then by Chapter II, (3B)(4), $q = p$; by Part (6), $Q_p(y, -x) = p$. By Chapter II, (3D), then $p = 3$, $y = 2$, $x = 1$.

Let $r > 1$, $n = ph$, $p$ being a prime, $p \nmid h$; so $h$ is the product of $r - 1$ distinct primes. From $a = (y^h + x^h)Q_p(y^h, -x^h)$, by (1) and the hypothesis, $y^h + x^h$ has $r - 1$ or $r$ distinct prime factors. By induction on $r$, if $y^h + x^h$ has exactly $r - 1$ distinct prime factors then $h = 3$, $y = 2$, $x = 1$. So $n = 3p$ (with $p > 3$), $r = 2$, and $2^{3p} + 1 = (2^p + 1)Q_3(2^p, -1)$.

If there exists a prime $q \neq 3$, such that $q \mid 2^p + 1$ then by Chapter II, (3B)(6), $v_q(Q_3(2^p, -1)) = v_q(3) = 0$, so $q \nmid Q_3(2^p, -1)$. Since $r = 2$ and $3 \mid 2^p + 1$ then $Q_3(2^p, -1)$ is a power of 3; again by Chapter II, (3B)(6), $Q_3(2^p, -1) = 3$, so by Chapter II, (3D), $2^p = 2$, which is impossible.

If $2^p + 1 = 3^s$ then $3^s \equiv 1 \pmod 8$, because $p > 3$. Thus $s = 2s'$. So $2^p = 3^s - 1 = (3^{s'} + 1)(3^{s'} - 1)$ hence $3^{s'} + 1 = 2^{p-c}$, $3^{s'} - 1 = 2^c$ with $p - c > c \geq 0$. Taking the difference $2 = 2^{p-c} - 2^c = 2^c(2^{p-2c} - 1)$, so $c = 1$, $p - 2c = 0$, i.e. $p = 2$, which is impossible.

If $y^h + x^h$ has exactly $r$ distinct prime factors, then each prime factor $q$ of $Q_p(y^h, -x^h)$ divides $y^h + x^h$. By Chapter II, (3B)(4), then $q = p$; $Q_p(y^h, -x^h) > 1$, therefore it is a power of $p$; by Part (6) of the same proposition, $Q_p(y^h, -x^h) = p$ and by Chapter II, (3D), $p = 3$, $y^h = 2$, $x^h = 1$, so $h = 1$, $n = 3$, concluding the proof.

(3) If $b = y^n - x^n$ has exactly $r$ distinct prime factors, by Chapter II, (3B)(2) and (3), $y^n - x^n$ is a multiple of $(y - x)\prod_{i=1}^r Q_{P_i}(y, x)$ and the integers $Q_{p_i}(y, x) > 1$ are pairwise relatively prime, hence by the hypothesis, they are prime powers. If $y - x > 1$ and $q$ is a prime dividing $y - x$ then there exists $i$ such that $q \mid Q_{p_i}(y, x)$. By Part (4) of the same proposition, $q = p_i$ and by Part (6), $Q_{p_i}(y, x) = p_i$. By Chapter II, (3D), $p_i = 3$, $y = 2$, $x = -1$, which is a contradiction. This proves that $y = x + 1$.  $\square$

As a corollary, we have:

**(3B)**    *Let $n \geq 3$ be a positive integer having $r$ distinct odd prime factors. If $0 < x < y < z$ are relatively prime integers such that*

$x^n + y^n = z^n$, then $z, y$ have at least $r + 1$ distinct prime factors and $x$ has at least $r$ distinct prime factors. Moreover, if $x$ has only $r$ such factors then $n$ is odd and $z - y = 1$.

PROOF. By Chapter I, $n$ is not a multiple of 3, nor a power of 2, so $r \geq 1$. Let $n = 2^u m$, with $u \geq 0$ and $m$ odd, having $r$ distinct prime factors. Let $x_1 = x^{2^u}$, $y_1 = y^{2^u}$, $z_1 = z^{2^u}$. Since $z^n = z_1^m = x_1^m + y_1^m$ and $y^n = y_1^m = z_1^m - y_1^m$ and since $m \neq 3$ and $z_1 - x_1 > 1$, it follows from (3A) that $z^n, y^n$, hence also $z, y$, have at least $r + 1$ distinct prime factors.

Similarly $x^n = x_1^m = z_1^m - y_1^m$, so $x^n$, hence also $x$, has at least $r$ distinct prime factors. If $x$ has only $r$ such factors then by (3A), $z_1 = y_1 + 1$, i.e., $z^{2^u} = y^{2^u} + 1$. So $u = 0, z = y + 1$ and $n$ odd.    □

More explicitly:

**(3C)**    Let $n > 2$, and let $0 < x < y < z$ be relatively prime integers such that $x^n + y^n = z^n$. Then:

(1) $y, z$ are not prime powers; and
(2) if $x$ is a prime power then $z = y + 1$ and $n$ is an odd prime.

PROOF. (1)    If $z$ or $y$ is a prime power, so is $z^n = x^n + y^n$, or $y^n = z^n - x^n$. By (3B), $n$ is a power of 2, $n \geq 4$, and this contradicts Fermat's theorem, which is true for such exponents.

(2)    If $x$ is a power of a prime $q$ then by (3B), $z = y + 1$ and $n = p^e$, $e \geq 1$, $p$ an odd prime. We show that $e = 1$.

If $e > 1$ then since $z^p - y^p > 1$ and $Q_{p^{e-1}}(z^p, y^p) = (z^{p^e} - y^{p^e})/(z^p - y^p) > 1$, it follows from $x^{p^e} = z^{p^e} - y^{p^e} = (z^p - y^p) \cdot Q_{p^{e-1}}(z^p, y^p)$ that both factors in the right-hand side are powers of $q$, greater than 1, hence multiples of $q$. By Chapter II, (3B)(4), $q$ divides $\gcd(z^p - y^p, Q_{p^{e-1}}(z^p, y^p)) = \gcd(p^{e-1}, z^p - y^p)$, hence $q = p$.

On the other hand, since $z = y + 1$ then $z^p - y^p = py^{p-1} + \binom{p}{2}y^{p-2} + \cdots + \binom{p}{p-1}y + 1$, so $p = q$ does not divide $z^p - y^p$, which is a contradiction.    □

Möller proved the following complement to (3A):

**(3D)**    If $m$ is odd, $m > 3$, with $r$ distinct prime factors, if $0 < x < y$ are relatively prime integers, then $b = y^{2m} - x^{2m}$ has at least $2r + 1$

*distinct prime factors.*

PROOF. $b = (y^m - x^m)(y^m + x^m)$ and since $\gcd(x, y) = 1$ then $d = \gcd(y^m - x^m, y^m + x^m) = 1$ or $2$. By (3A), $y^m - x^m$ and $y^m + x^m$ have at least $r$ distinct prime factors. Since $m > 3$, by (3A), $y^m + x^m$ has at least $r + 1$ distinct prime factors.

If $y^m - x^m$ has exactly $r$ distinct prime factors, by (3A), $y = x + 1$, thus $y^m - x^m$ is odd. Hence $d = 1$ and $b$ has at least $2r + 1$ distinct prime factors. If $y^m - x^m$ has at least $r + 1$ distinct prime factors, since $d = 1$ or $2$, then $b$ has at least $2(r + 1) - 1 = 2r + 1$ distinct prime factors. □

In particular, if $m$ is odd, $m > 3$ with $r$ distinct prime factors, if $0 < x < y < z$ are relatively prime integers such that $x^{2m} + y^{2m} = z^{2m}$ then $x, y$ have at least $2r + 1$ distinct prime factors. This was shown by Möller in 1955. Combining with (3B) it follows, with the above exponent $2m$, that $x, y, z$ cannot be prime powers.

Inkeri showed in 1946 that if $0 < x < y < z$, if $p$ is a prime number, $p \nmid xyz$ and $x^p + y^p = z^p$, then $z - y > 1$. Hence by (3C), $x$ is not a prime power. This provided a direct proof of Abel's conjecture in the case when $p \nmid xyz$. No direct proof is known, when $p \mid xyz$, that $x$ is not a prime power, so no direct proof that $z > y + 1$ has been devised.

In 1886, Catalan examined the implications of this eventuality.

**(3E)**   *Let $p$ be an odd prime number, and let $0 < x < y$ be integers such that $x^p + y^p = (y + 1)^p$. Then:*

   (1) $py(y + 1)$ *divides* $x^p - 1$.
   (2) $p \nmid x$, $p \mid x - 1$.
   (3) *If $q$ is a prime dividing $y + 1 - x$ then $q$ divides $x - 1$.*
   (4) $\gcd(x + y, y + 1 - x) = 1$.
   (5) $\gcd(2x - 1, 2y + 1) = 1$.
   (6) *$x$ is the only integer such that*

$$(py^{p-1})^{1/p} < x < (p(y + 1)^{p-1})^{1/p}.$$

PROOF. (1)   The polynomial $(Y + 1)^p - Y^p - 1$ is a multiple of $p, Y$, and $Y + 1$, hence $x^p - 1 = (y + 1)^p - y^p - 1 = py(y + 1)h$, $h$ an integer.

(2)   Since $p \mid x^p - 1$ then $p \nmid x$. From $x^p \equiv x \pmod{p}$ it follows that $p \mid x - 1$.

(3)   $y^p = (y + 1)^p - x^p = (y + 1 - x)k$, where $k$ is an integer; hence if $q \mid y + 1 - x$ then $q \mid y$, so $q \mid x - 1$.

(4)   If $q$ is a prime dividing $x + y$ and $y + 1 - x$ then $q \mid x - 1$ so $q \mid y$, hence $q \mid x$; but from $x^p + y^p = (y + 1)^p$ it follows that $\gcd(x, y) = 1$, a contradiction.

(5)   If $q$ is a prime dividing $2x - 1$, $2y + 1$, then it divides their sum $2(x + y)$ and their difference $2(y - x + 1)$; but $q$ is odd, so $q \mid y - x + 1$, $q \mid x + 1$, which is a contradiction.

(6)   We have

$$x^p = (y + 1)^p - y^p = \frac{(y + 1)^p - y^p}{(y + 1) - y}$$
$$= (y + 1)^{p-1} + (y + 1)^{p-2}y + \cdots + (y + 1)y^{p-2} + y^{p-1},$$

hence

$$py^{p-1} < x^p < p(y + 1)^{p-1},$$

and this yields the inequalities of the statement.

Finally, we note that if $x_1, x_2$ are integers such that

$$(py^{p-1})^{1/p} < x_1 < x_2 < (p(y + 1)^{p-1})^{1/p}$$

then

$$y < x_1 \left(\frac{x_1}{p}\right)^{1/(p-1)} \quad \text{and} \quad x_2 \left(\frac{x_2}{p}\right)^{1/(p-1)} < y + 1,$$

hence

$$(x_2 - x_1)\left(\frac{x_1}{p}\right)^{1/(p-1)} < x_2 \left(\frac{x_2}{p}\right)^{1/(p-1)} - x_1 \left(\frac{x_1}{p}\right)^{1/(p-1)} < 1.$$

But $1 \leq x_2 - x_1$ and $p = (p^p)^{1/p} < (py^{p-1})^{1/p} < x_1$ since by (2), $p \leq x - 1 < x < y$. Therefore $1 < (x_1/p)^{1/(p-1)}$ and so $1 < (x_2 - x_1)(x_1/p)^{1/(p-1)}$, which is a contradiction.   $\square$

We now shall use Barlow's relations of Chapter III, §1, with an obvious change of notation, since we consider the relation $x^p + y^p = z^p$ (with $0 < x < y < z$) instead of $x^p + y^p + z^p = 0$. In particular, we use the integers $r, s, t$ defined in those formulas.

In 1964, Dittmann proved the following facts:

**(3F)**    *Let $p$ be an odd prime, and let $0 < x < y$ be integers such that $x^p + y^p = (y+1)^p$ (hence $x, y$ are relatively prime). Then:*

(1) *If $p \mid y+1$ then $v_p(-s+1) = v_p(y+1) - 1$.*
(2) *If $p \mid y$ then $v_p(t-1) = v_p(y) - 1$.*

PROOF. (1)    Let $z = y+1$ and $z - x = -s^p$, so $s \neq -1$. By Chapter III, (1C), $n = v_p(y+1) \geq 2$ and $p^n$ divides $z$ and $x + y$. So $p^n$ divides $2z - (x+y) = z - x + 1 = -s^p + 1$. Since $s^p \equiv s \pmod{p}$ then $p \mid -s + 1$. Let $v_p(-s+1) = l \geq 1$. By Chapter II, (3B)(6), we have $n \leq v_p(-s^p + 1) = l + 1$, so $l \geq n - 1$.

If $v_p(-s+1) \geq n$ then by the fact just quoted above, $v_p(-s^p+1) \geq n + 1$, hence $p^{n+1}$ divides $-s^p + 1 = 2z - (x + y)$. By (1.2), $p^{pn-1}$ divides $x + y$, hence $p^{n+1}$ divides $2z$, so $p^{n+1} \mid z = y+1$, contrary to the hypothesis. This shows that $v_p(-s + 1) = n - 1$.

(2)    Let $v_p(y) = n$, so by Chapter II, (3B), $n \geq 2$ and by (1.2), $p^n$ divides $y$ and $y+1-x = 2y-(x+y-1)$. So $p^n$ divides $x+y-1 = t^p - 1$. Since $t^p \equiv t \pmod{p}$ then $p$ divides $t - 1$. Let $v_p(t-1) = l \geq 1$; then by Chapter II, (3B)(6), $n \leq v_p(t^p - 1) = l + 1$, hence $l \geq n - 1$.

If $v_p(t - 1) \geq n$ then $v_p(t^p - 1) \geq n + 1$ so $p^{n+1}$ divides $t^p - 1 = 2y - (y+1-x)$; but $p^{pn-1}$ divides $z - x = y + 1 - x$, therefore $p^{n+1}$ divides $2y$, so $p^{n+1} \mid y$, which is a contradiction.    □

The situation covered in the preceding result cannot yet be ruled out by a direct proof. With other methods, it may be shown that under the hypothesis of (3E), necessarily $p$ divides $y$ or $y + 1$.

In his thesis, Dittmann has also studied the possibility of a solution of Fermat's equation, with $y = x + 1$. He showed:

**(3G)**    *If $p$ is an odd prime number, if there exist positive integers $0 < x < z$ such that $x^p + (x + 1)^p = z^p$, then:*

(1) $p \mid x(x + 1)$.
(2) *If $p \mid x + 1$ then $-r = \left[-p^{n-1/p}s\right]$ and $r \equiv 1 \pmod{p}$.*
(3) *If $p \mid x$ then $-s = \left[-p^{n-1/p}r\right] + 1$ and $-s \equiv 1 \pmod{p}$.*

PROOF. (1)    If $p \nmid x(x + 1)$, by Chapter III, (1C), $z - x = -s^p$ and $z - (x + 1) = -r^p$ so $1 = r^p - s^p$, which is impossible, since $r, s$ are not zero.

(2)    If $p \mid x + 1$ then $p \nmid xz$, so by Chapter III, (1C), $z - x = -p^{pn-1}s^p$, $z - (x + 1) = -r^p$; hence $1 = r^p - p^{pn-1}s^p$, so $r \equiv 1$

(mod $p$) and $-r^p < -p^{pn-1}s^p$, so $-r < -p^{n-1/p}s$.

If $-r < -p^{n-1/p}s - 1$ then $r > p^{n-1/p}s + 1$, hence $r^p > p^{pn-1}s + 1$, which is a contradiction. This shows that $-r = [-p^{n-1/p}s]$.

(3)    If $p \mid x$ then $p \nmid (x + 1)z$, so by Chapter III, (1C), $z - x = -s^p$, $z - (x + 1) = -p^{pn-1}r^p$; hence $1 = p^{pn-1}r^p - s^p$, so $-s \equiv 1$ (mod $p$) and $-s^p > -p^{pn-1}r^p$, hence $-s > -p^{n-1/p}r$.

If $-s > -p^{n-1/p}r + 1$ then $s + 1 < p^{n-1/p}r$, hence $s^p + 1 < p^{pn-1}r^p$, which is a contradiction. This shows that $-s = [-p^{n-1/p}r] + 1$.    $\square$

## Bibliography

1881 Abel, N.H., Extraits de quelques lettres à Holmboe, *Oeuvres Complètes*, Vol. II, 2nd ed., pp. 254–255, Grondahl, Christiania, 1881.

1857 Talbot, W.H.F., *On Fermat's theorem*, Trans. Roy. Soc. Edinburgh, **21** (1857), 403–406.

1884 Jonquières, E. de, *Sur le dernier théorème de Fermat*, Atti Accad. Pont. Nuovi Lincei, **37** (1883/4), 146–149; reprinted in Sphinx-Oedipe, **5** (1910), 29–32.

1884 Jonquières, E. de, *Sur le dernier théorème de Fermat*, C. R. Acad. Sci. Paris, **98** (1884), 863–864.

1886 Catalan, E., *Sur le dernier théorème de Fermat*, Bull. Acad. Roy. Sci. Lett. Beaux-Arts Belgique, (3), **12** (1886), 498–500.

1886 Catalan, E., *Sur le dernier théorème de Fermat* (Mélanges Mathématiques, CCXV), Mém. Soc. Roy. Sci. Liège Sér. 2, **13** (1886), 387–397.

1887 Borletti, F., *Sopra il teorema di Fermat relativo all'equazione* $x^n + y^n = z^n$, Rend. Istit. Lombardo, (2), **20** (1887), 222–224.

1887 Mansion, P., *Sur le dernier théorème de Fermat*, Bull. Acad. Roy. Sci. Lett. Beaux-Arts Belgique, (3), **13** (1887), 16–17 and 225.

1891 Lucas, E., *Théorie des Nombres*, reprinted by A. Blanchard, Paris, 1961.

1895 Markoff, V., *Question 477*, L'Interm. Math. Sér. I, **2** (1895), 23.

1901 Gambioli, D., *Memoria bibliografica sull'ultimo teorema di Fermat*, Period. Mat., **16** (1901), 145–192.

1905 Sauer, R., *Eine polynomische Verallgemeinerung des Fer-*

*matschen Satzes*, Dissertation, Giessen, 1905.

1932 Mileikowsky, E.M., *Elementarer Beitrag zur Fermatschen Vermutung*, J. Reine Angew. Math., **166** (1932), 116–117.

1946 Inkeri, K., *Untersuchungen über die Fermatsche Vermutung*, Ann. Acad. Sci. Fenn. Ser. A, **33** (1946), 1–60.

1949 Izvekoff, J., *Sur une propriété des nombres premiers*, Bull. Soc. Math. Phys. Serbie, **1** (1949), 41–43.

1952 Bini, U., *La risoluzione delle equazioni $x^n \pm y^n = M$ e l'ultimo teorema di Fermat*, Archimede, **4** (1952), 50–57.

1955 Möller, K., *Untere Schranke für die Anzahl der Primzahlen, aus denen $x, y, z$ der Fermatschen Gleichung $x^n + y^n = z^n$ bestehen muss.*, Math. Nachr., **14** (1955), 25–28.

1964 Dittmann, G., *Untersuchungen über höhere Potenzen natürlicher Zahlen*, Thesis, Potsdam, 1964, 56 pp.

## VI.4. The First Case for Even Exponents

In this section, we give Terjanian's proof (1977; see also 1978) that the first case of Fermat's theorem holds for even exponents. It suffices to consider exponents $2p$, where $p$ is an odd prime. The proof will require only elementary considerations, so it is surprising that it was not found beforehand.

Several authors had considered Fermat's equation with even exponents, however their direct proofs are now all superseded by Terjanian's; yet we shall quote, and in a few cases prove, some of these statements.

To begin, in a letter to Gauss (1804), Sophie Germain stated without proof that if $p$ is a prime, $p \equiv 7 \pmod 8$, then $X^{p-1} + Y^{p-1} = Z^{p-1}$ has no solution in nonzero integers.

In the first of a long series of papers on Fermat's theorem, Kummer proved (1837) the following result, which was rediscovered by Niedermeier in 1943 and again by Griselle (1953) and Oeconomou (1956). We follow Griselle's proof:

**(4A)**    *Let $n \geq 2$ be an integer. If there exist nonzero integers $x, y, z$ such that $x^{2n} + y^{2n} = z^{2n}$ and $\gcd(n, xyz) = 1$, then $n \equiv 1 \pmod 8$.*

PROOF. By Fermat's theorem for the exponent 4, we may take $n$ odd, $n \geq 3$. We may also assume that $x, y, z$ are pairwise relatively prime positive integers, and that $x$ is even, while $y, z$ are odd, by the remark at the beginning of Chapter I, §1. Then

$$x^{2n} = z^{2n} - y^{2n}$$
$$= (z^2 - y^2)(z^{2(n-1)} + y^2 z^{2(n-2)} + \cdots + y^{2(n-2)} z^2 + y^{2(n-1)})$$
$$= (z^2 - y^2) \cdot Q_n(z^2, y^2),$$

with the notation of §1. By Chapter II, (3B)(4),

$$\gcd(z^2 - y^2, Q_n(z^2, y^2)) = \gcd(z^2 - y^2, n).$$

If $p$ is a prime dividing $n$ and $z^2 - y^2$ then a fortiori $p$ divides $x$, contrary to the hypothesis. Thus $z^2 - y^2$ and $Q_n(z^2, y^2)$ are relatively prime.

So $z^2 - y^2$ and $Q_n(z^2, y^2)$ are $2n$th powers. Moreover, $z^2 - y^2$ is even so $Q_n(z^2, y^2)$ is odd. Thus there exists an odd integer $k$ such that

$$z^{2(n-1)} + z^{2(n-2)} y^2 + z^{2(n-3)} y^4 + \cdots + y^{2(n-1)} = k^{2n}.$$

Each term of the above equality is an odd square, so it is of the form $(2a+1)^2 = 4a(a+1)+1 \equiv 1 \pmod 8$. Thus, we have the congruence $n \equiv 1 \pmod 8$. $\square$

From this result, it follows:

**(4B)**    *The set of primes $p$ such that the first case of Fermat's theorem holds for the exponent $2p$ is an infinite set.*

PROOF. According to Dirichlet's theorem for primes in arithmetic progressions, there exist infinitely many primes $p$ satisfying each of the congruences $p \equiv 3 \pmod 8$, $p \equiv 5 \pmod 8$, $p \equiv 7 \pmod 8$. For each such prime $p$ the first case of Fermat's theorem holds for the exponent $2p$, by virtue of (4A). $\square$

We note the following strengthening of (4A); statement (1) is due to Niedermeier (1944), while (2) was given by Grey (1954):

**(4C)**    *Let $p$ be an odd prime and assume that there exist nonzero relatively prime integers $x, y, z$ such that $x^{2p} + y^{2p} = z^{2p}$ :*

    (1) *If $3p$ does not divide $x, y, z$ then $p \equiv 1 \pmod 3$.*

(2) *If $2p$ does not divide $x, y, z$ then $p = 24a + 1$ (for some integer $a$) and $12a + 1$ has no factor congruent to 3 modulo 4.*

Another improvement over (4A) is the following ((1) was proved by Niedermeier in 1944, while (2) was given by Long in 1960):

**(4D)**    *Let $m \geq 3$ be an integer and assume that there exist nonzero integers $x, y, z$ such that $x^{2m} + y^{2m} = z^{2m}$ :*

(1) *If $m = p$ is a prime and if $5p$ does not divide $x, y, z$ then $p \equiv \pm 1 \pmod 5$.*
(2) *If $\gcd(m, xyz) = 1$ then $m \equiv 1$ or $49 \pmod{120}$.*

As a corollary:

**(4E)**    *If $n = 2m$ has last digit 4 or 6 (when written in decimal notation), then there exist no nonzero integers $x, y, z$ such that $\gcd(m, xyz) = 1$ and $x^n + y^n = z^n$.*

PROOF. We have $m \equiv \pm 2 \pmod 5$, hence $m \not\equiv 1$ or $49 \pmod{120}$ and the result follows from (4D). □

Long (1961) extended his method and proved:

**(4F)**    *If $p$ is a prime, $p \equiv -1 \pmod{10}$ and if there exist relatively prime integers $x, y, z$, prime to $p$ and such that $x^{2p} + y^{2p} = z^{2p}$, then $p$ is a square modulo 11, and consequently $p \equiv 49, 169, 289, 529, $ or $889 \pmod{1320}$.*

Oeconomou proved in 1956 the following results, involving the Legendre symbol:

**(4G)**    *Let $n > 1$ be an odd integer and assume that there exists an odd prime $q$ such that:*

(a) $\gcd(q - 1, n) = 1$.
(b) $(n/q) = -1$.
(c) *If $0 < m < (q - 1)/2$, with $n \equiv m \pmod{(q - 1)/2}$, if $a, b, c$ are integers such that $a^{2m} + b^{2m} \equiv c^{2m} \pmod q$, and*

$$\left( \frac{a^2 + b^2}{q} \right) = 1, \qquad \left( \frac{c^2 - a^2}{q} \right) = 1, \qquad \left( \frac{c^2 - b^2}{q} \right) = 1,$$

*then* $abc \equiv 0 \pmod{q}$.

*Then there exist no relatively prime nonzero integers* $x, y, z$ *such that* $\gcd(n, xyz) = 1$ *and* $x^{2n} + y^{2n} = z^{2n}$.

There are numerous possible choices of $q, m$ satisfying the conditions. For example: In any of the following cases, the first case of Fermat's theorem holds for the exponent $2n$:

    (a) $n \equiv -1 \pmod 3$.
    (b) $n \equiv \pm 2 \pmod 5$.
    (c) $(n/11) = -1$ and $n \equiv 4 \pmod 5$.
    (d) $(n/19) = -1$ and $n \equiv 4$ or $7 \pmod 9$.
    (e) $(n/23) = -1$ and $n \equiv 2$, or $3$, or $4$, or $10 \pmod{11}$.
    (f) $(n/29) = -1$ and $n \equiv 3$, or $5$, or $9$, or $11 \pmod{14}$ etc. ....

In this way, Oeconomou proved that the first case of Fermat's theorem holds for all even exponents less than $200\,000$ (with the possible exceptions of $108\,722$ and $188\,018$).

Gandhi proved in 1966:

**(4H)**    *Let* $p$ *be a prime,* $p \geq 5$. *If* $x, y, z$ *are pairwise relatively prime integers such that* $x^{p-1} + y^{p-1} = z^{p-1}$ *then* $z$ *is odd, and* $p$ *divides the one among the integers* $x, y$ *which is even, hence* $p$ *does not divide* $z$.

PROOF. Since $p-1$ is even, by the remark at the beginning of Chapter I, §1, $z$ is odd and $x, y$ have different parity, say $x$ is odd, $y$ is even.

If $p \nmid xy$ then $x^{p-1} \equiv y^{p-1} \equiv 1 \pmod p$ while $z^{p-1} \equiv 0 \pmod p$ or $\equiv 1 \pmod p$. Thus $2 \equiv x^{p-1} + y^{p-1} \equiv 0$ or $1 \pmod p$ which is impossible. Let us assume that $p \mid x$, $p \nmid y$ — this will lead to a contradiction. Let $m = (p-1)/2$. We have

$$x^{p-1} = z^{p-1} - y^{p-1} = (z^m - y^m)(z^m + y^m).$$

Then the two factors in the right-hand side are relatively prime, since they are both odd. Therefore each factor is a $(p-1)$th power:

$$\begin{cases} z^m - y^m = a^{p-1}, \\ z^m + y^m = b^{p-1}, \end{cases}$$

hence $x = ab$ and also $2z^m = a^{p-1} + b^{p-1}$. But $p \mid x$ and $\gcd(a, b) = 1$ so $p$ divides one and only one of the numbers $a, b$. Hence, $2z^m \equiv$

1 (mod $p$) and squaring this congruence, $4 \equiv 1$ (mod $p$), which is impossible.

This proves that $p$ divides $y$. Hence $p$ does not divide $z$.    □

In 1969, Raina proved:

**(4I)**    *Let $p$ be a prime, $p \geq 5$. If $x, y, z$ are positive pairwise relatively prime integers such that $x^{p-1} + y^{p-1} = z^{p-1}$ then $z$ is a quadratic residue modulo $p$.*

PROOF. Only one of $x, y$ is even, the other is odd. Suppose $y$ is even. By (4H), $z$ is odd and $p$ divides $y$.

Let $m = (p-1)/2$. From $(x^m)^2 + (y^m)^2 = (z^m)^2$, it follows by Chapter I, (1A), that there exist positive integers $a, b$, of different parity, such that $\gcd(a, b) = 1$ and

$$\begin{cases} x^m = a^2 - b^2, \\ y^m = 2ab, \\ z^m = a^2 + b^2. \end{cases}$$

Suppose first that $b$ is odd, so $a$ is even. Since $\gcd(2a, b) = 1$ then there exist integers $h, k$ such that

$$\begin{cases} 2a = h^m, \\ b = k^m. \end{cases}$$

Therefore $4z^m = h^{p-1} + 4k^{p-1}$. If $p \mid k$ then $p \nmid h$ so $4z^m \equiv 1$ (mod $p$). Hence, squaring this congruence, $16 \equiv 1$ (mod $p$) so $p = 5$, which is not possible because the equation $X^4 + Y^4 = Z^4$ has no solution in positive integers (Chapter I, (2C)).

So $p \nmid k$ and since $p \mid y$ then $p \mid h$. Therefore $h^{p-1} = 4a^2 \equiv 0$ (mod $16p$) and $z^{(p-1)/2} \equiv k^{p-1} \equiv 1$ (mod $p$).

Suppose now that $b$ is even, while $a$ is odd. Proceeding in the same way, we show that $z^{(p-1)/2} \equiv 1$ (mod $p$). Thus $z$ is a quadratic residue modulo $p$.    □

The next result in Raina's paper is vacuous since its hypothesis is never satisfied (namely, $z$ cannot be a prime, by (3C)).

In 1955, Becker published a paper in which he asserted that Fermat's theorem is true for all even exponents $2m > 2$. However, his proof is definitely wrong.

Now we shall give the proof of Terjanian's theorem, which contains all the above results as corollaries. Once more, we shall consider the quotient $Q_n(z, y) = (z^n - y^n)/(z - y)$ where $n$ is an odd natural number and $z, y$ are nonzero distinct integers (not necessarily positive).

For the convenience of the reader, we recall that if $m, n$ are nonzero odd integers, $n > 0$, $\gcd(m, n) = 1$, then the Jacobi symbol $(m/n)$ is defined. If $m > 0$ then $(n/m)$ is also defined, and the following reciprocity law is satisfied:

$$(4.1) \qquad \left(\frac{m}{n}\right) = (-1)^{(m-1)/2 \times (n-1)/2} \left(\frac{n}{m}\right).$$

Moreover,

$$(4.2) \qquad \left(\frac{-1}{n}\right) = (-1)^{(n-1)/2}.$$

**(4J)**    Let $y, z$ be distinct nonzero odd integers such that $y \equiv z$ (mod 4) and $\gcd(y, z) = 1$. Let $m, n$ be odd integers, $m \geq 1$, $n \geq 1$, $\gcd(m, n) = 1$. Then:

(1) $Q_m(z, y) \equiv m$ (mod 4), and in particular $Q_m(z, y)$ is odd.
(2) The Jacobi symbols $(Q_m(z, y)/Q_n(z, y))$ and $(m/n)$ are well defined and equal.

PROOF. (1)    Let $z = y + 4t$. Then

$$Q_m(z, y) = \frac{(y + 4t)^4 - y^m}{4t} = \binom{m}{1} y^{m-1} + \binom{m}{2} y^{m-2} 4t + \cdots$$
$$\equiv my^{m-1} \equiv m \pmod 4,$$

because $m - 1$ is even, $y$ is odd, so $y^{m-1} \equiv 1$ (mod 4).

(2)    First we note that since $\gcd(m, n) = 1$ then the Jacobi symbol $(m/n)$ is well defined. Similarly, from $\gcd(y, z) = 1$ and Chapter II, (3B)(2) it follows that $\gcd(Q_m(z, y), Q_n(z, y)) = 1$; since $Q_n(z, y) > 0$ then the Jacobi symbol $(Q_m(z, y)/Q_n(z, y))$ is also well defined.

The equality of the Jacobi symbols is proved by induction on $k = \min\{m, n\}$. It is trivial when $k = 1$. Let $k > 1$, so $m \neq n$, because $\gcd(m, n) = 1$.

If $m > n$ then there exist an odd integer $r$, $0 < r < n = k$, and an integer $q$ such that $m = qn + r$, or $m = qn - r$. If $m = qn + r$ then $m - r$ is even, so by Chapter II, (3B)(1),

$$\left(\frac{Q_m(z,y)}{Q_n(z,y)}\right) = \left(\frac{y^{m-r}Q_r(z,y)}{Q_n(z,y)}\right) = \left(\frac{Q_r(z,y)}{Q_n(z,y)}\right) = \left(\frac{r}{n}\right) = \left(\frac{m}{n}\right).$$

If $m = qn - r$ then $m - n$ and $n - r$ are even, so by Chapter II, (3B)(1), induction and the properties of the Jacobi symbol,

$$\left(\frac{Q_m(z,y)}{Q_n(z,y)}\right) = \left(\frac{-y^{m-n}z^{n-r}Q_r(z,y)}{Q_n(z,y)}\right)$$

$$= \left(\frac{-Q_r(z,y)}{Q_n(z,y)}\right) = \left(\frac{-1}{Q_n(z,y)}\right)\left(\frac{Q_r(z,y)}{Q_n(z,y)}\right)$$

$$= \left(\frac{-1}{Q_n(z,y)}\right)\left(\frac{r}{n}\right).$$

Since $Q_n(z,y) \equiv n \pmod 4$ by Part (1), then $(Q_n(z,y) - 1)/2 \equiv (n-1)/2 \pmod 2$, hence by (5.2), $(-1/Q_n(z,y)) = (-1/n)$. Thus

$$\left(\frac{Q_m(z,y)}{Q_n(z,y)}\right) = \left(\frac{-1}{n}\right)\left(\frac{r}{n}\right) = \left(\frac{m}{n}\right).$$

Now, if $m < n$, by the reciprocity law (4.1) for the Jacobi symbol and the above proof,

$$\left(\frac{Q_m(z,y)}{Q_n(z,y)}\right) = (-1)^{(Q_m(z,y)-1)/2 \times (Q_n(z,y)-1)/2}\left(\frac{Q_n(z,y)}{Q_m(z,y)}\right)$$

$$= (-1)^{(m-1)/2 \times (n-1)/2}\left(\frac{n}{m}\right) = \left(\frac{m}{n}\right). \quad \square$$

Terjanian's result now follows very easily:

**(4K)**    Let $p$ be an odd prime. If $x, y, z$ are nonzero integers such that $x^{2p} + y^{2p} = z^{2p}$ then $2p$ divides $x$ or $y$.

PROOF. There is no loss of generality to assume that $x, y, z$ are pairwise relatively prime. Also, $x, y$ cannot be both odd, since the exponent $2p$ is even (see remark at the beginning of Chapter I, §1). Let $x$ be even, so $y, z$ are odd. Then

$$x^{2p} = z^{2p} - y^{2p} = (z^2 - y^2)\frac{z^{2p} - y^{2p}}{z^2 - y^2}.$$

By Chapter II, (3B)(4),

$$\gcd\left(z^2 - y^2, \frac{z^{2p} - y^{2p}}{z^2 - y^2}\right) = p \quad \text{or } 1.$$

If the greatest common divisor is $p$ then $p$ divides $x^{2p}$ so $2p$ divides $x$.

We show now that it is not possible that $z^2 - y^2$ and $(z^{2p} - y^{2p})/(z^2 - y^2)$ be relatively prime. If they are, they must be squares. But

$$\frac{z^{2p} - y^{2p}}{z^2 - y^2} = \frac{z^p - y^p}{z - y} \times \frac{z^p + y^p}{z + y} = Q_p(z, y) \times Q_p(z, -y)$$

and these two factors are relatively prime because $\gcd(y, z) = 1$ and $y$ and $z$ are odd. So $Q_p(z, y)$, $Q_p(z, -y)$ are also squares. Since $p$ is not a square, there exists an odd prime $q$ such that $p$ is not a square modulo $q$.

Assume first that $z \equiv y \pmod{4}$. By (4.1),

$$-1 = \left(\frac{p}{q}\right) = \left(\frac{Q_p(z, y)}{Q_q(z, y)}\right),$$

which is an absurdity, because $Q_p(z, y)$ is a square. If $z \not\equiv y \pmod{4}$ then $z \equiv -y \pmod{4}$, hence again

$$-1 = \left(\frac{p}{q}\right) = \left(\frac{Q_p(z, -y)}{Q_q(z, -y)}\right),$$

which is again an absurdity. This concludes the proof.    □

In 1981, Rotkiewicz showed the following strengthening of Terjanian's result:

(4L)    Let $p$ be an odd prime. If $x, y, z$ are positive integers such that $x^{2p} + y^{2p} = z^{2p}$, then $8p^3$ divides $x$ or $y$.

PROOF. We may assume that $x, y, z$ are pairwise relatively prime and that $x$ is even, while $y, z$ are odd. Then by Terjanian's theorem, $2p$ divides $x$.

Now we show that 8 divides $x$. We have $\gcd(z^p - x^p, z^p + x^p) = 1$ because $\gcd(x, z) = 1$, $x$ is even and $z$ is odd. Since $y^{2p} = z^{2p} - x^{2p} = (z^p - x^p)(z^p + x^p)$ then $z^p - x^p = [(z^p - x^p)/(z - x)](z - x) = a^{2p}$, where $a$ is an odd positive integer. Since $p \mid x$ then $p \nmid z$ so $p \nmid z - x$.

By Chapter II, (3B)(5), $\gcd\left((z^p - x^p)/(z - x), z - x\right) = 1$, hence $(z^p - x^p)/(z - x) = b^2$, with $b$ an odd positive integer (because $z$ is odd). Hence $b^2 = (2b_1 + 1)^2 = 4b_1(b_1 + 1) \equiv 1 \pmod 8$ and $z^{p-1} + z^{p-2}x + z^{p-3}x^2 \equiv 1 \pmod 8$, because $x$ is even. But $z$ being odd, we have again $z^{p-1} \equiv 1 \pmod 8$, so $z^{p-3}x(z + x) \equiv 0 \pmod 8$. Hence $y \equiv 0 \pmod 8$, because $z, z + x$ are odd.

The proof that $p^3$ divides $x$ follows from the result of Vandiver, (1B). We have also $z^p + x^p = c^{2p}$, where $c$ is an odd positive integer. By Vandiver's theorem we have $x^p \equiv x \pmod{p^3}$. Since $p \mid x$ and $p \geq 3$ then $p^3$ divides $x$, showing that $8p^3$ divides $x$. $\square$

In 1950, Gut adapted ideas of Kummer and Mirimanoff and used methods of class field theory to derive a criterion, involving Euler numbers, for the first case of Fermat's theorem with exponent $2p$. Of course, this result is now obsolete.

In conclusion we mention explicitly the following easy fact, which will be used later:

**(4M)**     *If $x, y, z$ are nonzero relatively prime integer, $n \geq 3$ and $x^{2n} + y^{2n} = z^{2n}$, then $2 \mid xy$ and $3 \mid xy$.*

PROOF. If $x, y$ are odd then $x^{2n} + y^{2n} \equiv 2 \pmod 4$; now $z$ must be even and $z^{2n} \equiv 0 \pmod 4$, which is absurd. If $3 \nmid xy$ then $x^{2n} \equiv y^{2n} \equiv 1 \pmod 3$ but $z^{2n} \equiv 0$ or $1 \pmod 3$, so $x^{2n} + y^{2n} = z^{2n}$ is not true. $\square$

### Bibliography

1804 Germain, S., *Letter to Gauss, Nov. 21, 1804*; reprinted in *Oeuvres Philosophiques*, p. 298 (editées par H. Stupuy), Ritti, Paris, 1879.

1837 Kummer, E.E., *De aequatione $x^{2\lambda} + y^{2\lambda} = z^{2\lambda}$ per numeros integros resolvenda*, J. Reine Angew. Math., **17** (1837), 203–209; reprinted in *Collected Papers*, Vol. I, pp. 135–141.

1943 Niedermeier, F., *Ein elementarer Beitrag zur Fermatschen Vermutung*, J. Reine Angew. Math., **185** (1943), 111–112.

1944 Niedermeier, F., *Zwei Erweiterungen eines Kummerschen Kriteriums für die Fermatsche Gleichung*, Deutsche Math., **7** (1944), 518–519.

1950 Gut, M., *Eulersche Zahlen und großer Fermatscher Satz*, Comment. Math. Helv., **24** (1950), 73–99.

1953 Griselle, T., *Proof of Fermat's last theorem for* $n = 2(8a+1)$, Math. Mag., **26** (1953), 263.

1954 Grey, L.D., *A note on Fermat's last theorem*, Math. Mag., **27** (1954), 274–277.

1955 Becker, H.W., *Proof of F.L.T. for all even powers*, Math. Mag., **28** (1955), 297–298; and **29** (1956), 125.

1956 Oeconomou, G., *Sur le premier cas du théorème de Fermat pour les exposants pairs*, C. R. Acad. Sci. Paris, **243** (1956), 1588–1591.

1960 Long, L., *A note on Fermat's theorem*, Math. Gaz., **44** (1960), 261–262.

1961 Long, L., *On Fermat's last theorem*, Math. Gaz., **45** (1961), 319–321.

1966 Gandhi, J.M., *On Fermat's last theorem*, Math. Gaz., **50** (1966), 36–37.

1969 Raina, B.L., *On Fermat's last theorem*, Amer. Math. Monthly, **76** (1969), 49–51.

1977 Terjanian, G., *Sur l'équation* $x^{2p} + y^{2p} = z^{2p}$, C. R. Acad. Sci. Paris, **285** (1977), 973–975.

1978 Terjanian, G., *L'équation* $x^p - y^p = az^2$ *et le théorème de Fermat*, Sém. Th. des Nombres, Bordeaux, exposé no. 29, 1978, 7 pp.

1981 Rotkiewicz, A., *On Fermat's equation with exponent 2p*, Colloq. Math., **45** (1981), 101–102.

# VII
# Interludes 7 and 8

This chapter deals with polynomials which are intimately related with Fermat's equation.

## VII.1. Some Relevant Polynomial Identities

We give here some algebraic identities which are applicable in the study of Fermat's equation. Many of the early attempts to prove Fermat's theorem were based on some polynomial identities.

To begin, we indicate the following identity, which was used by Lamé in 1840; see also Lebesgue (1847), Mention (1847), and Catalan (1885); Gauss (1863) gave the special cases when $n = 3, 5, 7$; see also Rebout (1877), and Brocard (1878):

**(1A)**　　*If $X, Y, Z$ are indeterminates and $n$ is odd then*

$$(X+Y+Z)^n - (X+Y-Z)^n - (X-Y+Z)^n - (-X+Y+Z)^n$$

$$= 4nXYZ \sum_{\substack{i+j+k=(n-3)/2 \\ i,j,k \geq 0}} \frac{(n-1)!}{(2i+1)!\,(2j+1)!\,(2k+1)!} X^{2i} Y^{2j} Z^{2k}.$$

PROOF. By writing explicitly the $n$th powers of the left-hand side, we have:

$$(X+Y+Z)^n - (X+Y-Z)^n - (X-Y+Z)^n - (-X+Y+Z)^n$$

$$= \sum_{\substack{a+b+c=n \\ a,b,c \geq 0}} \frac{n!}{a!\,b!\,c!} X^a Y^b Z^c [1 - (-1)^c - (-1)^b - (-1)^a].$$

Since $n$ is odd and $a+b+c=n$ then one or three of the integers $a, b, c$ are odd. If only one is odd then $1 - (-1)^a - (-1)^b - (-1)^c = 0$. Thus, we have to consider only the summands with $a = 2i+1$, $b = 2j+1$, $c = 2k+1$, so $i+j+k = (n-3)/2$ and $1 - (-1)^a - (-1)^b - (-1)^c = 4$. We deduce at once the identity of the statement.    □

As a corollary, we have (see Werebrusow, 1908):

**(1B)**    *If $X, Y, Z$ are indeterminates and $n$ is odd then*

$$(X + Y + Z)^n - X^n - Y^n - Z^n = \frac{4n}{2^n}(X+Y)(Y+Z)(Z+X)$$

$$\times \sum_{\substack{i+j+k=(n-3)/2 \\ i,j,k \geq 0}} \frac{(n - 1)!\,(X + Y)^{2i}(Y + Z)^{2j}(Z + X)^{2k}}{(2i + 1)!\,(2j + 1)!\,(2k + 1)!}.$$

PROOF. We write the identity of (1A) with $U, V, W$ in place of $X, Y, Z$, respectively, where $U = (X + Y)/2$, $V = (Y + Z)/2$, $W = (Z + X)/2$. Then

$$U + V + W = X + Y + Z,$$
$$U + V - W = Y,$$
$$U - V + W = X,$$
$$-U + V + W = Z,$$

hence (1A) becomes

$$(X + Y + Z)^n - X^n - Y^n - Z^n$$

$$= \frac{4n}{2^n}(X + Y)(Y + Z)(Z + X)$$

$$\times \sum_{\substack{i+j+k=(n-3)/2 \\ i,j,k \geq 0}} \frac{(n - 1)!\,(X+Y)^{2i}(Y+Z)^{2j}(Z+X)^{2k}}{(2i + 1)!\,(2j + 1)!\,(2k + 1)!}.  \quad □$$

A special case is the following:

**(1C)**    *If $X, Y$ are indeterminates and $n$ is odd then*

$$(X + Y)^n - X^n - Y^n$$

$$= \frac{4n}{2^n}(X + Y)XY$$

$$\times \sum_{\substack{i+j+k=(n-3)/2 \\ i,j,k \geq 0}} \frac{(n-1)!}{(2i+1)!\,(2j+1)!\,(2k+1)!}(X + Y)^{2i}Y^{2j}X^{2k}.$$

PROOF. It suffices to replace $Z$ by 0 in the identity (1B).    □

Already in 1837, Kummer used the following identity for $(X + Y)^n - (X^n + Y^n)$, see also Mention (1847), Vachette (1861), Barisien (1906), Boutin and Gonzalez Quijano (1907), Bini (1907), Rose (1907), and Bachmann (1910):

**(1D)**    *If $X, Y$ are indeterminates and $n \geq 1$ then*

$$(X + Y)^n - (X^n + Y^n) = \sum_{i=1}^{\infty} (-1)^{i-1} \frac{n}{i} \binom{n-i-1}{i-1} X^i Y^i (X + Y)^{n-2i}$$

*(by convention, the terms in the above sum are zero when $2i > n$).*

PROOF. We show by induction that the identity

$$X^n + Y^n = (X + Y)^n + \sum_{i=1}^{\infty} (-1)^i \frac{n}{i} \binom{n-i-1}{i-1} X^i Y^i (X + Y)^{n-2i}$$

is true. For $n = 1, 2$ it is trivial. Then

$$X^{n+1} + Y^{n+1} = (X^n + Y^n)(X + Y) - XY(X^{n-1} + Y^{n-1})$$
$$= (X + Y)^{n+1}$$
$$+ \sum_{i=1}^{\infty} (-1)^i \frac{n}{i} \binom{n-i-1}{i-1} X^i Y^i (X + Y)^{n+1-2i}$$
$$- XY(X + Y)^{n-1}$$
$$- \sum_{i=1}^{\infty} (-1)^i \frac{n-1}{i} \binom{n-i-2}{i-1} X^{i+1} Y^{i+1} (X+Y)^{n-1-2i}$$

$$= (X+Y)^{n+1} + \sum_{i=1}^{\infty} (-1)^i c_i X^i Y^i (X+Y)^{n+1-2i},$$

where $c_1 = n+1$ and if $i \geq 2$, then

$$c_i = \frac{n}{i}\binom{n-i-1}{i-1} + \frac{n-1}{i-1}\binom{n-i-1}{i-2} = \frac{n+1}{i}\binom{n-i}{i-1}. \quad \square$$

In 1885 (and again in 1886) Catalan indicated another form of the identity (1B):

**(1E)**   *If $X, Y, Z$ are indeterminates and $n$ is odd, then*

$$\frac{(X+Y+Z)^n - X^n - Y^n - Z^n}{(X+Y)(Y+Z)(Z+X)} = P^{n-3} + H_1 P^{n-4} + H_2 P^{n-5}$$

$$+ \cdots + H_{n-3} + 2H^{(2)}_{(n-3)/2}$$

*where $P = X+Y+Z = H_1$; more generally, if $i \geq 1$, $H_i$ is the sum of all monomials of degree $i$, coefficients 1, in the indeterminates $X, Y, Z$, and $H^{(2)}_{(n-3)/2}$ is the sum of all monomials of degree $(n-3)/2$, coefficients 1, in $X^2, Y^2, Z^2$ (so it has degree $n-3$), that is, $H^{(2)}_{(n-3)/2} = H_{(n-3)/2}(X^2, Y^2, Z^2)$.*

PROOF. We have

$$Q_1 = \frac{P^n - (X^n + Y^n + Z^n)}{X+Y} = \frac{P^n - Z^n}{P-Z} - \frac{X^n + Y^n}{X+Y}$$
$$= (P^{n-1} + ZP^{n-2} + Z^2 P^{n-3} + \cdots + Z^{n-1})$$
$$- (X^{n-1} - YX^{n-2} + Y^2 X^{n-3} - \cdots + Y^{n-1}).$$

But, by Euclidean division, since $Y + Z = P - X$:

$$\frac{P^{n-1} + ZP^{n-2} + Z^2 P^{n-3} + \cdots + Z^{n-1}}{Y+Z}$$
$$= P^{n-2} + H_1(X,Z)P^{n-3} + H_2(X,Z)P^{n-4} + \cdots + H_{n-2}(X,Z)$$
$$+ \frac{X^{n-1} + ZX^{n-2} + \cdots + Z^{n-1}}{Y+Z},$$

where $H_i(X, Z)$ is the sum of all monomials of degree $i$, coefficient 1, in the indeterminates $X, Z$. Therefore

$$
\begin{aligned}
Q_2 &= \frac{Q_1}{Y + Z} \\
&= (P^{n-2} + H_1(X, Z)P^{n-3} + H_2(X, Z)P^{n-4} + \cdots + H_{n-2}(X, Z)) \\
&\quad + \frac{1}{Y + Z}[(Y + Z)X^{n-2} - (Y^2 - Z^2)X^{n-3} + (Y^3 + Z^3)X^{n-4} \\
&\qquad\qquad - \cdots - (Y^{n-1} - Z^{n-1})] \\
&= (P^{n-2} + H_1(X, Z)P^{n-3} + H_2(X, Z)P^{n-4} + \cdots + H_{n-2}(X, Z)) \\
&\quad + (X^{n-2} - (Y - Z)X^{n-3} + (Y^2 - ZY + Z^2)X^{n-4} - \cdots \\
&\qquad - (Y^{n-2} - ZY^{n-3} + Z^2 Y^{n-4} - \cdots - Z^{n-2}).
\end{aligned}
$$

By Euclidean division,

$$
\begin{aligned}
&\frac{P^{n-2} + H_1(X, Z)P^{n-3} + H_2(X, Z)P^{n-4} + \cdots + H_{n-2}(X, Z)}{Z + X} \\
&= P^{n-3} + H_1 P^{n-4} + H_2 P^{n-5} + \cdots + H_{n-3} \\
&\quad + \frac{Y^{n-2} + H_1(X, Z)Y^{n-3} + H_2(X, Z)Y^{n-4} + \cdots + H_{n-2}(X, Z)}{Z + X},
\end{aligned}
$$

where $H_i$ is as indicated in the statement. Then

$$
\begin{aligned}
Q_3 &= \frac{Q_2}{Z + X} \\
&= P^{n-3} + H_1 P^{n-4} + H_2 P^{n-5} + \cdots + H_{n-3} \\
&\quad + \frac{1}{Z + X}[Y^{n-2} + H_1(X, Z)Y^{n-3} + H_2(X, Z)Y^{n-4} \\
&\qquad\qquad + \cdots + H_{n-2}(X, Z) + X^{n-2} - (Y - Z)X^{n-3} \\
&\qquad\qquad + (Y^2 - ZY + Z^2)X^{n-4} - \cdots \\
&\qquad\qquad - (Y^{n-2} - ZY^{n-3} + Z^2 Y^{n-4} - \cdots - Z^{n-2})].
\end{aligned}
$$

But

$$
\begin{aligned}
\frac{H_1(X, Z)}{Z + X} &= 1, \\
\frac{H_3(X, Z)}{Z + X} &= X^2 + Z^2 = H_1^{(2)}(X, Z), \\
\frac{H_5(X, Z)}{Z + X} &= X^4 + X^2 Z^2 + Z^4 = H_2^{(2)}(X, Z),
\end{aligned}
$$

etc. Hence

$$Q_3 = P^{n-3} + H_1 P^{n-4} + H_2 P^{n-5} + \cdots + H_{n-3} + Y^{n-3}$$
$$+ H_1^{(2)}(X, Z)Y^{n-5} + H_2^{(2)}(X, Z)Y^{n-7} + \cdots + H_{(n-3)/2}^{(2)}(X, Z)$$
$$+ \frac{1}{Z+X}[Y^{n-2} + H_2(X, Z)Y^{n-4} + H_4(X, Z)Y^{n-6} +$$
$$\cdots + H_{n-3}(X, Z)Y + X^{n-2} - (Y - Z)X^{n-3}$$
$$+ (Y^2 - YZ + Z^2)X^{n-4} - \cdots$$
$$- (Y^{n-2} - ZY^{n-3} + Z^2 Y^{n-4} - \cdots - Z^{n-2})].$$

The expression in the brackets is equal to

$$YX^{n-3} + YZX^{n-4} + (YZ^2 + Y^3)X^{n-5}$$
$$+ (YZ^3 + Y^3 Z)X^{n-6} + (YZ^4 + Y^3 Z^2 + Y^5)X^{n-7} + \cdots$$
$$+ (YZ^{n-3} + Y^3 Z^{n-5} + \cdots + Y^{n-4} Z^2 + Y^{n-2})$$
$$+ X^{n-2} - (Y - Z)X^{n-3} + (Y^2 - YZ + Z^2)X^{n-4} - \cdots$$
$$- (Y^{n-2} - ZY^{n-3} + Z^2 Y^{n-4} - \cdots - Z^{n-2})$$
$$= X^{n-2} + ZX^{n-3} + (Y^2 + Z^2)X^{n-4} + (Y^2 Z + Z^3)X^{n-5}$$
$$+ (Y^4 + Y^2 Z^2 + Z^4)X^{n-6} + (Y^4 Z + Y^2 Z^3 + Z^5)X^{n-7}$$
$$+ \cdots + (Y^{n-3}Z + Y^{n-5}Z^3 + \cdots + Z^{n-2})$$
$$= X^{n-2} + ZX^{n-3} + H_1^{(2)}(Y, Z)X^{n-4} + H_1^{(2)}(Y, Z)ZX^{n-5}$$
$$+ H_2^{(2)}(Y, Z)X^{n-6} + H_2^{(2)}(Y, Z)ZX^{n-7} + \cdots + H_{(n-3)/2}^{(2)}(Y, Z)Z.$$

Hence

$$Q_3 = P^{n-3} + H_1 P^{n-4} + H_2 P^{n-5} + \cdots + H_{n-3} + Y^{n-3} +$$
$$H_1^{(2)}(X, Z)Y^{n-5} + H_2^{(2)}(X, Z)Y^{n-7} + \cdots + H_{(n-3)/2}^{(2)}(X, Z)$$
$$+ X^{n-3} + H_1^{(2)}(Y, Z)X^{n-5} + \cdots + H_{(n-3)/2}^{(2)}(Y, Z)$$
$$= P^{n-3} + H_1 P^{n-4} + H_2 P^{n-5} + \cdots + H_{n-3} + 2H_{(n-3)/2}^{(2)},$$

since each of the above last two lines is equal to $H_{(n-3)/2}^{(2)}$. $\square$

## Bibliography

1782 Waring, E., *Meditationes Algebraicae* (3rd ed.), Cambridge Univiversity Press, Cambridge, 1782.

1837 Kummer, E.E., *De aequatione $x^{2\lambda} + y^{2\lambda} = z^{2\lambda}$ per numeros integros resolvenda*, J. Reine Angew. Math., **17** (1837), 203–209; reprinted in *Collected Papers*, Vol. I, pp. 135–141, Springer-Verlag, Berlin, 1975.

1840 Lamé, G., *Mémoire d'analyse indéterminée démontrant que l'équation $x^7 + y^7 = z^7$ est impossible en nombres entiers*, J. Math. Pures Appl., **5** (1840), 195–211.

1847 Lamé, G., *Mémoire sur la résolution en nombres complexes de l'équation $A^n + B^n + C^n = 0$*, J. Math. Pures Appl., **12** (1847), 172–184.

1847 Lebesgue, V.A., *Sur la question $70^e$*, Nouv. Ann. Math., **6** (1847), 427–431.

1847 Mention, J., *Solution de la question 70*, Nouv. Ann. Math., **6** (1847), 399–400.

1861 Vachette, A., *Note sur certains développements et solution des questions 461, 468, et 479*, Nouv. Ann. Math., **20** (1861), 155–174.

1863 Gauss, C.F., *Zur Theorie der complexen Zahlen. (I) Neue Theorie der Zerlegung der Cuben*, Werke, Vol. II, pp. 387–391.

1877 Rebout, E., *Formation d'un cube entier qui soit égal à la somme de quatre cubes entiers*, Nouv. Ann. Math., (2), **16** (1877), 272–273.

1878 Brocard, H., *Sur divers articles de la Nouvelle Correspondance (Question 286)*, Nouv. Corr. Math., **4** (1878), 136–138.

1885 Serret, J.A., *Algèbre Supérieure*, Vol. I, p. 449 ($5^e$ édition), Gauthier-Villars, Paris, 1885.

1885 Catalan, E., *Sur le théorème de Fermat (1861)*, Mélanges Mathématiques XLVII, Mém. Soc. Roy. Sci. Liège, (2), **12** (1885), 179–187.

1886 Catalan, E., *Sur le dernier théorème de Fermat*, Mém. Soc. Roy. Sci. Liège, (2), **13** (1868), 387–397.

1906 Barisien, E.N., *Questions 3076 et 3077*, L'Interm. Math., **13** (1906), 142.

1907 Bini, U., *Sopra alcune congruenze*, Period. Mat., (3), **22** (1907), 180–183.

1907 Boutin, A. and Gonzalez Quijano, P.M., *Sur la question 3076 (de Barisien)*, L'Interm. Math., **14** (1907), 22–23.

1907 Rose, J., *Sur la question 3076* (*de Barisien*), L'Interm. Math., **14** (1907), 92–93.

1908 Werebrusow, A.S., *Question 3406*, L'Interm. Math., **15** (1908), 12.

1909 Dubouis, E., *Sur la question 3406* (*de Werebrusow*), L'Interm. Math., **16** (1909), 79–80.

1910 Bachmann, P., *Niedere Zahlentheorie*, Teubner, Leipzig, 1910; reprinted by Chelsea, New York, 1966.

1951 Perron, O., *Algebra*, Vol. I, W. de Gruyter, Berlin, 1951.

1959 Rédei, L., *Algebra*, Vol. I, Akademie Verlag, Geest & Portig, Leipzig, 1959.

## VII.2. The Cauchy Polynomials

In his proof of Fermat's theorem for the exponent 7, Lamé (1839, 1840) made use of a polynomial identity of degree 7. In the analysis of Lamé's paper, Cauchy and Liouville indicated a more general polynomial identity (1839), of which the following ones are special cases (Cauchy, 1841):

$$(X + Y)^5 - X^5 - Y^5 = 5XY(X + Y)(X^2 + XY + Y^2),$$
$$(X + Y)^7 - X^7 - Y^7 = 7XY(X + Y)(X^2 + XY + Y^2)^2.$$

In this way the study of the polynomial $(X + Y)^n - X^n - Y^n$ was initiated.

If $n \geq 3$, $n$ odd, then this polynomial is a multiple of $X, Y, X + Y$; moreover, if $n = p$ is an odd prime, then it is also a multiple of $p$.

**(2A)**    Let $n \equiv \pm 1 \pmod 6$. *The exact power of* $X^2 + XY + Y^2$ *dividing* $(X + Y)^n - (X^n + Y^n)$ *has exponent*

$$e = \begin{cases} 1 & \text{when} \quad n \equiv -1 \pmod 6, \\ 2 & \text{when} \quad n \equiv 1 \pmod 6. \end{cases}$$

PROOF. We shall show that $(X + 1)^n - (X^n + 1) = (X^2 + X + 1)^e H_n(X)$ where $e$ is as indicated, $H_n(X) \in \mathbb{Z}[X]$ and $X^2 + X + 1$ does not divide $H_n(X)$. By homogenization, we deduce the statement (2A).

Let $G_n(X) = (X+1)^n - (X^n+1)$ and let $\omega = (-1 + \sqrt{-3})/2$ be a primitive cubic root of 1, hence $\omega^2 + \omega + 1 = 0$. Then $G_n(\omega) = (\omega+1)^n - (\omega^n+1) = -(\omega^{2n} + \omega^n + 1) = -(\omega^{3n} - 1)/(\omega^n - 1) = 0$.

So $G_n(X)$ is a multiple of the minimal polynomial $X^2 + X + 1$ of $\omega$. Explicitly, by division, $G_n(X) = F(X)(X^2 + X + 1) + (aX + b)$ where $F(X) \in \mathbb{Z}[X]$ and $a, b \in \mathbb{Z}$. So $0 = G_n(\omega) = F(\omega)(\omega^2 + \omega + 1) + a\omega + b = a\omega + b$, hence $a = 0$ (otherwise $\omega = -b/a \in \mathbb{Q}$, which is not true) and also $b = 0$.

Next, $(X^2 + X + 1)^2$ divides $G_n(X)$ if and only if $\omega$ is a double root of $G_n(X)$, that is, $\omega$ is a root of the derivative $G_n{}'(X) = n((X+1)^{n-1} - X^{n-1})$, i.e., $n[(\omega+1)^{n-1} - \omega^{n-1}] = 0$. Since $\omega + 1 = -\omega^2$, this is equivalent to $(\omega^{n-1} - 1)\omega^{n-1} = 0$, that is, $\omega^{n-1} = 1$. Finally, this holds if and only if 3 divides $n - 1$, so $n \equiv 1 \pmod 6$.

We show that $(X^2 + X + 1)^3$ does not divide $G_n(X)$. If it did, then $n \equiv 1 \pmod 6$ and moreover $X^2 + X + 1$ divides $G_n{}''(X) = n(n-1)((X+1)^{n-2} - X^{n-2})$. So $G_n{}''(\omega) = 0$, hence $(\omega+1)^{n-2} = \omega^{n-2}$ and $-\omega^{2(n-2)} = \omega^{n-2}$. Therefore $\omega^{n-2} = -1$ and $\omega^{2(n-2)} = 1$. This implies that 3 divides $2(n-2)$, so $n \equiv 2 \pmod 3$, hence $n \equiv -1 \pmod 6$, which is a contradiction.  □

We may therefore write: if $n \equiv \pm 1 \pmod 6$ then

(2.1)
$$(X+Y)^n - (X^n + Y^n) = XY(X+Y)(X^2 + XY + Y^2)^e E_n(X, Y),$$

and if $n = p > 3$ is a prime then

(2.2)
$$(X+Y)^p - (X^p + Y^p) = pXY(X+Y)(X^2 + XY + Y^2)^e C_p(X, Y),$$

where $E_p(X, Y) = pC_p(X, Y)$ and $e = 1$, or 2, according to $n \equiv -1$, or 1 $\pmod 6$.

There are numerous proofs of the above result (or variants of it) in the literature, to wit by Cayley (1878), Glaisher (1878, 1879), Muir (1878), Catalan (1884, 1885, 1886), Lucas (1888, 1891), Barisien (1906), Taupin and Retall (1907), Ursus and Grigorieff (1907), Candido (1907), and Brčić-Kostić (1952).

In 1878, Glaisher expressed the above result in the following form:

(2B)   *If $n$ is odd then $(X - Y)^n + (Y - Z)^n + (Z - X)^n$ is divisible* by

$$\tfrac{1}{3}[(X - Y)^3 + (Y - Z)^3 + (Z - X)^3].$$

If $n \equiv -1 \pmod 6$ then the above polynomial is also divisible by

$$\tfrac{1}{2}[(X - Y)^2 + (Y - Z)^2 + (Z - X)^2].$$

If $n \equiv 1 \pmod 6$, then it is divisible by

$$\tfrac{1}{2}[(X - Y)^4 + (Y - Z)^4 + (Z - X)^4].$$

PROOF. Let $A, B$ be indeterminates and let $n \geq 3$ be odd. Then $AB(A + B)$ divides $(A + B)^n - A^n - B^n$.

Let $A = Z - Y$, $B = X - Z$ so $A + B = X - Y$. Then $AB(A+B) = (Z - Y)(X - Z)(X - Y) = \tfrac{1}{3}[(Y - Z)^3 + (Z - X)^3 + (X - Y)^3]$ divides $(X - Y)^n + (Y - Z)^n + (Z - X)^n$. Next,

$$\begin{aligned} A^2 + AB + B^2 &= (Z - Y)^2 + (Z - Y)(X - Z) + (X - Z)^2 \\ &= \tfrac{1}{2}[(Y - Z)^2 + (Z - X)^2 + (X - Y)^2] \end{aligned}$$

and

$$\begin{aligned} (A^2 + AB + B^2)^2 &= \tfrac{1}{4}[(Y - Z)^2 + (Z - X)^2 + (X - Y)^2]^2 \\ &= \tfrac{1}{2}[(Y - Z)^4 + (Z - X)^4 + (X - Y)^4]. \end{aligned}$$

Hence by (2A) if $n \equiv -1 \pmod 6$ then $\tfrac{1}{2}[(X - Y)^2 + (Y - Z)^2 + (Z - X)^2]$ divides the given polynomial while if $n \equiv 1 \pmod 6$ then $\tfrac{1}{2}[(X - Y)^4 + (Y - Z)^4 + (Z - X)^4]$ divides the given polynomial. □

The following special result was explicitly given by Catalan (1884, 1885), Gérono (1885), Nester (1907), Welsch (1909), and Brocard (1910):

**(2C)**   Let $p$ be a prime number.

   (1) If $(X+Y)^p - X^p - Y^p = pXY(X+Y)P^2$, where $P \in \mathbb{Z}[X, Y]$ then $p = 3$, $P = 1$, or $p = 7$, $P = X^2 + XY + Y^2$.
   (2) If $2^{p-1} - 1 = pN^2$ where $N$ is an integer then $p = 3$, $N = 1$ or $p = 7$, $N = 3$.

PROOF. We establish the two statements simultaneously. First we observe that $p \neq 2$. Taking $X = Y = 1$ we have $2^p - 2 = 2pN^2$, where $N = P(1,1) \in \mathbb{Z}$.

If $p = 3$ then $N = 1$. We assume now $p \neq 3$. We have $(2^{(p-1)/2} + 1)(2^{(p-1)/2} - 1) = pN^2$. The two factors in the left-hand side are odd, hence they are relatively prime. So, one of the factors is a square, while the other one is $p$ times a square.

But $2^{(p-1)/2} - 1 \equiv 3 \pmod 4$ since $(p-1)/2 \geq 2$, so $2^{(p-1)/2} - 1$ is not a square. Hence $2^{(p-1)/2} + 1 = M^2$, $M$ being an integer. Since $2^{(p-1)/2} = M^2 - 1 = (M-1)(M+1)$, it follows that $M - 1$, $M + 1$ are powers of 2. But $(M+1) - (M-1) = 2$, so $M - 1 = 2$, $M = 3$ and necessarily $p = 7$, $N = 3$. Therefore

$$P^2 = \frac{(X+Y)^7 - X^7 - Y^7}{7XY(X+Y)} = (X^2 + XY + Y^2)^2. \quad \square$$

The polynomial $C_p(X,1)$ will be simply denoted by $C_p(X)$ and called the *Cauchy polynomial* for the prime $p \geq 5$. If $p = 6k \pm 1$ then $C_p(X)$ has degree $6(k-1)$. We note the following special cases:

$$C_5(X) = 1,$$
$$C_7(X) = 1,$$
$$C_{11}(X) = X^6 + 3X^5 + 7X^4 + 9X^3 + 7X^2 + 3X + 1$$
$$= (X^2 + XY + Y^2)^3 + [XY(X+Y)]^2,$$
$$C_{13}(X) = X^6 + 3X^5 + 8X^4 + 11X^3 + 8X^2 + 3X + 1$$
$$= (X^2 + XY + Y^2)^3 + 2[XY(X+Y)]^2.$$

These expressions in terms of $XY(X+Y)$ and $X^2 + XY + Y^2$ will soon be generalized for arbitrary values of $p$.

The Cauchy polynomials satisfy the following properties, where $p = 6k \pm 1$ (see Mirimanoff (1903), Klösgen (1970)):

**(2D)**

(1) $C_p(X) = X^{6(k-1)}C_p(1/X)$,
    $C_p(X) = C_p(-1 - X)$.
(2) $C_p(0) = C_p(-1) = 1$.
(3) $C_p(X)$ *has no real roots.*
(4) *All the (imaginary) roots of $C_p(X)$ are simple and belong to*

$k - 1$ *disjoint sets, each composed of six distinct roots*

$$\left\{ z, \frac{1}{z}, -(1+z), -\frac{1}{1+z}, -\frac{z}{1+z}, -\frac{1+z}{z} \right\}.$$

PROOF. (1) Since

$$C_p(X) = \frac{(X+1)^p - X^p - 1}{pX(X+1)(X^2+X+1)^e}$$

(with $e = 1$ or 2) then $C_p(1/X) = X^{-p+2e+3}C_p(X)$. From

$$e = \begin{cases} 1 & \text{when} \quad p = 6k-1, \\ 2 & \text{when} \quad p = 6k+1, \end{cases}$$

it follows that $C_p(X) = X^{6(k-1)}C_p(1/X)$. Similarly $C_p(X) = C_p(-1-X)$.

(2)    Since $C_p(X)$ is a monic and symmetric polynomial then $C_p(0) = 1$. Also by (1), $C_p(-1) = C_p(0) = 1$.

(3)    If $z > 0$ is a real root of $C_p(X)$ then $(z+1)^p = z^p + 1$, which is impossible. If $z < -1$ is a real root of $C_p(X)$ then $-(1+z) > 0$ would be a positive real root of $C_p(-1-X) = C_p(X)$, which is a contradiction. Similarly, if $-1 < z < 0$ is a real root of $C_p(X)$ then $1/z < -1$ would be a real root of $C_p(X)$, which is impossible.

(4)    Let $z$ be any root of $C_p(X)$. By (1) it follows that $1/z$ and $-(1+z)$ are roots of $C_p(X)$, hence $-(1+1/z) = -(z+1)/z$, $-1/(1+z)$ and $-z/(z+1)$ are also roots of $C_p(X)$. The sets of roots considered above are either equal or disjoint. Indeed, if

$$t \in M_z = \left\{ z, \frac{1}{z}, -(1+z), -\frac{1}{1+z}, -\frac{z}{1+z}, -\frac{1+z}{z} \right\}$$

it is easy to verify that $M_t = M_z$.

Let $z$ be an imaginary root of $C_p(X)$, and suppose it is a double root; this happens if and only if $z$ is a root of $C_p'$. Since

$$C_p'(X) = 6(k-1)X^{6(k-1)-1}C_p\left(\frac{1}{X}\right) - X^{6(k-1)-2}C_p'\left(\frac{1}{X}\right)$$

and $C_p'(X) = -C_p'(-1-X)$, it follows that $1/z$ and $-1-z$ are also double roots, hence each element of the set $M_z$ is also a double root.

From $(X+1)^p - X^p - 1 = pX(X+1)(X^2+X+1)^e C_p(X)$, taking derivatives we have

$$(X+1)^{p-1} - X^{p-1} = (X(X+1)(X^2+X+1)^e)' C_p(X)$$
$$+ X(X+1)(X^2+X+1)^e C_p'(X).$$

Hence $(z+1)^{p-1} = z^{p-1}$, so $(1+1/z)^{p-1} = 1$, that is, $1 + 1/z$ is a $(p-1)$th root of 1, so $|1 + 1/z| = 1$.

Beginning with the double root $-z/(1+z)$ (instead of $z$), we deduce that $1 + (-(1+z)/z) = -1/z$ is also a $(p-1)$th root of 1, and $|1/z| = 1$. Thus, the triangle with vertices $0, 1, \xi = -1/z$ has sides of length 1, so it is equilateral, therefore $\xi$ is a primitive sixth root of 1. So $\xi^2 - \xi + 1 = 0$, and $\xi^3 = -1$. Then $-(z+1)/z = -1 - 1/z = -1 + \xi$ satisfies

$$(-1+\xi)^3 = -1 + 3\xi - 3\xi^2 + \xi^3 = -1 + 3\xi - 3\xi + 3 - 1 = 1,$$

so its minimal polynomial is $X^2 + X + 1$. Since $-(z+1)/z$ is also a root of $C_p(X)$ then $X^2 + X + 1$ divides $C_p(X)$, contradicting (2A). $\quad\square$

Mirimanoff also investigated in 1903 whether each polynomial $C_p(X)$ is irreducible, and he conjectured this to be true. Klösgen (1970) has verified with a computer that if $p \leq 31$, the Cauchy polynomial $C_p(X)$ is an irreducible polynomial of $\mathbb{Z}[X]$. Since the maximum of the coefficients of $C_p(X)$ grows very fast with $p$, the test was not continued any further.

We report without proof the following results, due to Helou, 1997:

If $n \geq 3$, $n$ odd, then for every prime $p$, the polynomial $C_n$ modulo $p$ is reducible over the field with $p$ elements. Moreover, if $n$ is an odd prime and for some prime $p$, $C_n$ modulo $p$ has at most three factors, then $C_n$ is irreducible. Helou attributes to Filaseta the proof that $C_{2p}$ is irreducible for every odd prime $p$; a proof is given by Helou in his paper (1997). More results about the Cauchy polynomials are in the paper by Terjanian (1989).

We shall need another expression for the Cauchy polynomials. More generally, we shall express $S_n(X, Y) = (X+Y)^n + (-1)^n(X^n + Y^n)$ as a polynomial in $U = X^2 + XY + Y^2$ and $V = XY(X+Y)$. There are two methods to achieve this result.

We may consider the polynomial of degree 3 in the indeterminate $T$, with coefficients in $\mathbb{Z}[X, Y]$, whose roots are $X + Y, -X, -Y$:

$$F(T) = (T + X)(T + Y)(T - X - Y)$$
$$= T^3 + a_1 T^2 + a_2 T + a_3,$$

with

$$\begin{cases} a_1 = X + Y - (X + Y) = 0, \\ a_2 = XY - X(X + Y) - Y(X + Y) = -(X^2 + XY + Y^2), \\ a_3 = -XY(X + Y). \end{cases}$$

Then $S_n(X, Y) = (X + Y)^n + (-1)^n(X^n + Y^n)$ is the sum of the $n$th powers of the roots of $F(T)$. We shall use the following classical expression, due to Waring (1782), for the sum

$$p_n = x_1^n + \cdots + x_k^n \qquad (n = 0, 1, \dots)$$

of the roots of a polynomial

$$T^k + a_1 T^{k-1} + \cdots + a_k = \prod_{i=1}^k (T - x_i).$$

The special cases $n = 1, 2, 3, 4$ were known to Girard (1629); see also Saalschütz (1906). The proof may be found in Serret (1885), Lucas (1891), Perron (1951). A modern algebraic proof was given by Rédei (1952, 1959). Another proof, using power series, is given by Cesàro and Kowalewski (1904).

LEMMA 2.1. *With above notations:*

$$p_n = n \prod_{i=1}^n \frac{(-1)^i}{i} \{(a_1 + \cdots + a_k)^i\}_n,$$

*where $\{(a_1 + \cdots + a_k)^i\}_n$ is the sum of all monomials $(i!/(i_1! \cdots i_k!)) \times a_1^{i_1} \cdots a_k^{i_k}$ (with $i_1 + \cdots + i_k = i$, $0 \le i_1, \dots, i_k$) having weight equal to $n$, that is, $i_1 + 2i_2 + \cdots + ki_k = n$.*

The next result was proved for even exponents by Ferrers and Jackson (1852; see reference in Dickson's *History of the Theory of Numbers*, Vol. II, p. 747); the proof for arbitrary exponent appears in Todhunter's book (1861). It was proved again by Muir (1879), using the above lemma. Kapferer (1949) rediscovered it, with the same proof, when $n$ is a prime. In 1969, Carlitz and Hunter gave a proof with the method of power series.

**(2E)**   *Let $n \geq 2$, let $U = X^2 + XY + Y^2$, and let $V = XY(X+Y)$. Then*

$$(X+Y)^n + (-X)^n + (-Y)^n$$

$$= \sum_r \frac{n}{n-3-2r} \binom{r}{n-3-2r} U^{3r-n+3} V^{n-2r-2}$$

*(summation for $\max\{0, (n-3)/3\} \leq r \leq (n-2)/2$, with the convention that if $n$ is even and $r = (n-2)/2$ then the coefficient of the summand is equal to 2).*

PROOF. FIRST PROOF (MUIR). By the above lemma

$$(X+Y)^n + (-X)^n + (-Y)^n$$

$$= \sum_{i=1}^{n} \frac{(-1)^i n}{i} \{(a_2 + a_3)^i\}_n$$

$$= n \sum_{i=1}^{n} \frac{(-1)^i}{i} \sum_{\substack{i_2+i_3=1 \\ 2i_2+3i_3=n}} \frac{i!}{i_2! \, i_3!} a_2^{i_2} a_3^{i_3}$$

$$= n \sum_{i=1}^{n} (-1)^i \sum_{\substack{i_2+i_3=1 \\ 2i_2+3i_3=n}} \frac{(i_2+i_3-1)!}{i_2! \, i_3!} a_2^{i_2} a_3^{i_3}$$

$$= n \sum \frac{(i_2+i_3-1)!}{i_2! \, i_3!} (-a_2)^{i_2} (-a_3)^{i_3}$$

(sum for all $i_2, i_3$, such that $0 \leq i_2, i_3$; $1 \leq i_2+i_3$; $2i_2+3i_3 = n$). But $-a_2 = U$, $-a_3 = V$ and

$$\frac{(i_2+i_3-1)!}{i_2! \, i_3!} = \frac{1}{i_3} \binom{i_2+i_3-1}{i_3-1}$$

(valid when $i_2 \neq 0$ or $i_3 \neq 1$). Let $i_2 + i_3 - 1 = r$, hence $n - 2 - 2r = (2i_2 + 3i_3) - 2(i_2 + i_3 - 1) - 2 = i_3$. Therefore $i_2 = 3r - n + 3$ and

$$(X+Y)^n + (-X)^n + (-Y)^n$$

$$= \sum_r \frac{n}{n-2-2r} \binom{r}{n-3-2r} U^{3r-n+3} V^{n-2r-2},$$

the sum being extended for $\max\{0, (n-3)/3\} \leq r \leq (n-2)/2$. Moreover, if $n$ is even and $r = (n-2)/2$ then the corresponding summand has coefficient 2 (since $i_2 = n/2$, $i_3 = 0$).   □

SECOND PROOF (CARLITZ & HUNTER). This proof does not require Waring's lemma. We consider the identity

$$\frac{Z}{1-ZW} + \frac{X}{1-XW} + \frac{Y}{1-YW}$$
$$= \frac{(X+Y+Z) - 2(XY+YZ+ZX)W + 3XYZW^2}{1 - (X+Y+Z)W + (XY+YZ+ZX)W^2 - XYZW^3}.$$

Let $Z = -(X+Y)$, so $XY + YZ + ZX = -U$, $XYZ = -V$ and changing signs:

(2.3)
$$\frac{(X+Y)}{1+(X+Y)W} - \frac{X}{1-XW} - \frac{Y}{1-YW} = \frac{-2UW + 3VW^2}{1 - UW^2 + VW^3}.$$

The formal power series expansion of $1/(1 - UW^2 + VW^3)$ is

$$\frac{1}{1 - UW + VW^3} = \frac{1}{1 - W^2(U - VW)}$$
$$= \sum_{r=0}^{\infty} W^{2r}(U - VW)^r$$
$$= \sum_{r=0}^{\infty} W^{2r} \left[ \sum_{s=0}^{r} (-1)^s \binom{r}{s} U^{r-s} V^s W^s \right]$$
$$= \sum_{n=0}^{\infty} (-1)^n W^n \sum_{r=0}^{\infty} \binom{r}{n-2r} U^{3r-n} V^{n-2r}$$

(this last sum is actually for $n/3 \le r \le n/2$; by convention the other summands are zero). The left-hand side of (2.3) is equal to

$$\sum_{n=0}^{\infty} [X^{n+1} + Y^{n+1} - (-1)^n (X+Y)^{n+1}] W^n.$$

The right-hand side of (2.3) is equal to

$$\sum_{n=0}^{\infty} (-1)^{n+1} W^{n+1} \sum_{r=0}^{\infty} 2 \binom{r}{n-2r} U^{3r+1-n} V^{n-2r}$$
$$+ \sum_{n=0}^{\infty} (-1)^{n+2} W^{n+2} \sum_{r=0}^{\infty} 3 \binom{r}{n-2r} U^{3r-n} V^{n+1-2r}$$

$$= \sum_{n=0}^{\infty}(-1)^{n+1}W^{n+1}\sum_{r=0}^{\infty}2\binom{r}{n-2r}U^{3r+1-n}V^{n-2r}$$

$$+ \sum_{n=1}^{\infty}(-1)^{n+1}W^{n+1}\sum_{r=0}^{\infty}3\binom{r}{n-1-2r}U^{3r+1-n}V^{n-2r}$$

$$= -2UW + 3VW^2 - 2U^2W^3$$

$$+ \sum_{n=3}^{\infty}(-1)^{n+1}W^{n+1}\sum_{r=0}^{\infty}\frac{n+1}{n-2r}\binom{r}{n-1-2r}U^{3r+1-n}V^{n-2r}$$

$$= -2UW + 3VW^2 - 2U^2W^3$$

$$+ \sum_{n=4}^{\infty}(-1)^{n}W^{n}\sum_{r=0}^{\infty}\frac{n+1}{n-1-2r}\binom{r}{n-2-2r}U^{3r+2-n}V^{n-1-2r},$$

because

$$2\binom{r}{n-2r} + 3\binom{r}{n-1-2r} = \frac{n+2}{n-2r}\binom{r}{n-1-2r}$$

(for $n \geq 3$). On the other hand, the left-hand side of (2.3) is equal to

$$\sum_{n=0}^{\infty}[(-1)^{n}(X+Y)^{n+1} - X^{n+1} - Y^{n+1}]W^{n}$$

$$= \sum_{n=2}^{\infty}[(X+Y)^{n} + (-X)^{n} + (-Y)^{n}](-1)^{n-1}W^{n-1}.$$

Comparing the two sides of (2.3), as computed above, yields the statement.  □


We note the following special cases. Taking an odd exponent,

$$(2.4) \ (X+Y)^{2k+1} - X^{2k+1} - Y^{2k+1}$$

$$= \sum_{\frac{2}{3}(k-1)\leq r\leq k-1}\frac{2k+1}{2k-1-2r}\binom{r}{2k-2-2r}U^{3r-2k+2}V^{2k-2r-1}.$$

If $p$ is a prime, $p \neq 2, 3$, letting $p = 6k \pm 1$, we have (see F. Lucas,

1897, and Kapferer, 1949):

$$(2.5) \qquad (X+Y)^p - X^p - Y^p = pXY(X+Y)(X^2+XY+Y^2)^2$$

$$\times \sum_{i=0}^{(p-7)/6} \binom{\dfrac{p-3}{2}-i}{2i} \frac{1}{2i+1} U^{(p-7)/2-3i}V^{2i},$$

when $p \equiv 1 \pmod 6$, and

$$(2.6) \qquad (X+Y)^p - X^p - Y^p = pXY(X+Y)(X^2+XY+Y^2)$$

$$\times \sum_{i=0}^{(p-5)/6} \binom{\dfrac{p-3}{2}-i}{2i} \frac{1}{2i+1} U^{(p-5)/2-3i}V^{2i},$$

when $p \equiv -1 \pmod 6$. If the exponent is even, we have

$$(2.7) \ (X+Y)^{2k} + X^{2k} + Y^{2k}$$

$$= \sum_{(2k-3)/3 \le r \le k-1} \frac{k}{k-1-r} \binom{r}{2k-3-2r} U^{3r-2k+3}V^{2k-2r-2}.$$

In 1879, elaborating on previous papers of Glaisher (1878, 1879), Muir found the following recurrence relation and divisibility properties of the polynomials $S_n(X,Y)$:

**(2F)**

(1) $VS_n(X,Y) + US_{n+1}(X,Y) = S_{n+3}(X,Y)$ *for* $n \ge 0$.
(2) $6S_n(X,Y) = 3S_2(X,Y)S_{n-2}(X,Y) + 2S_3(X,Y)S_{n-3}(X,Y)$
  *for* $n \ge 3$.
(3) *If* $n \equiv 0 \pmod 6$ *then* $U,V \nmid S_n(X,Y)$.
  *If* $n \equiv 1 \pmod 6$ *then* $U^2V \mid S_n(X,Y)$.
  *If* $n \equiv 2 \pmod 6$ *then* $U \mid S_n(X,Y)$.
  *If* $n \equiv 3 \pmod 6$ *then* $V \mid S_n(X,Y)$.
  *If* $n \equiv 4 \pmod 6$ *then* $U^2 \mid S_n(X,Y)$.
  *If* $n \equiv 5 \pmod 6$ *then* $UV \mid S_n(X,Y)$.

PROOF. (1)   Let $n = 2m$. By (2.4) and (2.7) and noting that

$$\frac{2m}{2m-2r} \binom{r-1}{2m-1-2r} + \frac{2m+1}{2m+1-2r} \binom{r-1}{2m-2r}$$

$$= \frac{2m+3}{2m+1-2r} \binom{r}{2m-2r}$$

it follows that

$$VS_{2m}(X,Y) + US_{2m+1}(X,Y) = S_{2m+3}(X,Y).$$

Taking $n = 2m + 1$ and proceeding similarly, we deduce that

$$VS_{2m+1}(X,Y) + US_{2m+2}(X,Y) = S_{2m+4}(X,Y)$$

and this proves (1).

(2)   Noting that $S_2(X,Y) = 2U$ and $S_3(X,Y) = 3V$, it follows from (1) that

$$3S_2(X,Y)S_{n-2}(X,Y) + 2S_3(X,Y)S_{n-3}(X,Y)$$
$$= 6\left[VS_{n-3}(X,Y) + US_{n-2}(S,Y)\right] = 6S_n(X,Y).$$

(3)   This follows immediately from (2.4) and (2.7).   $\square$

Using an extension of Waring's formula, MacMahon indicated in 1884 some more algebraic identities of the same family.

## Bibliography

1629 Girard, A., *Invention nouvelle en l'algèbre*, Amsterdam, 1629; reprinted by B. de Haan, Leyden, 1884.

1782 Waring, E., *Meditationes Algebraicae* (3rd ed.), Cambridge University Press, Cambridge, 1782.

1839 Lamé, G., *Mémoire sur le dernier théorème de Fermat*, C. R. Acad. Sci. Paris, **9** (1839), 45–46.

1839 Cauchy, A. and Liouville, J., *Rapport sur un mémoire de M. Lamé relatif au dernier théorème de Fermat*, C. R. Acad. Sci. Paris, **9** (1839), 359–363; reprinted in J. Math. Pures Appl., **5** (1840), 211–215; also in *Oeuvres Complètes*, (1), Vol. IV, pp. 499–504, Gauthier-Villars, Paris, 1911.

1840 Lamé, G., *Mémoire d'analyse indéterminée démontrant que l'équation $x^7 + y^7 = z^7$ est impossible en nombres entiers*, J. Math. Pures Appl., **5** (1840), 195–211.

1841 Cauchy, A., *Exercices d'analyse et de physique mathématique*, **2** (1841), 137–144 (Notes sur quelques théorèmes d'algèbre); *Oeuvres Complètes*, (2), Vol. XII, pp. 157–166, Gauthier-Villars, Paris, 1916.

1848 Ferrers, N.M. and Jackson, J.S., *Solutions of the Cambridge Senate-House Problems for 1848–1851*, pp. 83–85.

1861 Todhunter, I., *Theory of Equations*, Macmillan, London, 1861, pp. 173–176.

1878 Cayley, A., *An algebraic identity*, Messenger Math., (2), **8** (1878), 45–46.

1878 Glaisher, J.W.L., *Note on the above paper of Cayley*, Messenger Math., (2), **8** (1878), 46–47.

1878 Glaisher, J.W.L., *On a class of algebraic identities*, Messenger Math., (2), **8** (1878), 53–56.

1878 Glaisher, J.W.L., *Note on Cayley's theorem*, Messenger Math., (2), **8** (1878), 121.

1878 Glaisher, J.W.L., *Note on Cauchy's theorem relating to the factors of $(x + y)^n - x^n - y^n$*, Quart. J. Pure Appl. Math., **15** (1878), 365–366.

1878 Muir, T., *Cauchy's theorem regarding the divisibility of $(x + y)^n + (-x)^n + (-y)^n$*, Messenger Math., (2), **8** (1878), 119–120.

1879 Glaisher, J.W.L., *On Cauchy's theorem relating to the factors of $(x+y)^n - x^n - y^n$*, Quart. J. Pure Appl. Math., **16** (1879), 89–98.

1879 Muir, T., *On an expansion of $(x+y)^n + (-x)^n + (-y)^n$*, Quart. J. Pure Appl. Math., **16** (1879), 9–14.

1884 Catalan, E., *Question 1489*, Nouv. Ann. Math., (3), **3** (1884), 351.

1884 MacMahon, P.A., *Algebraic identities arising out of an extension of Waring's formula*, Messenger Math., **14** (1884), 8–11.

1885 Catalan, E., *Solution de la question 1489*, Nouv. Ann. Math., (3), **4** (1885), 520–524.

1885 Gérono, G.C., *Note du rédacteur sur l'article de Catalan*, Nouv. Ann. Math., (3), **4** (1885), 523–524.

1885 Serret, J.A., *Algèbre Supérieure*, Vol. I, p. 449 (5$^e$ édition), Gauthier-Villars, Paris, 1885.

1886 Catalan, E., *Sur le dernier théorème de Fermat*, Mém. Soc. Roy. Sci. Liège, (2), **13** (1886), 387–397.

1886 Matthews, G.B., *A note in connexion with Fermat's last theorem*, Messenger Math., **15** (1886), 68–74.

1888 Lucas, E., *Sur un théorème de Cauchy*, Assoc. Française Avanc. Sci., Oran, 1888, pp. 29–31.

1891 Lucas, E., *Théorie des Nombres*, reprinted by A. Blanchard,

Paris, 1961.

1897 Lucas, F., *Note relative à la théorie des nombres*, Bull. Soc. Math. France, **25** (1897), 33–35.

1903 Mirimanoff, D., *Sur l'équation* $(x + 1)^l - x^l - 1 = 0$, Nouv. Ann. Math., (4), **3** (1903), 385–397.

1904 Cesàro, E. and Kowalewski, S., *Elementares Lehrbuch der algebraischen Analysis und der Infinitesimalrechnung*, Teubner, Leipzig, 1904.

1906 Barisien, E.N., *Questions 3076 et 3077*, L'Interm. Math., **13** (1906), page 142.

1906 Saalschütz, L., *Albert Girard und die Waringsche Formel*, Arch. Math. Phys., (3), **12** (1906), 205–207.

1907 Nester, *Question 3230*, L'Interm. Math., **14** (1907), 126.

1907 Candido, G., *Sur la question 3230 (de Nester)*, L'Interm. Math., **14** (1907), 2.

1907 Taupin and Retall, V., *Sur la question 3077 (de Barisien)*, L'Interm. Math., **14** (1907), 36–39.

1907 Ursus and Grigorieff, E., *Sur la question 3077 (de Barisien)*, L'Interm. Math., **14** (1907), 93–95.

1909 Welsch, *Solution de la question 3230 (de Nester)*, L'Interm. Math., **16** (1909), 14–15.

1910 Brocard, H., *Sur la question 3230 (de Nester)*, L'Interm. Math., **17** (1910), 278–279.

1949 Kapferer, H., *Über ein Kriterium zur Fermatschen Vermutung*, Comment. Math. Helv., **23** (1949), 64–75.

1951 Perron, O., *Algebra*, Vol. I, W. de Gruyter, Berlin, 1951.

1952 Brčić-Kostić, M., *L'extension de la congruence* $(a+b)^n - a^n - b^n \equiv 0 \pmod{n}$ *(n un nombre premier)* (in Croatian, French summary), Hrvatsko Prirodoslovno Društvo, Glas. Mat. Ser. II, **7** (1952), 7–11.

1952 Rédei, L., *Kurzer Beweis der Waringschen Formel*, Acta Math. Acad. Sci. Hungar., **3** (1952), 151–153.

1959 Rédei, L., *Algebra*, Vol. I, Akademie Verlag, Geest & Portig, Leipzig, 1959.

1969 Carlitz, L. and Hunter, J.A.H., *Sums of powers of Fibonacci and Lucas numbers*, Fibonacci Quart., **7** (1969), 467–473.

1970 Klösgen, W., *Untersuchungen über Fermatsche Kongruenzen*, Gesellschaft Math. Datenverarbeitung, Bonn, Nr. 36, 1970, 124 pp.

1989 Terjanian, G., *Sur le loi de réciprocité des puissances l-èmes*, Acta Arith., **54** (1989), 87–125.

1997 Helou, C., *Cauchy–Mirimanoff polynomials*, C. R. Math. Rep. Acad. Sci. Canada, **19** (1997), 51–57.

# VIII
# Reformulations, Consequences, and Criteria

In this chapter we give a variety of results, a good indication of the wide search for solutions of Fermat's problem. There are reformulations into equivalent problems, also a number of consequences of the truth of Fermat's theorem as well as statements which follow from the assumption that the theorem is false for some exponent.

## VIII.1. Reformulation and Consequences of Fermat's Last Theorem

In this section, we shall indicate some propositions which may be proved if we assume the truth of Fermat's last theorem. Among these propositions, some imply, conversely, the truth of Fermat's last theorem.

### A. Diophantine Equations Related to Fermat's Equation.

There have been many instances where certain diophantine equations were compared to Fermat's equation. We describe in succession the results originated with Lebesgue, Christilles, Perrin, Hurwitz, and Kapferer. We also discuss briefly the equations of Frey which played a central role in the recent approach and solution of Fermat's problem by Wiles.

## A1.    Lebesgue

The first result of this kind in the literature is due to Lebesgue in 1840. It was proved again by Terquem in 1846 and rediscovered by Pocklington in 1913:

**(1A)**    *If Fermat's last theorem is true for the exponent $n \geq 3$ then the equation $X^{2n} + Y^{2n} = Z^2$ has only trivial solutions.*

PROOF. Suppose that $x, y, z$ are nonzero positive integers such that $x^{2n} + y^{2n} = z^2$. It is easily seen that we may assume without loss of generality that $x, y, z$ are pairwise relatively prime. Moreover, $x, y$ cannot be both odd, otherwise $z^2 \equiv 2 \pmod 4$, which is not true. So, for example, $x$ is even, $y$ is odd, hence $z$ is odd.

Let $x = 2^a x'$, with $a \geq 1$, $x'$ odd, $x' > 0$. Then $(z + y^n)(z - y^n) = z^2 - y^{2n} = x^{2n} = 2^{2an} x'^{2n}$. We note that $\gcd(z + y^n,\, z - y^n) = 2$, so

$$\begin{cases} z \pm y^n = 2r^{2n}, \\ z \mp y^n = 2^{2an-1} s^{2n}, \end{cases}$$

with $r, s$ odd, positive, $\gcd(r, s) = 1$.

Adding and subtracting, we obtain

$$\begin{cases} z = r^{2n} + 2^{2an-2} s^{2n}, \\ \pm y^n = r^{2n} - 2^{2an-2} s^{2n} = (r^n - 2^{an-1} s^n)(r^n + 2^{an-1} s^n). \end{cases}$$

Since $\gcd(r^n - 2^{an-1} s^n, r^n + 2^{an-1} s^n) = 1$ then

$$\begin{cases} r^n + 2^{an-1} s^n = t^n, & t > 0, \\ r^n - 2^{an-1} s^n = \pm u^n, & u \geq 0. \end{cases}$$

Therefore

$$t^n \mp u^n = 2^{an} s^n = (2^a s)^n.$$

By hypothesis, we must have $u = 0$, hence $r^n = 2^{an-1} s^n$ is odd, so $an = 1$, which is a contradiction.    $\square$

As a corollary, Liouville proved in 1840 (see also Terquem, 1846):

**(1B)**    *If Fermat's last theorem is true for the exponent $n \geq 3$ then the equation $X^{2n} - Y^{2n} = 2Z^n$ has only trivial solutions.*

PROOF. Suppose that $x, y, z$ are nonzero integers such that $x^{2n} - y^{2n} = 2z^n$. Let $t = y^{2n} + z^n$, so $t - z^n = y^{2n}$ and also $t + z^n = x^{2n}$. Hence $t^2 - z^{2n} = (xy)^{2n}$, so $z^{2n} + (xy)^{2n} = t^2$.

By (1A) we must have $t = 0$, and then necessarily $z = 0, xy = 0$, which is a contradiction.  $\square$

## A2.  Christilles

The next result was proved by Christilles in 1967. We begin with a lemma:

LEMMA 1.1.  *The equation* $X^3 + Y^3 + Z^3 = 3XYZ$ *has a solution in nonzero integers* $x, y, z$ *if and only if* $x + y + z = 0$ *or* $x = y = z$.

PROOF. We have the identity

$$(1.1) \quad X^3 + Y^3 + Z^3 - 3XYZ$$
$$= (X + Y + Z)(X^2 + Y^2 + Z^2 - XY - YZ - ZX).$$

If $x, y, z$ are nonzero integers then $x^3 + y^3 + z^3 = 3xyz$ if and only if $x + y + z = 0$ or $x^2 + y^2 + z^2 = xy + yz + zx$.

However, for any integers $x, y$ the equation $Z^2 - (x + y)Z + (x^2 + y^2 - xy) = 0$ has solutions $z = ((x + y) \pm (x - y)\sqrt{-3})/2$, which are integers exactly when $x = y$, and in this case $x = y = z$.  $\square$

**(1C)**   *Let* $n \geq 3$. *The following statements are equivalent:*

(1) *Fermat's last theorem is true for the exponent* $n$.
(2) *The only solutions in nonzero integers* $x, y, z$ *of* $X^{3n} + Y^{3n} + Z^{3n} = 3X^n Y^n Z^n$ *are* $x = y = z$.

PROOF. (1) $\to$ (2)  Assume that $x, y, z$ are nonzero integers which are not all equal and such that $x^{3n} + y^{3n} + z^{3n} = 3x^n y^n z^n$. By the lemma we have $x^n + y^n + z^n = 0$.

(2) $\to$ (1)  Suppose that $x, y, z$ are nonzero integers such that $x^n + y^n + z^n = 0$; then $x, y, z$ cannot be all equal and by the lemma, $x^{3n} + y^{3n} + z^{3n} = 3x^n y^n z^n$.  $\square$

Christilles has also indicated in the same paper the following sufficient condition for Fermat's theorem:

**(1D)**    Let $n \geq 3$. If $X^{5n} + Y^{5n} + Z^{5n} = 5X^n Y^n Z^n (Z^{2n} - X^n Y^n)$ has no solution in nonzero integers, then Fermat's last theorem is true for the exponent $n$.

PROOF. Suppose that $x, y, z$ are nonzero integers such that $x^n + y^n + z^n = 0$. Then

$$-z^{5n} = (x^n + y^n)^5 = x^{5n} + y^{5n} + 5x^n y^n (x^{3n} + y^{3n}) + 10x^{2n} y^{2n} (x^n + y^n).$$

So

$$x^{5n} + y^{5n} + z^{5n} = -5x^n y^n [x^{3n} + y^{3n} - 3x^n y^n z^n + x^n y^n z^n].$$

By (1.1) and the hypothesis $x^{3n} + y^{3n} + z^{3n} = 3x^n y^n z^n$, hence $x^{5n} + y^{5n} + z^{5n} = 5x^n y^n z^n [z^{2n} - x^n y^n]$ with $x, y, z \neq 0$, proving the result. □

A3.    Perrin

Perrin showed in 1885 the following fact concerning the cubic Fermat equation:

**(1E)**    The following statements are equivalent and true:

(1) Fermat's last theorem is true for the exponent 3.
(2) For every $n \geq 1$ the equation

(1.2)                    $X^3 + Y^3 + 3^{3n-1} Z^3 = 2 \times 3^n XYZ$

has no solution in nonzero integers $x, y, z$, not multiples of 3.

PROOF. (1) $\rightarrow$ (2)    Suppose that there exists an integer $n \geq 1$ and nonzero integers $x, y, z$, $3 \nmid xyz$, satisfying equation (1.2). Let

$$\begin{cases} a = x(3^n yz - x^2), \\ b = y(3^n xz - y^2), \\ c = 3^n z(xy - 3^{2n-1} z^2). \end{cases}$$

Since $x, y, z \neq 0$ and $3 \nmid xyz$ then $a, b, c \neq 0$. From (1.2) we have, taking the square and the cube

$$x^6 + y^6 + 3^{6n-2} z^6 + 2x^3 y^3 + 2 \times 3^{3n-1} x^3 z^3 + 2 \times 3^{3n-1} y^3 z^3 = 4 \times 3^{2n} x^2 y^2 z^2$$

and

$$x^9 + y^9 + 3^{9n-3}z^9 + 3x^6y^3 + 3x^3y^6 + 3^{3n}x^6z^3$$
$$+2 \times 3^{3n}x^3y^3z^3 + 3^{3n}y^6z^3 + 3^{6n-1}x^3z^6 + 3^{6n-1}y^3z^6$$
$$= 8 \times 3^{3n}x^3y^3z^3.$$

Next we have

$$a^3 = x^3(3^{3n}y^3z^3 - 3^{2n+1}x^2y^2z^2 + 3^{n+1}x^4yz - x^6),$$
$$b^3 = y^3(3^{3n}x^3z^3 - 3^{2n+1}x^2y^2z^2 + 3^{n+1}xy^4z - y^6),$$
$$c^3 = 3^{3n}z^3(x^3y^3 - 3^{2n}x^2y^2z^2 + 3^{4n-1}xyz^4 - 3^{6n-3}z^6).$$

Adding these equalities, taking into account (1.2) and the preceding relations, leads in a straightforward manner to

$$a^3 + b^3 + c^3 = 0.$$

$(2) \rightarrow (1)$   Suppose that there exist nonzero integers $a, b, c$ such that $a^3 + b^3 + c^3 = 0$; we may assume $a, b, c$ pairwise relatively prime. From the identity

$$(X + Y + Z)^3 = X^3 + Y^3 + Z^3 + 3(X + Y)(Y + Z)(Z + X)$$

it follows that

$$(a + b + c)^3 = 3(a + b)(b + c)(c + a).$$

But $a+b$, $b+c$, $c+a$ are also pairwise relatively prime; for example, if a prime $p$ divides $a + b$ and also $b + c$ then since $p$ divides $a + b + c$ it would divide $c$ and $a$, contrary to the hypothesis.

Hence, one of the factors, say $a + b$, is a multiple of 3, and necessarily there exist integers $n \geq 1$, $x, y, z$ such that

$$\begin{cases} a + b = 3^{3n-1}z^3, \\ b + c = x^3, \end{cases}$$

and $3^n xyz = a + b + c$. Hence

$$x^3 + y^3 + 3^{3n-1}z^3 = 2(a + b + c) = 2 \times 3^n xyz.$$

Finally, $x, y, z \neq 0$ (if, for example, $z = 0$ then $a = -b$ so $a = -b = \pm 1$ which implies $c = 0$) and also $3 \nmid xyz$ because $a + b$, $b + c$, $c + a$ are pairwise relatively prime.  $\square$

### A4.  Hurwitz

In 1908, Hurwitz considered the diophantine equation

(1.3)  $$X^m Y^n + Y^m Z^n + Z^m X^n = 0,$$

where $m > n > 0$ and $\gcd(m, n) = 1$, without loss of generality. He proved:

**(1F)**   *The above equation has only the trivial solution if and only if Fermat's theorem is true for the exponent* $m^2 - mn + n^2$.

PROOF. Let $x, y, z$ be nonzero integers such that $x^m y^n + y^m z^n + z^m x^n = 0$. We may assume $\gcd(x, y, z) = 1$.

Let $u = \gcd(y, z)$, $v = \gcd(z, x)$, $w = \gcd(x, y)$. Then $u, v, w$ are pairwise relatively prime. Hence $vw$ divides $x$ and we may write $x = x_1 vw$, $x_1$ an integer. Similarly $y = y_1 wu$, $z = z_1 uv$ with $y_1, z_1$ integers. Moreover, $x_1, y_1, z_1$ are pairwise relatively prime. Substituting in the equation and dividing by $u^n v^n w^n$ we have

$$x_1^m y_1^n v^{m-n} w^m + y_1^m z_1^n w^{m-n} u^m + z_1^m x_1^n u^{m-n} v^m = 0.$$

Thus $u^{m-n}$ divides $x_1^m y_1^n v^{m-n} w^n$. But $\gcd(u, v) = 1$, $\gcd(u, w) = 1$ and $\gcd(u, x) = 1$, hence $\gcd(u, x_1) = 1$. So $u^{m-n}$ divides $y_1^n$. On the other hand, $y_1^n$ divides $z_1^m x_1^n u^{m-n} v^m$. Since $y_1$ is relatively prime to $x_1, z_1$, and $v$ then $y_1^n$ divides $u^{m-n}$. Hence $y_1^n = \pm u^{m-n}$.

In the same way $x_1^n = \pm w^{m-n}$, $z_1^n = \pm v^{m-n}$. Since $\gcd(m, n) = 1$ then $\gcd(m, m - n) = 1$, hence $x_1 = w_1^{m-n}$, $y_1 = u_1^{m-n}$, $z_1 = v_1^{m-n}$ for some integers $u_1, v_1, w_1$ such that $u = \pm u_1^n$, $v = \pm v_1^n$, $w = \pm w_1^n$. It follows that

$$w_1^{(m-n)m} u_1^{(m-n)n} v_1^{(m-n)n} w_1^{mn} \pm u_1^{(m-n)m} v_1^{(m-n)n} w_1^{(m-n)n} u_1^{mn}$$
$$\pm v_1^{(m-n)m} w_1^{(m-n)n} u_1^{(m-n)n} v_1^{mn} = 0.$$

Multiplying with $u_1^{n^2 - mn} v_1^{n^2 - mn} w_1^{n^2 - mn}$, and noting that $m^2 - mn + n^2$ is odd, we obtain the equation

$$w_1^{m^2 - mn + n^2} + (\pm u_1)^{m^2 - mn + n^2} + (\pm v_1)^{m^2 - mn + n^2} = 0.$$

Conversely, if the nonzero integers $u, v, w$ satisfy the relation

$$u^{m^2 - mn + n^2} + v^{m^2 - mn + n^2} + w^{m^2 - mn + n^2} = 0,$$

multiplying with $u^{mn}v^{mn}w^{mn}$, we obtain

$$u^{m^2+n^2}v^{mn}w^{mn} + v^{m^2+n^2}u^{mn}w^{mn} + w^{m^2+n^2}u^{mn}v^{mn} = 0.$$

Letting

$$\begin{cases} x = v^n w^m, \\ y = w^n u^m, \\ z = u^n v^m, \end{cases}$$

we obtain $x^m y^n + y^m z^n + x^m x^n = 0$.  $\square$

As a special case, letting $m = 3$, $n = 1$, it follows that

$$X^3 Y + Y^3 Z + Z^3 Y = 0$$

has only the trivial solution.

### A5.   Kapferer

In 1933, Kapferer published a proof that Fermat's last theorem is true for the exponent $n \geq 3$ if and only if the equation

$$(1.4) \qquad\qquad Z^3 - Y^2 = 3^2 \times 2^{2n-2} X^{2n}$$

has no solution in nonzero pairwise relatively prime integers $x, y, z$.

The proof of Kapferer contained a flaw, partly corrected by Ribenboim (communicated to Inkeri). Inkeri found and proved (correctly) the results which we give below. Comments by Gandhi and Stuff (1975) on this matter were inaccurate.

We begin with the special case $n = 3$, which was given by Fueter (1930):

**(1G)**    *The equation*

$$(1.5) \qquad\qquad Z^3 - Y^2 = 3^3 \times 2^4 X^6$$

*has no solution in nonzero integers* $x, y, z$ *with* $\gcd(y, z) = 1$.

PROOF. Assume that $x, y, z$ are nonzero integers such that $\gcd(y, z) = 1$ and

$$z^3 - y^2 = 3^3 \times 2^4 x^6.$$

Let

$$\begin{cases} u = 6^2 x^3 + y, \\ v = 6^2 x^3 - y, \\ w = 6xz. \end{cases}$$

Then $u, v, w \in \mathbb{Z}$, $w \neq 0$. We show that $u \neq 0$, $v \neq 0$. If $u = 0$ then $6^2 x^3 = -y$, so $y \neq \pm 1$. If $p$ is a prime dividing $y$, then $p \mid 6x$. So $p \mid z$, which is contrary to the hypothesis that $\gcd(y, z) = 1$. Similarly, $v = 0$.

Finally, $u^3 + v^3 = (6^2 x^3 + y)^3 + (6^2 x^3 - y)^3 = 2 \times 6^6 x^p + 6^3 x^3 y^2 = 6^3 x^3 (2 \times 6^3 x^6 + y^2) = 6^3 x^3 z^3 = w^3$. This is, however, impossible. $\square$

It may be shown in the same way that there are no integers $x, y, z, \neq 0$ with $\gcd(y, z) = 1$ such that $z^3 - 3y^2 = 2^4 x^6$.

We give now Inkeri's equivalence, which is a corrected form of the one previously given by Kapferer (1933). It also involves the related equation

$$(1.6) \qquad\qquad Z^3 - 3Y^2 = 2^{2n-2} X^{2n}.$$

**(1H)**    *Let $n \geq 3$ be an odd integer. The following statements are equivalent:*

(1) *Fermat's last theorem is true for the exponent $n$.*
(2) *The equations (1.4) and (1.6) have no solution in nonzero integers $x, y, z$ with $\gcd(y, z) = 1$.*

PROOF. $(1) \Rightarrow (2)$    We assume that there exist nonzero integers $x, y, z$ such that $\gcd(y, z) = 1$ and $z^3 - y^2 = 3^3 \times 2^{2n-2} x^{2n}$. Let

$$\begin{cases} a = y, \\ b = 3 \times 2^{n-1} x^n, \\ c = z, \end{cases}$$

so $a^2 + 3b^2 = c^3$. Then $c$ is odd and $\gcd(a, b) = 1$ because $\gcd(a, c) = \gcd(y, z) = 1$.

By Chapter I, Lemma 4.7, there exist integers $r, s \neq 0$, $\gcd(r, 3s) = 1$, $r \not\equiv s \pmod{2}$, such that

$$\begin{cases} a = r(r^2 - 9s^2), \\ b = 3s(r^2 - s^2), \\ c = r^2 + 3s^2. \end{cases}$$

Then

$$2^{n-1} x^n = \frac{b}{3} = s(r^2 - s^2) = s(r - s)(r + s).$$

Since $x \neq 0$, then $s$, $r - s$, $r + s$ are nonzero. Also $s$ is even, because $r \not\equiv s \pmod 2$. Since $\gcd(r, s) = 1$ then $s$, $r - s$, $r + s$ are pairwise relatively prime. So there exist nonzero integers $u, v, w$ such that $2s = w^n$, $r - s = u^n$, $-r - s = v^n$. So $u^n + v^n + w^n = 0$. We note that $u, v, w \neq 0$ and $\gcd(u, v, w) = 1$.

Now we assume that there exist nonzero integers $x, y, z$ such that $\gcd(y, z) = 1$ and $z^3 - 3y^2 = 2^{2n-1}x^{2n}$. Let

$$\begin{cases} a = 2^{n-1}x^n, \\ b = y, \\ c = z, \end{cases}$$

so $a^2 + 3b^2 = c^3$. Again $c$ is odd and $\gcd(a, b) = 1$.

By Chapter I, Lemma 4.7, there exist nonzero integers $r, s$, $\gcd(r, 3s) = 1$, $r \not\equiv s \pmod 2$ such that

$$\begin{cases} a = r(r^2 - 9s^2), \\ b = 3s(r^2 - s^2), \\ c = r^2 + 3s^2. \end{cases}$$

Then

$$2^{n-1}x^n = a = r(r - 3s)(r + 3s).$$

We note that $r$ is necessarily even, since $r - 3s$, $r + 3s$ are odd. So $r$, $r - 3s$, $r + 3s$ are pairwise relatively prime. Hence there exist nonzero integers $u, v, w$ such that $2r = w^n$, $-r + 3s = u^n$, $-r - 3s = v^n$ and so $u^n + v^n + w^n = 0$, with $u, v, w \neq 0$, $\gcd(u, v, w) = 1$.

$(2) \Rightarrow (1)$   We assume that $u, v, w$ are nonzero integers such that $u^n + v^n + w^n = 0$. Without loss of generality, we may assume that $u, v, w$ are pairwise relatively prime, $w$ is even, and $u, v$ are odd. Let $s = w^n/2$ and $r = u^n + s$. Hence $2s = w^n$, $r - s = u^n$, $-r - s = v^n$. Let

$$\begin{cases} a = r(r^2 - 9s^2), \\ b = 3s(r^2 - s^2), \\ c = r^2 + 3s^2. \end{cases}$$

By substitution, we obtain $a^2 + 3b^2 = c^3$ and $b = 3s(r^2 - s^2) = -3 \times 2^{n-1}\left(\frac{1}{2}uvw\right)^n$. Let

$$\begin{cases} x = \frac{1}{2}uvw, \\ y = a, \\ z = c, \end{cases}$$

so $z^3 - y^2 = 3^3 \times 2^{2n-2}x^{2n}$. Since $\gcd(u,v) = 1$ then $\gcd(r,s) = 1$ and $x \neq 0$ by assumption. Now $w$ is even, then $s$ is even, so $y$ is odd, hence $y \neq 0$. Next, $3 \nmid uvw$ if and only if $3 \mid r$. Indeed, if $3 \nmid uvw$ then $3 \nmid s(r^2 - s^2)$, so necessarily $3 \mid r$, and conversely, if $3 \mid r$, from $\gcd(r,s) = 1$ then $3 \nmid s(r^2 - s^2)$, so $3 \nmid uvw$.

Now we show that $\gcd(x,y) = 1$ which will be done in several steps:

(a) Since $s$ is even and $\gcd(r,s) = 1$ then $r$ is odd. Then $\gcd(r^2 - s^2, r^2 - 9s^2) = 1$. Indeed, if $p \mid r^2 - s^2$, and $p \mid r^2 - 9s^2$ then $p \neq 2$ and $p \mid 8r^2$, $p \mid 8s^2$, so $p \mid \gcd(r,s) = 1$, which is impossible.

(b) $\gcd(r, r^2 - s^2) = \gcd(s, r^2 - 9s^2) = 1$ because $\gcd(r,s) = 1$.

(c) Then $\gcd(a, b/3) = 1$. Since $x \mid b/3$ and $y = a$ then $\gcd(x,y) = 1$. Since $y$ is odd, we also have $\gcd(2x, y) = 1$. Next we have: $\gcd(a,b) = 1$ or $3$. More precisely $\gcd(a,b) = 1$ when $3 \nmid r$ and $\gcd(a,b) = 3$ when $3 \mid r$. We note $3 \mid r$ if and only if $3 \mid r(r^2 - 9s^2) = a$ and this is equivalent to $3 \nmid uvw$.

Also $\gcd(y,z) = \gcd(a,c) = 1$ or $3$ according to $3 \nmid r$ or $3 \mid r$. Indeed, if $p$ is a prime, $e \geq 1$ and $p^e \mid \gcd(a,c)$ then from $a^2 + 3b^2 = c^3$, $p^2 \mid 3b^2$ so $p \mid b$, hence $p = 3$, $\gcd(a,b) = 3$ and $3^2 \nmid c$, showing that if $\gcd(a,c) \neq 1$, then $\gcd(a,c) = 3$ and $3 \mid r$. Conversely, if $3 \mid r$ then $\gcd(a,b) = 3$, so $3^2 \nmid c$ and by the above, $\gcd(a,c) = 3$.

If $\gcd(y,z) = 3$ then $9 \mid y$. Let $x_1 = x$, $y_1 = y/9$, $z_1 = z/3$, then $z_1^3 - 3y_1^2 = 2^{2n-2}x_1^{2n}$ with $x_1, y_1, z_1 \neq 0$, $\gcd(y_1, z_1) = 1$. This concludes the proof. $\square$

For further use, we note that (with the above notations), $\gcd(2x, y) = 1$. Here is a related result:

(1I)    Let $n \geq 3$. The following statements are equivalent:

(1) Fermat's last theorem is true for the exponent $n$.

(2) The equation (1.4) has no solution in nonzero integers $x, y, z$ such that $\gcd(2x, y) = 1$.

PROOF. $(1) \Rightarrow (2)$ Assume that $x, y, z \neq 0$, $\gcd(2x, y) = 1$ and $z^3 - y^2 = 3^3 \times 2^{2n-1}x^{2n}$. If $p$ is a prime and $p \mid \gcd(y, z)$ then $p \nmid 2x$, so $p = 3$, so $3^3 \mid y^2$, hence $9 \mid y$. Let $x_1 = x$, $y_1 = y/9$, $z_1 = z/3$, so $z_1^3 - 3y_1^2 = 2^{2n-1}x_1^{2n}$ with $x_1, y_1, z_1 \neq 0$, $\gcd(y_1, z_1) = 1$. By the implication $(1) \Rightarrow (2)$ of (1H) then there exist $u, v, w \neq 0$ such that $u^n + v^n + w^n = 0$.

$(2) \Rightarrow (1)$ Assume that $u, v, w \neq 0$, $\gcd(u, v, w) = 1$, and $u^n + v^n + w^n = 0$. By the proof of the implication $(2) \Rightarrow (1)$ of (1H) and the remark following this proof, there exist $x, y, z \neq 0$, $\gcd(2x, y) = 1$ and $z^3 - y^2 = 3^3 \times 2^{2n-1} x^{2n}$. $\square$

We examine in more detail the situation:

**(1J)**   *Assume that* (1.4) *has no solution in nonzero relatively prime integers. If $u, v, w \neq 0$, $\gcd(u, v, w) = 1$, and $u^n + v^n = w^n$, then:*

(1) $n$ *is odd,* $u \equiv v \equiv -w \pmod{3}$, $3 \nmid uvw$.
(2) *If $p$ is any prime dividing $n$, then $p \equiv 1 \pmod 6$.*

PROOF. (1)   By $(2) \Rightarrow (1)$ in (1H), there exist nonzero integers $x, y, z$ such that $z^3 - y^2 = 3^3 \times 2^{2n-1} x^{2n}$ with $\gcd(y, z) = 3$ and also $3 \nmid uvw$.

By Chapter VI, (4L), $n$ is odd, hence $u^n + v^n + (-w)^n = 0$. Let $r = u^n$, $s = -v^n$, so $r, s \neq 0$, $r \neq s$ (since $w \neq 0$). We have $r \not\equiv s \pmod{3}$, since $3 \nmid w$. Then $u \equiv u^n = r \not\equiv s = -v^n \equiv -v \pmod{3}$. Since $3 \nmid uv$ then $u \equiv v \pmod{3}$. By symmetry, $u \equiv v \equiv -w \pmod{3}$.

(2)   Let $p$ be a prime dividing $n$, $n = pt$, $u_1 = u^t$, $v_1 = v^t$, $w_1 = w^t$, so $u_1^p + v_1^p + (-w_1)^p = 0$ with $\gcd(u_1, v_1, w_1) = 1$, $u_1 \equiv v_1 \equiv -w_1 \pmod{3}$. By Chapter VI, (2A), we have $p \equiv 1 \pmod 6$. $\square$

It was shown by Inkeri, using nonelementary methods (namely class field theory), that under the above circumstances, $3^p \equiv 3 \pmod{p^2}$; this congruence is very seldom satisfied by a prime $p$.

It is appropriate to indicate now some facts about the equation

$$(1.7) \qquad\qquad Z^3 - Y^2 = c.$$

(For more details, see my book, Ribenboim, 1994.) Euler proved in 1738 that if $c = \pm 1$ the only solution of (1.7) in integers greater than 1 is $y = 3$, $z = 2$. It was shown by Siegel in 1929, using deep analytical methods, that for each given $c$ there exist at most finitely many solutions for equation (1.7).

Using (1H) we may show:

**(1K)**   *If $c = 3^3 \times 2^{2n-2} x^{2n}$ where $n \geq 3$, $x = 1$, $p$ or $pq$ (for* *distinct primes $p, q$), then equation (1.4) has no solution in nonzero relatively prime integers $y, z$.*

PROOF. Assume that $y, z$ are nonzero relatively prime integers such that $z^3 - y^2 = 3^3 \times 2^{2n-2} x^{2n}$. By the proof of implication (1) $\Rightarrow$ (2) of (1H), we have $2^{n-1} x^n \, r(r^2 - s^2)$ with $r, s \neq 0$, $r \not\equiv s \pmod{2}$, $\gcd(r, s) = 1$. Then $\gcd(r, r^2 - s^2) = 1$ and clearly $r^2 - s^2 \neq 1$.

If $x = 1$ then $r = \pm 1$, $r^2 - s^2 = \pm 2^{n-1}$, so $s^2 = 2^{n-1} + 1$, which is impossible.

If $x = p$ then $2^{n-1} p^n = r(r^2 - s^2)$. If $p = 2$, the above argument leads to a contradiction. If $p \neq 2$, then either $r = 2^{n-1}$, $r^2 - s^2 = p^n$, or vice versa. In the first case, $r - s = \pm 1$, $r + s = \pm p^n$ (or $r - s = \pm p^n$, $r + s = \pm 1$); in all cases, we reach a contradiction like $2^n = \pm (p^n + 1)$ or similar relations.

If $x = pq$ and $p$ or $q$ is 2, we use the preceding argument. If $p, q$ are odd then $2^{n-1} p^n q^n = r(r - s)(r + s)$. If $r = \pm p^n$ then $r - s = \pm 2^{n-1}$, $r + s = \pm q^n$ (or $\mp q^n$) or vice versa, and $2p^n = \pm (2^{n-1} \pm q^n)$, which is impossible. Similarly, $r \neq \pm q^n$, so $r = \pm 2^{n-1}$, $r - s = \pm p^n$, $r + s = \pm q^n$ (with appropriate sign) hence $2^n = \pm p^n \pm q^n$. By Chapter VI, (3B), this is impossible. □

The paper of Yahya (1973), where he published a proof of Fermat's last theorem, contained flaws of which one was the use of the incorrect result of Kapferer. Yahya has also related Fermat's equation to another diophantine equation. Inkeri examined this relationship and proved correctly the following statement:

**(1L)**    *Let $n \geq 3$. The following statements are equivalent:*

    (1) *Fermat's last theorem is true for the exponent $n$.*
    (2) *The equation*

(1.8)                    $$2^{2n-2} X^3 - Y^2 X - Z^n = 0$$

*has no solution in nonzero integers $x, y, z$ such that $\gcd(z, y) = \gcd(y, 2z) = 1$.*

## A6.    Frey

We find in the thesis of Hellegouarch (1972) an elliptic curve associated to a hypothetical solution of Fermat's equation with exponent

$2p^n$ (where $n \geq 1$ and $p$ is a prime); the aim was to show the non-existence of points of certain orders in elliptic curves.

In 1986, Frey had, independently, the same idea to associate to each solution of Fermat's equation an elliptic curve. Explicitly, if $a, b$ are nonzero relatively prime integers, $n > 3$ and $a$ is even, the *Frey curve* is the curve with equation

$$(1.9) \qquad Y^2 = X(X - a^n)(X + b^n).$$

It is an elliptic curve, whose properties were studied by Frey. Ribet and Wiles studied this curve under the assumption that there exist a nonzero integer $c$ such that $a^n + b^n = c^n$. The use of advanced theories of elliptic curves, modular forms, Galois representations, led ultimately to a contradiction, thus showing that Fermat's last theorem is true. We shall discuss this matter further in the Epilogue.

## B. Reformulations of Fermat's Last Theorem.

The following statements, which are equivalent to Fermat's theorem, were given by Pérez-Cacho in 1946. The equivalence between (1), (2), (3), and (4) was first proved by Bendz in 1901 and was rediscovered by Krasner, who published his paper in 1939 (see also Rivoire (1968)), and by Chowla in 1978 (see also Inkeri (1984)). Lind indicated some partial results in 1909.

**(1M)**    *Let $m \geq 2$, $n = 2m - 1$. The following statements are equivalent:*

  (1) *The equation $X^n + Y^n = Z^n$ has only the trivial solutions in integers.*

  (2) *The equation $X(1 + X) = T^n$ has only the trivial solutions in $\mathbb{Q}$.*

  (3) *The equation $X^2 = 4Y^n + 1$ has only the trivial solutions in $\mathbb{Q}$.*

  (4) *The equation $X^2 = Y^{n+1} - 4Y$ has only the trivial solutions in $\mathbb{Q}$.*

  (5) *For every nonzero rational number $a$, the polynomial $Z^2 - a^m Z + a$ is irreducible over $\mathbb{Q}$.*

  (6) *The equation $(XY)^m = X + Y$ has only the trivial solutions in $\mathbb{Q}$.*

  (7) *The equation $X^m = X/Y + Y$ has only the trivial solutions in $\mathbb{Q}$.*

(8) *If $u_1, r$ are nonzero rational numbers, and if $u_1, u_2, \ldots$ is a geometric progression of ratio $r$, then $u_m^2 - u_1 + r \neq 0$.*

(9) *If $\triangle$ is a triangle with vertices $A, B, C$, if the angle $\widehat{CAB} = 90°$, if $|\overline{AB}| = 2$, $|\overline{AB}| + |\overline{BC}|$ is an nth power of a rational number then $|\overline{AC}|$ is not rational.*

*Moreover, these conditions imply:*

(10) *The tangents to the parabola $Y^2 = 4X$ at every rational point distinct from the origin, cut the curve $Y = X^m$ at irrational points.*

PROOF. (1) $\rightarrow$ (2)    Let $a, b, c, d$ be nonzero integers, such that $b > 0$, $d > 0$, $\gcd(a, b) = \gcd(c, d) = 1$, and $a/b\,(1 + a/b) = (c/d)^n$. Then $a(a + b)/b^2 = c^n/d^n$. Hence $a(a + b) = c^n$, $b^2 = d^n$. Since $n$ is odd, then $b = y^n$ for some integer $y$. Since $\gcd(a, a + b) = 1$, then $a = x^n$, $a + b = z^n$, for nonzero integers $y, z$. Thus $x^n + y^n = z^n$.

(2) $\rightarrow$ (3)    Let $x, y$ be nonzero rational numbers such that $x^2 = 4y^n + 1$. Then $x \neq \pm 1$ and $((x - 1)/2)((x - 1)/2 + 1) = y^n$.

(3) $\rightarrow$ (4).    If $x, y$ are nonzero rational numbers such that $x^2 = y^{n+1} - 4y$, dividing by $y^{n+1} = y^{2m}$, we deduce that

$$\left(\frac{x}{y^m}\right)^2 = 1 + 4\left(\frac{-1}{y}\right)^n.$$

(4) $\rightarrow$ (5)    The discriminant of $Z^2 - a^m Z + a$ is $a^{n+1} - 4a \neq 0$ (since $n$ is odd). By hypothesis, $a^{n+1} - 4a$ cannot be a square, hence $Z^2 - a^m Z + a$ has no root in $\mathbb{Q}$.

(5) $\rightarrow$ (6)    If $x, y$ are nonzero rational numbers such that $(xy)^m = x + y$, let $xy = a$. Then $Z^2 - a^m Z + a$ has a solution in $\mathbb{Q}$.

(6) $\rightarrow$ (7)    Let $x, y$ be nonzero rational numbers such that $x^m = x/y + y$. Let $t = x/y$ so $t + y = (ty)^m$.

(7) $\rightarrow$ (1)    Let $a, b, c$ be nonzero integers such that $a^n + b^n = c^n$. Let $x = c^2/ab$ and $y = cb^{m-1}/a^m$. Then

$$\frac{x}{y} + y = \frac{ca^{m-1}}{b^m} + \frac{cb^{m-1}}{a^m} = \frac{c(a^n + b^n)}{a^m b^m} = \left(\frac{c^2}{ab}\right)^m = x^m.$$

(3) $\rightarrow$ (8)    Assume that $u_m^2 = u_1 - r$. Since $u_m = u_1 r^{m-1}$ then the equation $r^{2(m-1)} Z^2 - Z + r$ has the rational root $u_1$, therefore its discriminant is a square, that is, $1 - 4r^{2m-1} = s^2$. Hence $s$ and $-r$ satisfy $X^2 = 4Y^n + 1$.

(8) $\rightarrow$ (3)    The proof of the converse is similar.

If $x, y$ are nonzero rational numbers such that $4y^n + 1 = x^2$ then the equation $y^{2(m-1)}Z^2 - Z - y = 0$ has a rational root $u_1$. Let $u_i = u_1(-y)^{i-1}$; then $u_m^2 = u_1^2 y^{2(m-1)} = u_1 + (-y)$.

(5) $\leftrightarrow$ (9)   The condition (5) is equivalent to the fact that $a^{2m} - 4a$ is not a square in $\mathbb{Q}$, that is, $a^{2n} - 4a^n$ is not a square in $\mathbb{Q}$, or still, $(a^{m-1}Z)^2 = a^{2n} - 4a^n$ has no rational solution, for every rational number $a \neq 0$. Adding 4, this is equivalent to the non-existence of rational solutions for

$$4 + (a^{m-1}Z)^2 = (a^n - 2)^2,$$

which is in turn equivalent to (9).

Now we show that:

(7) $\rightarrow$ (10)   Let $(x_1, y_1) \neq (0, 0)$ be a rational point of the parabola $Y^2 = 4X$. The tangent to the parabola at this point has equation

$$y_1 Y = 2(X + x_1),$$

that is,

$$y_1 Y = 2\left(X + \frac{y_1^2}{4}\right),$$

so

$$Y = \frac{2}{y_1}X + \frac{y_1}{2}.$$

The intersections of the tangent with the curve $Y = X^m$ are the points $(x, y)$ such that $y = (2/y_1)x + y_1/2$, $y = x^m$. Thus $(x, y_1/2)$ is a solution of the equation $X^m = X/Y + Y$. By (7), $x$ must be irrational.   $\square$

In 1958, Pérez-Cacho showed:

(1N)   *Let $n \geq 2$. There is a bijection between the following sets:*

(F)   *The set of solutions $(x, y, z)$ of $X^n + Y^n = Z^n$, where $x, y, z$ are nonzero pairwise relatively prime natural numbers; and*

(F')   *the set of solutions $(u, v, w, t)$ of $U^n + V^{2n} = W^n + T^{2n}$ where $u, v, w, t$ are nonzero natural numbers, $\gcd(u, v) = \gcd(w, t) = \gcd(v, t) = 1$,   and $w = v \gcd(u, w)$, $u = t \gcd(u, w)$, $t \neq v$.*

PROOF. Let $(x, y, z) \in F$, let $u = xz$, $v = y$, $w = yz$, $t = x$. Then $v \neq t$, since $x \neq y$ because $2x^n$ is not an $n$th power. Also $u^n + v^{2n} =$

$$x^n z^n + y^{2n} = x^n(x^n + y^n) + y^{2n} = (x^n + y^n)y^n + x^{2n} = y^n z^n + x^{2n} = w^n + t^{2n}.$$

We have $u, v, w, t \neq 0$, $\gcd(u, v) = \gcd(w, t) = \gcd(v, t) = 1$, $z = \gcd(u, w)$ and $u = t \gcd(u, w)$, $w = v \gcd(u, w)$. Thus $(u, v, w, t) \in F'$.

Conversely, let $(u, v, w, t) \in F'$. Let $x = t$, $y = v$, $z = \gcd(u, w)$. Then $w = vz = yz$, $u = tz = xz$ and $x^{2n} - y^{2n} = t^{2n} - v^{2n} = u^n - w^n = (x^n - y^n)z^n$. Since $t \neq v$ then $x^n + y^n = z^n$.

Clearly the above correspondence between (F) and (F') is a bijection.    □

In 1979, Vrănceanu indicated a less interesting property equivalent to Fermat's last theorem for the exponent $n$.

We give now an equivalent combinatorial (!) formulation of Fermat's last theorem. It appeared in a short note by Quine (1989).

Consider a set of $n \geq 3$ balls, which are to be arranged into $z$ bins which are white, red, or blue. Let:

$W = $ number of white bins;
$B = $ number of blue bins; and
$R = $ number of red bins.

Then $z = W + B + R$. Let:

$(r'b) = $ number of arrangements of the $n$ balls into the bins, such that red bins receive no balls, but at least a blue bin has a ball.

$(rb') = $ same, but with no balls in blue bins and some ball in some red bin.

$(rb) = $ same, with at least a ball in some red bin and at least a ball in a blue bin.

$(w) = $ same, with all balls in the white bins.

We have:

**(1O)**    *Fermat's last theorem is true for the exponent $n$ if and only if $(w) \neq (rb)$.*

PROOF. We have

$z^n = $ number of all arrangements of the $n$ balls in the $z$ bins.

So

$$z^n = (w) + (r'b) + (rb') + (rb).$$

Let $x = R+W$ and $y = B+W$. Then the number of all arrangements of the $n$ balls into the bins which are red or white is

$$x^n = (w) + (rb').$$

Similarly, the number of arrangements of the $n$ balls which are blue or white is

$$y^n = (w) + (r'b).$$

If Fermat's last theorem is true for $n \geq 3$ then $z^n \neq x^n + y^n$ and, comparing, $(rb) \neq (w)$.

On the other hand, if Fermat's last theorem is false for $n \geq 3$, let $x, y, z$ be positive integers such that $x^n + y^n = z^n$. Let $W, B, R$ be given by

$$\begin{cases} B = z - x, \\ R = z - y, \\ W = x + y - z. \end{cases}$$

By the above argument, $(w) = (rb)$. $\square$

## Bibliography

1738 Euler, L., *Theorematum qourundam arithmeticorum demon-strationes*, Comm. Acad. Sci. Petrop., **10** (1738), 125–146; reprinted in *Opera Omnia*, Commentationes Arithmeticae, Vol. I, pp. 38–58, Teubner, Leipzig, 1915.

1840 Lebesgue, V.A., *Sur un théorème de Fermat*, J. Math. Pures Appl., **5** (1840), 184–185.

1840 Liouville, J., *Sur l'équation $Z^{2n} - Y^{2n} = 2X^n$*, J. Math. Pures Appl., **5** (1840), 360.

1846 Terquem, O., *Théorèmes sur les puissances des nombres*, Nouv. Ann. Math., **5** (1846), 70–78.

1885 Perrin, R., *Sur l'équation indéterminée $x^3 + y^3 = z^3$*, Bull. Soc. Math. France, **13** (1885), 194–197.

1901 Bendz, T.R., *Öfver Diophantiska Ekvationen $x^n + y^n = z^n$*, Almqvist & Wiksells Boktryckeri, Uppsala, 1901, 35 pp.

1908 Hurwitz, A., *Über die diophantische Gleichung $x^3y + y^3z + z^3x = 0$*, Math. Ann., **65** (1908), 428–430; reprinted in *Math. Werke*, Vol. II, pp. 427–429, Birkhäuser, Basel, 1963.

1909 Lind, B., *Einige zahlentheoretische Sätze*, Arch. Math. Phys., (3), **15** (1909), 368–369.

1913 Pocklington, H.C., *Some diophantine impossibilities*, Proc. Cambridge Philos. Soc., **17** (1913), 108–121.

1929 Siegel, C.L., *Über einige Anwendungen diophantischer Approximationen*, Abh. Preuss. Akad. Wiss. Berlin, Phys. Math. Kl., **1** (1929), 57 pp.

1930 Fueter, R., *Über kubische diophantische Gleichungen*, Comment. Math. Helv., **2** (1930), 69–89.

1933 Kapferer, H., *Über die diophantischen Gleichungen $Z^3 - Y^2 = 3^3 \cdot 2^\lambda \cdot X^{\lambda+2}$ und deren Abhängigkeit von der Fermatschen Vermutung*, Heidelberger Akad., Math. Naturwiss. Klasse, Abh., **2** (1933), 32–37.

1939 Krasner, M., *Sur le théorème de Fermat*, C. R. Acad. Sci. Paris, **208** (1939), 1468–1471.

1946 Pérez-Cacho, L., *El último teorema de Fermat y los números de Mersenne*, Rev. Real Acad. Cienc. Exact. Fiis. y Natur., Madrid, **40** (1946), 39–57.

1958 Pérez-Cacho, L., *Sobre algunas cuestiones de la teoria de números*, Rev. Mat. Hisp.-Amer., (4), **18** (1958), 10–27 and 113–124.

1967 Christilles, W.E., *A note concerning Fermat's conjecture*, Amer. Math. Monthly, **74** (1967), 292–294.

1968 Rivoire, P., *Dernier Théorème de Fermat et Groupes de Classes dans Certains Corps Quadratiques Imaginaires*, Thèse, Université Clermont-Ferrand, 1968, 59 pp.; reprinted in Ann. Sci. Univ. Clermont-Ferrand II, **68** (1979), 1–35.

1972 Hellegouarch, Y., *Courbes Elliptiques et Équation de Fermat*, Thesis, Univiversité Besançon, 1972.

1973 Yahya, Q.A.M.M., *On general proof of Fermat's last theorem*, Portugal. Math., **32** (1973), 157–170.

1975 Gandhi, J.M. and Stuff, M., *Comments on certain results about Fermat's last theorem*, Notices Amer. Math. Soc., **22** (1975), A-502.

1978 Chowla, S., *L-series and elliptic curves*, Part 4, On Fermat's last theorem. *Number Theory Day*, Springer Lecture Notes, No. 626, pp. 19–24, Springer-Verlag, New York, 1978.

1979 Vrănceanu, G., *Une interprétation géométrique du théorème de Fermat*, Rev. Roumaine Math. Pures Appl., **24** (1979), 1137–1140.

1984 Inkeri, K., *On certain equivalent statements for Fermat's last theorem – with requisite corrections*, Ann. Univ. Turkuenis, Ser. AI, **186** (1984) 12–22; reprinted in *Collected Papers of Kustaa Inkeri* (editor, P. Ribenboim), Queen's Papers in Pure and Applied Mathematics, Vol. 91, Kingston, Ontario, 1992.

1986 Frey, G., *Elliptic curves and solutions of $A - B = C$*, in: Sém. Th. Nombres, Paris, 1985–1986 (editor, C. Goldstein), Progress in Math., Birkhäuser, Boston, 1986, pp. 39–51.

1987 Frey, G., *Links between elliptic curves and solutions of $A - B = C$*, J. Indian Math. Soc., **51** (1987), 117–145.

1989 Quine, W.V., *Fermat's last theorem in combinatorial form*, Amer. Math. Monthly, **95** (1989), 626.

1994 Ribenboim, P., *Catalan's Conjecture*, Academic Press, Boston, 1994.

## VIII.2. Criteria for Fermat's Last Theorem

In this section we gather various results of a different nature, all proved with elementary methods. For the convenience of the reader, we classify them into subsections:

   A. Connection with Euler's totient function.
   B. Connection with the Möbius function.
   C. Proof that a nontrivial solution cannot be in arithmetical progression.
   D. Criterion with a Legendre symbol.
   E. Criterion with a discriminant.
   F. Connection with a cubic congruence.
   G. Criterion with a determinant.
   H. Connection with a binary quadratic form.
   I. The non-existence of algebraic identities yielding solutions of Fermat's equation.
   J. Criterion with second-order linear recurrences.
   K. Perturbation of one exponent.
   L. Divisibility condition for Pythagorean triples.

**A. Connection with Euler's Totient Function.** The first result was proved by Pérez-Cacho in 1928, in a slightly weaker form:

**(2A)**    *Let $x, y$ be nonzero relatively prime integers, let $n \geq 2$ be an integer, and let $p$ be any prime factor of $n$, $n = pm$. Let $z$ be any integer such that $z \geq 3$, $z$ divides $x^n \pm y^n$, but $z$ does not divide $x^m \pm y^m$. Then $p$ divides $\varphi(z)$.*

PROOF. If $p = 2$, it is true since $z \geq 3$, so $\varphi(z)$ is even.

Let $p$ be an odd prime, $n = pm$, and suppose that $p \nmid \varphi(z)$. Then there exist integers $r, s$ such that $rp - s\varphi(z) = 1$. Since $\varphi(z)$ is even, then $r$ is odd.

If we assume $x^{pm} \equiv -y^{pm} \pmod{z}$ but $x^m \not\equiv -y^m \pmod{z}$, then $x^{rpm} \equiv -y^{rpm} \pmod{z}$ hence $x^{(s\varphi(z)+1)m} \equiv -y^{(s\varphi(z)+1)m} \pmod{z}$. We have $\gcd(z, x) = \gcd(z, y) = 1$. Indeed, for example, if a prime $q$ divides $x$ and $z$, it divides $x^n + y^n$, hence also $y$, contrary to the hypothesis. By Euler's theorem, $x^{\varphi(z)} \equiv y^{\varphi(z)} \equiv 1 \pmod{z}$, so $x^m \equiv -y^m \pmod{z}$, which contradicts the hypothesis.

The proof is similar when $z \mid x^n - y^n$, $z \mid x^m - y^m$.  □

In particular, if $\gcd(x, y) = 1$ and $p \mid n$ and taking $z = x^n \pm y^n$, then $p \mid \varphi(x^n \pm y^n)$. As a matter of fact, as shown by Pérez-Cacho, this last assertion also holds when $\gcd(x, y) = d \neq 1$. Indeed, let $x = dx_1$, $y = dy_1$, so $\gcd(x_1, y_1) = 1$; then $p \mid \varphi(x_1^n \pm y_1^n)$; but $x^n \pm y^n = d^n(x_1^n \pm y_1^n)$ hence $p \mid \varphi(x^n \pm y^n)$.

As a corollary, we have the result proved by Swistak in 1969:

**(2B)**    *If $p$ is an odd prime and $0 < x < y < z$ are positive integers such that $x^p + y^p = z^p$ then $p$ divides $\varphi(x)$, $\varphi(y)$, and $\varphi(z)$.*

PROOF. We may assume without loss of generality that $x, y, z$ are pairwise relatively prime, because if, for example, $x = dx_1$ and $p \mid \varphi(x_1)$ then also $p \mid \varphi(x)$.

We have $3 \leq z$, $z \mid x^p + y^p$ and $z \nmid x + y$; indeed, $z^p = x^p + y^p < (x + y)^p$, so $z < x + y < 2z$. By (2A), $p \mid \varphi(z)$.

Similarly, $x \mid z^p - y^p$, $x \nmid z - y$ since $z - y < x$. Also

$$x^p = (z - y)\left(z^{p-1} + pz^{p-2}y + \binom{p}{2}z^{p-3}y^2 + \cdots + y^{p-1}\right)$$
$$> 2^{p-1}p > 2^p,$$

so $x \geq 3$. By (2A), $p \mid \varphi(x)$.

Finally, $y \mid z^p - x^p$, $y \nmid z - x$ since $z - x < y$ and $3 \leq x < y$, so again $p \mid \varphi(y)$. $\square$

Bussi indicated in 1943 the following corollary:

**(2C)**   If $x, y, z$ are pairwise relatively prime positive integers, if $p$ is a prime, $p \nmid xyz$, and $x^p + y^p = z^p$, then there exist primes $q, r, s$, such that $q \equiv r \equiv s \equiv 1 \pmod{p}$ and $q \mid x$, $r \mid y$, $s \mid z$.

PROOF. By (2B), $p$ divides $\varphi(x)$, $\varphi(y)$, $\varphi(z)$. Since $p \nmid xyz$ then there exist primes $q, r, s$, such that $p \mid q - 1$, $p \mid r - 1$, $p \mid s - 1$, and $q \mid x$, $r \mid y$, $s \mid z$. $\square$

This result is also a corollary of Chapter III, (1B). Another remark made by Bussi in 1932 is the following:

**(2D)**   Let $n > 2$, and let $x, y, z$ be pairwise relatively prime positive integers such that $x^n + y^n = z^n$. If $k$ is any integer such that $\varphi(k) = n$ then $\gcd(k, xyz) > 1$.

PROOF. If $\gcd(k, xyz) = 1$ then $x^{\varphi(k)} \equiv y^{\varphi(k)} \equiv z^{\varphi(k)} \equiv 1 \pmod{k}$. But $\varphi(k) = n$, so $1 \equiv z^n = x^n + y^n \equiv 2 \pmod{k}$, which is absurd. $\square$

**B. Connection with the Möbius Function.**   Rameswar Rao proved in 1969:

**(2E)**   If $n \geq 3$ is an odd integer, and if $x, y, z$ are positive integers such that $x^n + y^n = z^n$, then the Möbius function $\mu$ vanishes for $x + y$.

PROOF. Since $n$ is odd $x + y$ divides $x^n + y^n = z^n$. Any prime factor $p$ of $x + y$ divides $z^n$ hence divides $z$.
   If $\mu(x+y) \neq 0$ then $x+y$ has no square factor, hence $x+y$ divides $z$, in particular $x+y \leq z$. This is a contradiction since $z < x+y$. $\square$

**C. Proof that a Nontrivial Solution Cannot be in Arithmetical Progression.** The next result is due to Bottari (1907); it was rediscovered by Goldziher (1913), by Mihaljinec (1952), and by

Rameswar Rao (1969). In 1908, Cattaneo gave the following simple proof, which is exactly the same as Rameswar Rao's:

**(2F)**  *If $n > 2$ and $x, y, z$ are positive integers such that $x^n + y^n = z^n$ then $x, y, z$ cannot be in arithmetic progression.*

PROOF. Let us assume the contrary, so there exists a positive integer $a$ such that $x = y - a$, $z = y + a$. Then

$$(2.1) \qquad\qquad (y - a)^n + y^n = (y + a)^n.$$

Dividing, if necessary, by the greatest common divisor of $a, y$, we may assume that $\gcd(a, y) = 1$. We also see that $y$ cannot be odd. The relation (2.1) implies that $y^n = am$, for some integer $m$. Since $\gcd(a, y) = 1$ then $a = 1$, so $(y - 1)^n + y^n = (y + 1)^n$.

We see that $n$ cannot be odd, otherwise

$$y^n = 2\left[\binom{n}{1}y^{n-1} + \binom{n}{3}y^{n-3} + \cdots + \binom{n}{n-2}y^2 + 1\right]$$

and since $y$ is even then $2^{n-1}$ would divide the odd number in brackets.

Since $n$ is even then dividing by $y$:

$$y^{n-1} = 2\left[\binom{n}{1}y^{n-2} + \binom{n}{3}y^{n-4} + \cdots + \binom{n}{n-1}\right],$$

so $y^{n-1} = 2l$. Hence $y/2$ divides $l$. Since $y/2$ divides each summand in the bracket but the last one, it follows also that $y/2$ divides $\binom{n}{n-1} = n$, so $y \leq 2n$.

Hence $y^{n-1} > 2\binom{n}{1}y^{n-2} \geq y^{n-1}$, a contradiction.  □

**D. Criterion with a Legendre Symbol.** In 1958, Pérez-Cacho indicated the following criterion:

**(2G)**  *Let $p$ be an odd prime. Assume that if $x, y, z$ are any nonzero relatively prime integers then there exists a prime $q$, $q \neq p$, $q$ dividing $z^2 - pxy$ and*

$$\left(\frac{p^{p+1} - 4p}{q}\right) = -1.$$

*Then Fermat's last theorem is true for the exponent $p$.*

PROOF. We assume that there exist nonzero relatively prime integers $x, y, z$ such that $x^p + y^p = z^p$. By hypothesis, there exists a prime $q$, $q \neq p$, such that $q$ divides $z^2 - pxy$ and $((p^{p+1} - 4p)/q) = -1$.

From $z^{2p} = x^{2p} + 2x^p y^p + y^{2p}$ it follows that

$$z^{2(p+1)} - 4z^2 x^p y^p = z^2(x^{2p} - 2x^p y^p + y^{2p}) = [z(x^p - y^p)]^2.$$

Let $k$ be defined by $z^2 = pxy + kq$. Then

$$
\begin{aligned}
(p^{p+1} - 4p)(xy)^{p+1} &= (z^2 - kq)^{p+1} - 4(xy)^p(z^2 - kq) \\
&\equiv z^{2(p+1)} - 4z^2 x^p y^p \\
&\equiv [z(x^p - y^p)]^2 \pmod{q}.
\end{aligned}
$$

Since $p + 1$ is even, then $p^{p+1} - 4p$ is a square modulo $q$, which is a contradiction.  $\square$


**E. Criterion with a Discriminant.** In 1949, Kapferer gave a criterion involving the discriminant of a certain polynomial. In the proof of this result we shall require some facts about the resultant and the discriminant of binary forms, which have been gathered in Chapter II, §4.

The following lemma was explicitly used by Kapferer; a proof is given in his paper.

LEMMA 2.1. *Let $F(X, Y)$, $G(X, Y)$ be forms of degree $n, m$, respectively, let $L(X, Y)$, $M(X, Y)$ be forms of degree $k$, and let*

$$\Phi(X, Y) = F(L(X, Y), M(X, Y))$$

*(form of degree $kn$),*

$$\Gamma(X, Y) = G(L(X, Y), M(X, Y))$$

*(form of degree $km$). Then*

$$R(\Phi, \Gamma) = [R(F, G)]^k [R(L, M)]^{mn}.$$

PROOF. The result is trivial if $n = 0$ or $m = 0$, so we assume $n \geq 1$, $m \geq 1$.

Let $F(X, Y) = \prod_{i=1}^{n}(\alpha_i' X - \alpha_i Y)$ where $\alpha_i, \alpha_i'$ are not both zero (for each $i = 1, \ldots, n$). Thus $\Phi(X, Y) = F(L, M) = \prod_{i=1}^{n}(\alpha_i' L - \alpha_i M)$. Similarly, let $G(X, Y) = \prod_{j=1}^{m}(\beta_j' X - \beta_j Y)$, so $\Gamma(X, Y) = \prod_{j=1}^{m}(\beta_j' L - \beta_j M)$.

Then by Chapter II, (4A),

$$R(\Phi, \Gamma) = \prod_{i=1}^{n} \prod_{j=1}^{m} R(\alpha_i' L - \alpha_i M, \beta_j' L - \beta_j M).$$

For example, let $\beta_j \neq 0$; then

$$
\begin{aligned}
R(\alpha_i' L &- \alpha_i M, \beta_j' L - \beta_j M) \\
&= \frac{1}{\beta_j^k} R(\beta_j \alpha_i' L - \beta_j \alpha_i M, \beta_j' L - \beta_j M) \\
&= \frac{1}{\beta_j^k} R\left((\beta_j \alpha_i' - \alpha_i \beta_j')L, \beta_j' L - \beta_j M\right) \\
&= \frac{(\beta_j \alpha_i' - \alpha_i \beta_j')^k}{\beta_j^k} R(L, \beta_j' L - \beta_j M) \\
&= \frac{(-1)^k (\beta_j' \alpha_i - \beta_j \alpha_i')^k}{\beta_j^k} R(L, -\beta_j M) \\
&= (\beta_j' \alpha_i - \beta_j \alpha_i')^k R(L, M).
\end{aligned}
$$

Hence

$$
\begin{aligned}
R(\Phi, \Gamma) &= \prod_{i=1}^{n} \prod_{j=1}^{m} (\beta_j' \alpha_i - \beta_j \alpha_i')^k R(L, M) \\
&= [R(F, G)]^k [R(L, M)]^{mn}. \quad \square
\end{aligned}
$$

Now we give Kapferer's criterion:

**(2H)**    *Let $p$ be a prime number, $p > 7$. If there exist nonzero integers $x, y, z$ such that*

$$p \nmid xyz(x - y)(y - z)(z - x)(x^2 + y^2 + z^2)$$

*and $x^p + y^p + z^p \equiv 0 \pmod{p^2}$, then $p$ divides the discriminant of the homogeneous polynomial*

$$K_p(X, Y) = \sum_{i=0}^{(p-7)/6} \binom{\dfrac{p-3}{2} - i}{2i} \frac{1}{2i+1} X^{(p-7)/6-i} Y^i$$

*when* $p \equiv 1 \pmod 6$,

$$K_p(X,Y) = \sum_{i=0}^{(p-5)/6} \binom{\frac{p-3}{2} - i}{2i} \frac{1}{2i+1} X^{(p-5)/6-i} Y^i$$

*when* $p \equiv -1 \pmod 6$.

PROOF. Indeed,

$$x^p \equiv x \pmod p, \qquad y^p \equiv y \pmod p, \qquad z^p \equiv z \pmod p,$$

hence $x + y + z \equiv 0 \pmod p$ and so $z^p \equiv -(x+y)^p \pmod{p^2}$. Thus $(x+y)^p - x^p - y^p \equiv 0 \pmod{p^2}$.

We note that $y - z \equiv x + 2y \pmod p$, $z - x \equiv -(2x+y) \pmod p$, $x^2 + y^2 + z^2 \equiv 2(x^2 + xy + y^2) \pmod p$, hence $p \nmid xy(x+y)(x - y)(2x+y)(2y+x)(x^2 + xy + y^2)$.

We have seen in Chapter VII, (2.2), that

$$(X + Y)^p - X^p - Y^p = pXY(X + Y)(X^2 + XY + Y^2)^e C_p(X,Y),$$

where $C_p(X,Y) \in \mathbb{Z}[X,Y]$ is the homogenized Cauchy polynomial and $e = 1$ or $2$, according to whether $p \equiv -1$, or $1 \pmod 6$. Thus $p$ divides $C_p(x,y)$.

We show that $p$ divides the discriminant of $C_p(X,Y)$. Indeed, let $Q(X,Y) = XY(X+Y)(X^2 + XY + Y^2)^e$, so

$$\frac{1}{p}[(X+Y)^p - X^p - Y^p] = Q(X,Y)C_p(X,Y).$$

Hence taking the partial derivatives:

$$(X+Y)^{p-1} - X^{p-1} = Q(X,Y)\frac{\partial C_p}{\partial X}(X,Y) + \frac{\partial Q}{\partial X}(X,Y)C_p(X,Y),$$

$$(X+Y)^{p-1} - Y^{p-1} = Q(X,Y)\frac{\partial C_p}{\partial Y}(X,Y) + \frac{\partial Q}{\partial Y}(X,Y)C_p(X,Y).$$

Since $p \nmid xy(x+y)$ then $x^{p-1} \equiv y^{p-1} \equiv (x+y)^{p-1} \equiv 1 \pmod p$; on the other hand, $p \nmid Q(x,y)$. Therefore, since $p$ divides $C_p(x,y)$ then $p \mid (\partial C_p/\partial X)(x,y)$ and $p \mid (\partial C_p/\partial Y)(x,y)$. So,

$$\text{Discr}(C_p(X,Y)) = R\left(\frac{\partial C_p}{\partial X}(X,Y), \frac{\partial C_p}{\partial Y}(X,Y)\right) \equiv 0 \pmod p.$$

Now we use the expression of $C_p(X, Y)$ as a homogeneous polynomial in

$$L(X, Y) = (X^2 + XY + Y^2)^3 = U^3,$$
$$M(X, Y) = X^2 Y^2 (X + Y)^2 = V^2,$$

which was derived in Chapter VII, (2.5) and (2.6): $C_p(X, Y) = K_p(L, M)$ where

$$K_p(L, M) = \sum_{i=0}^{(p-7)/6} \binom{\frac{p-3}{2} - i}{2i} \frac{1}{2i + 1} L^{(p-7)/6-i} M^i,$$

when $p \equiv 1 \pmod 6$,

$$K_p(L, M) = \sum_{i=0}^{(p-5)/6} \binom{\frac{p-3}{2} - i}{2i} \frac{1}{2i + 1} L^{(p-5)/6-i} M^i,$$

when $p \equiv -1 \pmod 6$. We have

$$\frac{\partial C_p}{\partial X} = \frac{\partial K_p}{\partial L} \cdot \frac{\partial L}{\partial X} + \frac{\partial K_p}{\partial M} \cdot \frac{\partial M}{\partial X},$$
$$\frac{\partial C_p}{\partial Y} = \frac{\partial K_p}{\partial L} \cdot \frac{\partial L}{\partial Y} + \frac{\partial K_p}{\partial M} \cdot \frac{\partial M}{\partial Y},$$

where

$$\frac{\partial L}{\partial X} = 3(X^2 + XY + Y^2)^2 (2X + Y),$$

$$\frac{\partial M}{\partial X} = 2XY^2 (X + Y)(2X + Y),$$

$$\frac{\partial L}{\partial Y} = 3(X^2 + XY + Y^2)^2 (X + 2Y),$$

$$\frac{\partial M}{\partial Y} = 2X^2 Y (X + Y)(X + 2Y).$$

Letting $L(x, y) = r$, $M(r, s) = s$, then $p \nmid rs$ and from

$$\frac{\partial C_p}{\partial X}(x, y) \equiv \frac{\partial C_p}{\partial Y}(x, y) \equiv 0 \pmod p,$$

it follows that either

$$\frac{\partial K_p}{\partial L}(r, s) \equiv \frac{\partial K_p}{\partial M}(r, s) \equiv 0 \pmod p$$

or

$$\det \begin{pmatrix} \dfrac{\partial L}{\partial X}(x,y) & \dfrac{\partial M}{\partial X}(x,y) \\[2mm] \dfrac{\partial L}{\partial Y}(x,y) & \dfrac{\partial M}{\partial Y}(x,y) \end{pmatrix} \equiv 0 \pmod{p}.$$

Computing the determinant, this condition becomes

$$6(x^2 + xy + y^2)^2(2x + y)(x + 2y)xy(x + y)(x - y) \equiv 0 \pmod{p}.$$

However, by hypothesis this determinant is not a multiple of $p$. Therefore

$$\frac{\partial K_p}{\partial L}(r, s) \equiv \frac{\partial K_p}{\partial M}(r, s) \equiv 0 \pmod{p},$$

that is, $p$ divides the resultant of the binary forms $(\partial K_p/\partial L)(X, Y)$, $(\partial K_p \partial M)(X, Y)$, having degree $m - 1$, where

$$m = \begin{cases} \dfrac{p - 7}{6} & \text{when} \quad p \equiv 1 \pmod 6, \\[3mm] \dfrac{p - 5}{6} & \text{when} \quad p \equiv -1 \pmod 6. \end{cases}$$

Since $L(X, Y)$, $M(X, Y)$ have degree 6, by Chapter V, (1A),

$$R\left(\frac{\partial K_p}{\partial L}(X, Y), \frac{\partial K_p}{\partial M}(X, Y)\right)$$

$$= \left[R\left(\frac{\partial K_p}{\partial L}(L, M), \frac{\partial K_p}{\partial M}(L, M)\right)\right]^6 R(L(X, Y), M(X, Y))^{(m-1)^2}.$$

But

$$R(L(X, Y), M(X, Y)) = R((X^2 + XY + Y^2)^3, X^2 Y^2(X + Y)^2)$$
$$= \left[R(X^2 + XY + Y^2, XY(X + Y))\right]^6 = 1.$$

Hence $p$ divides the resultant

$$R\left(\frac{\partial K_p}{\partial L}(L, M), \frac{\partial K_p}{\partial M}(L, M)\right) = \mathrm{Discr}(K_p(L, M)). \quad \square$$

We note that we have actually shown that if $r = (x^2 + xy + y^2)^3 \not\equiv 0$ (mod $p$) and $s = x^2 y^2(x + y)^2 \not\equiv 0$ (mod $p$) then $(r, s)$ is a multiple root of the congruence $K_p(L, M) \equiv 0 \pmod{p}$.

From (2G) we obtain the following criterion:

**(2I)**    *Let $p$ be a prime number. If there exist nonzero relatively prime integers $x, y, z$ such that $x^p + y^p + z^p = 0$ then either*

$$p \mid xyz(x-y)(y-z)(z-x)(x^2 + y^2 + z^2)$$

*or $p$ divides the discriminant of the polynomial*

$$F_p(T) = F(T) = \sum_i \binom{m-i}{2i} \frac{1}{2i+1} T^i,$$

*where $m = (p-3)/2$ and the summation extends from $i = 0$ to*

$$i = \begin{cases} \dfrac{p-7}{6} & when \quad p \equiv 1 \pmod 6, \\ \dfrac{p-5}{6} & when \quad p \equiv -1 \pmod 6. \end{cases}$$

PROOF. Suppose that $x, y, z$ are nonzero relatively prime integers such that $p \nmid xyz(x-y)(y-z)(z-x)(x^2+y^2+z^2)$ and $x^p+y^p+z^p = 0$, hence $p > 7$.

Then $x^p + y^p + z^p \equiv 0 \pmod{p^2}$, hence by (2G), $p$ divides the discriminant of $K_p(X, Y)$, and more precisely, $(r, s)$ with

$$r = (x^2 + xy + y^2)^3 \not\equiv 0 \pmod p,$$
$$s = x^2 y^2 (x+y)^2 \not\equiv 0 \pmod p,$$

is a multiple root of the congruence $K_p(X, Y) \equiv 0 \pmod p$.

We write $T = Y/X$, so

$$K_p(1, Y/X) = F(T), \qquad K_p(X, Y) = X^m F(T),$$

where

$$m = \begin{cases} \dfrac{p-7}{6} & when \quad p \equiv 1 \pmod 6, \\ \dfrac{p-5}{6} & when \quad p \equiv -1 \pmod 6. \end{cases}$$

But

$$\frac{\partial K_p}{\partial X} = mX^{m-1}F(T) - X^{m-2}YF'(T),$$
$$\frac{\partial K_p}{\partial Y} = X^{m-1}F'(T),$$

and since $(r, s)$ is a common root of the congruences

$$\frac{\partial K_p}{\partial X}(r, s) \equiv 0 \pmod{p},$$

$$\frac{\partial K_p}{\partial Y}(r, s) \equiv 0 \pmod{p},$$

letting $t \in \mathbb{Z}$ be such that $tr \equiv s \pmod{p}$, then

$$\begin{cases} 0 \equiv mr^{m-1}F(t) - r^{m-2}sF'(t) \pmod{p}, \\ 0 \equiv r^{m-1}F'(t) \pmod{p}, \end{cases}$$

so $F(t) \equiv F'(t) \equiv 0 \pmod{p}$. This implies that $p$ divides the discriminant of $F(T)$.    □

**F. Connection with a Cubic Congruence.** In 1944, Pierre used a method somewhat similar to the one of Kapferer, to reduce Fermat's equation to two systems of congruences.

To begin, we need a lemma (part of these assertions were proved by Mirimanoff in 1907 and Skolem in 1937; see also Skolem (1941)).

LEMMA 2.2. *Let $p = 6k \pm 1$ be a prime number, let $a, b \in \mathbb{Z}$, $p \nmid b$, and consider the congruence*

$$X^3 + aX + b \equiv 0 \pmod{p}.$$

*Let $t \in \mathbb{Z}$ be such that $tb^2 \equiv a^3 \pmod{p}$, and*

$$\Psi_p(T) = T^{k-1} - 3\frac{(p-5)(p-7)}{2^2 \times 3!}T^{k-2} + \cdots$$

$$+ (-1)^{s-1}(2s-1)\frac{[p-(2s+1)][p-(2s+3)]\cdots[p-(4s-1)]}{2^{2(s-1)}(2s-1)!}T^{k-s}$$

$$+ \cdots$$

$$+ (-1)^{k-1}(2k-1)\frac{[p-(2k+1)][p-(2k+3)]\cdots[p-(4k-1)]}{2^{2(k-1)}(2k-1)!}.$$

(1) *If $-(4t + 27)$ is not a quadratic residue modulo $p$, the congruence has a unique solution $x$, $0 < x < p$.*
(2) *If $4t + 27 \equiv 0 \pmod{p}$, the congruence has two distinct solutions $x_1, x_2$, $0 < x_1, x_2 < p$.*
(3) *If $-(4t + 27)$ is a quadratic residue modulo $p$ and if $\Psi_p(t) \equiv 0 \pmod{p}$ then the congruence has three distinct solutions $x_1, x_2, x_3$, $0 < x_1, x_2, x_3 < p$.*

(4) *If* $-(4t + 27)$ *is a quadratic residue modulo* $p$, *but* $\Psi_p(t) \not\equiv 0$ (mod $p$), *then the congruence has no solution.*

(5) *There are* $k$ *values of* $t$ *such that the congruence has several distinct solutions and* $2k$ *values of* $t$ *for which the congruence has no solutions.*

In this respect, Cailler (1908) indicated a relation between the resolution of congruences of third degree and linear recurrences of second order; see also Mirimanoff (1909).

Let $F_p(T)$ be the polynomial defined in (2H). With the above lemma, Pierre proved:

**(2J)**    *Let* $x, y, z$ *be relatively prime integers such that* $p \not| xyz(x - y)(y - z)(z - x)(x^2 + y^2 + z^2)$ *and* $x^p + y^p + z^p = 0$. *Let* $r = (x^2 + xy + y^2)^3$, $s = x^2y^2(x + y)^2$.

(1) *If* $t$ *satisfies the congruence* $rt \equiv s \pmod{p^3}$ *then* $F_p(t) \equiv 0 \pmod{p^3}$.

(2) *If* $t$ *satisfies the congruence* $r + ts \equiv 0 \pmod{p}$ *then*

$$\left(\frac{-4t - 27}{p}\right) = 1, \qquad \Psi_p(t) \equiv 0 \pmod{p}.$$

PROOF. (1)    By Chapter VI, (1C), $x + y + z \equiv 0 \pmod{p^3}$. Hence $x^p + y^p = -z^p \equiv (x + y)^p \pmod{p^4}$. Thus $p^4$ divides $(x + y)^p - x^p - y^p = pxy(x + y)(x^2 + xy + y^2)^e C_p(x, y)$ with $e = 2$ when $p \equiv 1 \pmod 6$, $e = 1$ when $p \equiv -1 \pmod 6$. Since $p \not| z$ then $p \not| x + y$.

Similarly, since $p \not| x^2 + y^2 + z^2$ then $p \not| x^2 + xy + y^2$. Thus $p^3$ divides $C_p(x, y)$. With the previous notations, $C_p(x, y) = K_p(r, s) \equiv r^m F_p(t) \pmod{p^3}$, with $m = (p - 7)/6$ when $p \equiv 1 \pmod 6$, $m = (p - 5)/6$ when $p \equiv -1 \pmod 6$. Since $p \not| r$ then $F_p(t) \equiv 0 \pmod{p^3}$.

(2)    We observe: the congruence

$$(x^2 + xy + y^2)^3 + Tx^2y^2(x + y)^2 \equiv 0 \pmod{p}$$

has a solution $t$ if and only if

$$\begin{cases} d^3x^3 + tdx + t \equiv 0 \pmod{p}, \\ d^3y^3 + tdy + t \equiv 0 \pmod{p}, \end{cases}$$

with some $d$, not multiple of $p$.

Indeed, if $dx, dy$ are solutions of $X^3 + tX + t \equiv 0 \pmod{p}$, then the other solution is congruent to $-d(x + y)$ modulo $p$. Hence

$$\begin{cases} d^2[xy - x(x + y) - y(x + y)] \equiv t \pmod{p}, \\ -d^3 xy(x + y) \equiv -t \pmod{p}. \end{cases}$$

Therefore

$$\begin{cases} d^2(x^2 + xy + y^2) \equiv -t \pmod{p}, \\ d^3 xy(x + y) \equiv t \pmod{p}, \end{cases}$$

so $(x^2 + xy + y^2)^3 + tx^2 y^2 (x + y)^2 \equiv 0 \pmod{p}$.

Conversely, let $d$ be such that $dxy(x + y) \equiv -(x^2 + xy + y^2)$ $\pmod{p}$. Consider the congruence $X^3 + aX + b \equiv 0 \pmod{p}$ with roots $dx, dy, -d(x + y)$. Then

$$\begin{cases} d^2[xy - x(x + y) - y(x + y)] \equiv a \pmod{p}, \\ -d^3 xy(x + y) \equiv -b \pmod{p}. \end{cases}$$

Thus

$$\begin{cases} d^2(x^2 + xy + y^2) \equiv -a \pmod{p}, \\ d^3 xy(x + y) \equiv b \pmod{p}. \end{cases}$$

So $b \equiv a \equiv t \pmod{p}$, hence

$$\begin{cases} d^3 x^3 + tdx + t \equiv 0 \pmod{p}, \\ d^3 y^3 + tdy + t \equiv 0 \pmod{p}. \end{cases}$$

It follows from the above considerations that the congruence $X^3 + tX + t \equiv 0 \pmod{p}$ has three incongruent solutions $dx, dy$, and $-d(x + y)$. Indeed, if, for example, $dx \equiv -d(x + y) \pmod{p}$ then $x \equiv -x(x + y) \equiv z \pmod{p}$, contrary to the hypothesis. By the preceding lemma, $-(4t + 27)$ is a quadratic residue modulo $p$ and $\Psi_p(t) \equiv 0 \pmod{p}$.

This concludes the proof.  $\square$

It is worthwhile to recall that if $p \equiv -1 \pmod{6}$ then $p \nmid x^2 + xy + y^2$, hence $p \nmid x^2 + y^2 + z^2$ (see Chapter VI, Lemma 1.2).

But Pollaczek has also shown in 1917 (see Chapter VI, (2J)) that even if $p \equiv 1 \pmod{6}$, from $x^p + y^p + z^p = 0$, $p \nmid xyz$, it follows that $p \nmid x^2 + xy + y^2$, or equivalently, $p \nmid x^2 + y^2 + z^2$.

Similarly, if $x \equiv y \pmod{p}$ then $x^p \equiv y^p \pmod{p^2}$, $z \equiv -2x$ $\pmod{p}$, $z^p \equiv -2^p x^p \pmod{p^2}$, so $2x^p \equiv 2^p x^p \pmod{p^2}$. Thus if $p \nmid x$ then $2^p \equiv 2 \pmod{p^2}$. Actually, Inkeri showed in 1946 that

$2^p \equiv 2 \pmod{p^4}$ (see his paper). Hence, if $p$ is such that $2^p \not\equiv 2$ $\pmod{p^4}$ then necessarily $p \nmid (x-y)(y-z)(z-x)$.

Wieferich showed in 1909 that if the first case of Fermat's last theorem is assumed false for the exponent $p$ then $2^p \equiv 2 \pmod{p^2}$. This was the first of a series of criteria of similar type, discovered by Mirimanoff, Vandiver, Pollaczek, Rosser, and Granville. The proofs of these results are lengthy and nonelementary. For a fuller discussion, see my book *13 Lectures on Fermat's Last Theorem*.

**G. Criterion with a Determinant.** In 1907, Bini used a classical recurrence relation to obtain an expression in terms of a determinant:

**(2K)**    *Let $x, y, z$ be any numbers, and let*

$$\begin{cases} a = x + y + z, \\ b = xy + yz + zx, \\ c = xyz, \end{cases}$$

*and for every $n \geq 1$ let $S_n = x^n + y^n + z^n$. Then:*

(1) $S_n - aS_{n-1} + bS_{n-2} + cS_{n-3} = 0.$
(2)

$$S_n = \det \begin{pmatrix} 0 & -a & b & -c & 0 & \cdots & 0 & 0 \\ 0 & 1 & -a & b & -c & \cdots & 0 & 0 \\ 0 & 0 & 1 & -a & b & \cdots & 0 & 0 \\ 0 & 0 & 0 & 1 & -a & \cdots & 0 & 0 \\ \vdots & \vdots & \vdots & \vdots & \vdots & \ddots & \vdots & \vdots \\ 0 & 0 & 0 & 0 & 0 & \cdots & b & -c \\ 3c & 0 & 0 & 0 & 0 & \cdots & -a & b \\ -2b & 0 & 0 & 0 & 0 & \cdots & 1 & -a \\ a & 0 & 0 & 0 & 0 & \cdots & 0 & 1 \end{pmatrix}.$$

PROOF. (1)    $x, y, z$ satisfy the equation

$$X^3 - aX^2 + bX - c = 0.$$

Multiplying with $X^{n-3}$, we have

$$X^n - aX^{n-1} + bX^{n-2} - cX^{n-3} = 0.$$

Replacing $X$ by $x, y, z$ and adding the relations so obtained, we have

$$S_n - aS_{n-1} + bS_{n-2} - cS_{n-3} = 0.$$

(2)    We write the above relations for $k = 1, 2, \ldots, n$. Thus $S_1, S_2, \ldots, S_n$ satisfy a system of $n$ linear equations:

$$
\begin{cases}
S_n - aS_{n-1} + bS_{n-2} - cS_{n-3} = 0, \\
\qquad\qquad\vdots \\
S_4 - aS_3 + bS_2 - cS_1 = 0, \\
S_3 - aS_2 + bS_1 = 3c, \\
S_2 - aS_1 = -2b, \\
S_1 = a.
\end{cases}
$$

The determinant of the coefficients of the system is equal to 1. Applying Cramer's rule, it follows that $S_n$ is given by the determinant indicated. $\square$

It should be pointed out here that Bini proved that this implies: if $p$ is an odd prime and $x + y + z = 0$ then $xyz$ divides $x^p + y^p + z^p$. But, as a matter of fact, this statement follows at once from the remark preceding (2A) of Chapter VII, noting that $-z = x + y$.

**H. Connection with a Binary Quadratic Form.** In 1963, Pignataro linked Fermat's equation with the representation of a $p$th power by means of a binary quadratic form.

We first recall the following well-known fact. Fermat considered in 1657 the equation $X^2 - dY^2 = 1$, where $d$ is a positive integer, but not a square (this equation has been erroneously named after Pell). Fermat stated that he had proved by the method of descent the existence of infinitely many solutions in integers. However, the first published proof is due to Lagrange, around 1766. Explicitly the result is the following:

LEMMA 2.3. *Let $p$ be a positive integer, but not a square. There exists a solution in positive integers $(x_1, y_1)$ for the equation $X^2 - dY^2 = 1$ such that: $(x, y)$ is a solution in integers if and only if there exists an integer $m$ such that $x + y\sqrt{d} = (x_1 + y_1\sqrt{d})^m$. In particular, for different values of $m$ one obtains different solutions and therefore $X^2 - dY^2 = 1$ has infinitely many solutions in positive integers.*

The proof of Lagrange used continued fractions. Another proof may be found, for example, in Ribenboim (1999).

We introduce the following notation. If $b, c$ are nonzero integers, let $\langle b, c \rangle = bX^2 + cY^2$.

**(2L)**    *Let $p$ be an odd prime, and let $x$ be an odd positive integer which is minimal such that there exist positive integers $y, z$ satisfying $x^p + y^p = z^p$. Then:*

(1) *$yz$ is a quadratic residue modulo $x$; and*
(2) *$x^p$ is represented by the quadratic form $\langle z, -y \rangle$ in infinitely many ways.*

PROOF. (1)   We first note that since $x^p + y^p = z^p$ and $x$ is minimal then $x, y, z$ are pairwise relatively prime. We write

$$x^p = z^p - y^p = z \left( z^{(p-1)/2} \right)^2 - y \left( y^{(p-1)/2} \right)^2,$$

so $\langle z, -y \rangle$ represents $x^p$. Also, $z \left( z^{(p-1)/2} \right)^2 \equiv y \left( y^{(p-1)/2} \right)^2 \pmod{x}$, hence $yz \left( z^{(p-1)/2} \right)^2 \equiv y^{p+1} \pmod{x}$, so $yz$ is a quadratic residue modulo $x$.

(2)   Now we show that $y, z$ cannot be both squares. Otherwise $y = y_1^2$, $z = z_1^2$ with $y_1, z_1 > 0$, so

$$x^p = z_1^{2p} - y_1^{2p} = \left( z_1^p - y_1^p \right) \left( z_1^p + y_1^p \right).$$

Since $x$ is odd and $\gcd(y, z) = 1$ then $\gcd \left( z_1^p - y_1^p, z_1^p + y_1^p \right) = 1$ and clearly $z_1^p + y_1^p > 1$. Hence $z_1^p - y_1^p$ is a $p$th power of an odd positive integer $x'$, $0 < x' < x$, which contradicts the minimality of $x$.

As seen above, $x^p = za^2 - yb^2$ where $a = z^{(p-1)/2}$, $b = y^{(p-1)/2}$. Consider the equation $X^2 - yzY^2 = 1$. Since $\gcd(y, z) = 1$ and $y, z$ are not both squares, then $yz$ is not a square.

By the above lemma, there exist positive integers $u_1, v_1$ such that $1 = u_1^2 - yzv_1^2$ and moreover for every integer $m$ if $u_m + v_m \sqrt{yz} = (u_1 + v_1 \sqrt{yz})^m$ then $1 = u_m^2 - yzv_m^2$. We note that if $1 \leq m$ then

$u_m < u_{m+1}$ and $v_m < v_{m+1}$. Then

$$x^p = (za^2 - yb^2)(u_m^2 - yzv_m^2)$$

$$= z\left(a^2 - \frac{y}{z}b^2\right)\left(u_m^2 - \frac{y}{z}z^2v_m^2\right)$$

$$= z\left(a - b\sqrt{\frac{y}{z}}\right)\left(u_m - v_m z\sqrt{\frac{y}{z}}\right)\left(a + b\sqrt{\frac{y}{z}}\right)\left(u_m + v_m z\sqrt{\frac{y}{z}}\right)$$

$$= z\left[(au_m + bv_m y) - (av_m z + bu_m)\sqrt{\frac{y}{z}}\right]$$

$$\cdot \left[(au_m + bv_m y) + (av_m z + bu_m)\sqrt{\frac{y}{z}}\right]$$

$$= z(au_m + bv_m y)^2 - y(av_m z + bu_m)^2.$$

This shows that $x^p$ is represented in infinitely many ways by $\langle z, -y \rangle$.
□

## I. The Non-Existence of Algebraic Identities Yielding Solutions of Fermat's Equation.

In 1895, Jonquières investigated whether there would be algebraic relations connecting hypothetical solutions of Fermat's equation.

If $n = 2$, we have the algebraic relation

$$(X^2 + Y^2)^2 - (X^2 - Y^2)^2 = (2XY)^2$$

and, as was indicated in Chapter I, (1A), this yields all primitive solutions of the Pythagorean equation.

We shall see that nothing of the kind exists if $n > 2$.

LEMMA 2.4. *Let* $F = F_0 + F_1 + F_2 + \cdots \in \mathbb{Z}[X, Y]$ *where* $F_i$ *is the homogeneous part of degree* $i$ *of* $F$.

*If* $n \geq 1$, *there exist homogeneous polynomials of degree* $n$, $P_2 \in \mathbb{Z}[X_0, X_1]$, $P_3 \in \mathbb{Z}[X_0, X_1, X_2], \ldots$ *(depending only on* $n$) *such that the homogeneous parts of* $F^n$ *are:*

$$(F^n)_0 = (F_0)^n,$$
$$(F^n)_1 = n(F_0)^{n-1}F_1,$$
$$(F^n)_2 = n(F_0)^{n-1}F_2 + P_2(F_0, F_1),$$
$$(F^n)_3 = n(F_0)^{n-1}F_3 + P_3(F_0, F_1, F_2),$$
$$\vdots$$
$$(F^n)_k = n(F_0)^{n-1}F_k + P_k(F_0, F_1, \ldots, F_{k-1}).$$

PROOF. By raising $F$ to the $n$th power, we note that the homogeneous part of $(F^n)_k$ consists of $n(F_0)^{n-k}F_k$ plus a contribution involving the homogeneous parts of degree less than $k$. This is a polynomial expression, with multinomial coefficients, independent of $F$, depending only on $n$.    □

**(2M)**    *If $n > 2$, there do not exist polynomials $F, G \in \mathbb{Z}[X, Y]$ such that $X^n Y^n = F^n - G^n$.*

PROOF. Let $F = F_0 + F_1 + F_2 + \cdots$, $G = G_0 + G_1 + G_2 + \cdots$ where $F_k, G_k$ are the homogeneous parts of degree $k$ of $F, G$, respectively. If $F^n - G^n = X^n Y^n$, then $(F^n)_k = (G^n)_k$ for $k = 0, 1, \ldots, 2n - 1$, $(F^n)_{2n} - (G^n)_{2n} = X^n Y^n$. By the lemma, $F_0 = G_0, \ldots, F_{2n-1} = G_{2n-1}$ and hence also

$$
\begin{aligned}
n(F_0)^{n-1}F_{2n} &= (F^n)_{2n} - P_{2n}(F_0, F_1, \ldots, F_{2n-1}) \\
&= (G^n)_{2n} + X^n Y^n - P_{2n}(G_0, G_1, \ldots, G_{2n-1}) \\
&= n(G_0)^{n-1}G_{2n} + X^n Y^n,
\end{aligned}
$$

so $n(F_0)^{n-1}(F_{2n} - G_{2n}) = X^n Y^n$. This implies that $n = 1$, which is a contradiction.    □

We recall from Chapter VI, (3C), that if $n > 2$, if $0 < x < y < z$ are integers such that $x^n + y^n = z^n$, then $y$ is not a prime power. So $y = ab$, with $a, b > 1$, $\gcd(a, b) = 1$. The above result tells that it is impossible to find polynomials $F, G \in \mathbb{Z}[X, Y]$, such that for all $a, b \in \mathbb{Z}$ $a^n b^n = [F(a, b)]^n - [G(a, b)]^n$.

**J. Criterion with Second-Order Linear Recurrences.** We rephrase the results in Chapter III, (1A), (1B), (1C) in terms of second-order linear recurring sequences (see Kiss (1980)). We recall that if $m > 1$, if $(R_k)_{k \geq 0}$ is a sequence of integers, then $r(m)$ is the smallest index $r$ such that $m$ divides $R_r$.

**(2N)**    *Let $p$ be an odd prime, let $x, y, z$ be relatively prime integers such that $x^p + y^p + z^p = 0$. Let $A = x - y$, $B = -xy$, $R_0 = 0$, $R_1 = 1$, and for every $k \geq 2$ let $R_k = AR_{k-1} - BR_{k-2}$, and $D = A^2 - 4B = (x + y)^2$.*

(1) *If $p \nmid z$ then $R_p = d^p$, where $d$ is an integer, $R_p \equiv D^{(p-1)/2} \equiv 1 \pmod{p^2}$ and $d \equiv 1 \pmod{p}$, $r(d) = r(d^2) = \cdots = r(d^p) = p$ (with the notation of the lemma).*

(2) *If $p \mid z$ then $R_p/p = d^p$, where $d$ is an integer, $R_p \equiv p \pmod{p^4}$, $D \equiv 0 \pmod{p^{4p-2}}$ and $r(d) = r(d^2) = \cdots = r(d^p) = p$, $r(d^{p+1}) \neq p$.*

PROOF. (1)    The roots of $X^2 - AX + B$ are $x, -y$. By Chapter V, Lemma 2.1, $R_p = (x^p + y^p)/(x + y) = d^p$ (with $d > 1$), $x + y = c^p$, as follows from Chapter III, (1A), where $c = t$, $d = t_1 \equiv 1 \pmod{p}$, by Chapter III, (1B), (with the notation previously used). Then $R_p = d^p \equiv 1 \pmod{p^2}$. Since $p \nmid z$ and $x + y + z \equiv 0 \pmod{p}$ then $p \nmid x + y$ so $p \nmid c$. Hence $D^{(p-1)/2} = (x+y)^{p-1} = c^{p(p-1)} \equiv 1 \pmod{p^2}$.

Finally, since $d^p \mid R_p$ then by Chapter V, Lemma 2.2, $r(d), r(d^2), \cdots, r(d^p)$ divide $p$; since $d > 1$ then $r(d^i) \neq 1$, so $r(d) = r(d^2) = \cdots = r(d^p) = p$.

(2)    Assuming that $p \mid z$, it follows as before from Chapter III, (1B) and (2C), that $R_p = (x^p + y^p)/(x + y) = pd^p$, where $d = t_1 \equiv 1 \pmod{p^2}$, $x + y = p^{np-1}c^p$ where $n \geq 2$, $c = t$. Then $R_p/p = d^p \equiv 1 \pmod{p^3}$ so $R_p \equiv p \pmod{p^4}$. Also, $D = (x+y)^2 \equiv 0 \pmod{p^{4p-2}}$.

Finally, since $d^p \mid R_p$ but $d^{p+1} \nmid R_p$ then $r(d) = \cdots = r(d^p) = p$, $r(d^{p+1}) \neq p$.    □

And now we give a result of Kiss and Phong (1979), containing an interpretation of the congruence $q^{p-1} \equiv 1 \pmod{p^2}$ in terms of an appropriate recurring sequence.

**(2O)**    *Let $p, q$ be distinct primes such that $p \neq 2$ and $p \nmid q - 1$. Let $A = q + 1$, $B = q$, $R_0 = 0$, $R_1 = 1$ and every $k \geq 2$ let $R_k = AR_{k-1} - BR_{k-2}$. Let $D = A^2 - 4B = (q - 1)^2$. Then the following statements are equivalent:*

(a) *$q^{p-1} \equiv 1 \pmod{p^2}$;*
(b) *$p^2 / R_{p-1}$; and*
(c) *$r(p) = r(p^2)$.*

PROOF. (a) $\leftrightarrow$ (b)    The roots of $X^2 - AX + B = X^2 - (q+1)X + q$ are $\alpha = q$, $\beta = 1$. By Chapter V, Lemma 2.1(1), $R_{p-1} = (q^{p-1} - 1)/(q - 1)$. Since $p \nmid q - 1$ then $p^2 \mid R_{p-1}$ if and only if $p^2 \mid q^{p-1} - 1$.

(b) $\leftrightarrow$ (c)   By Chapter V, Lemma 2.2(5), $r(p)$ divides $p-(D/p) = p-1$, since $D$ is a square. Let $p-1 = sr(p)$. By Chapter V, Lemma 2.2(8), we have $R_{p-1} = R_{sr(p)} \equiv SR_{r(p)}R_{r(p)+1}^{s-1}$ (mod $R_{r(p)}^2$).

Since $p \mid R_{r(p)}$ then $p^2 \mid R_{r(p)}^2$; also by Chapter V, Lemma 2.2(2), $p \mid R_{sr(p)}$, $p \nmid R_{r(p)+1}$ and $p \nmid s$ (since $s \le p-1$). Hence $p^2 \mid R_{p-1}$ if and only if $p^2 \mid R_{r(p)}$. Let $v_p(R_{r(p)}) = k \ge 2$.

By Part (6) of Lemma 2.2 of Chapter V, $r(p^k) = r(p)$. But since $p \mid R_{r(p^2)}$ then $r(p) \mid r(p^2)$. On the other hand, $2 \le k$, hence $p^2 \mid R_{r(p^k)} = R_{r(p)}$, hence $r(p^2) \mid r(p)$, showing the equality. And conversely, if $r(p) = r(p^2)$ then $p^2 \mid R_{r(p)}$.   $\square$

## K. Perturbation of One Exponent. 

The following result (see Schaumberger (1973), Klamkin (1974)) is rather a curiosity; it tells that if one "perturbs" even slightly one of the exponents in Fermat's equation, the new equation has infinitely many solutions in integers. We first show:

**(2P)**    *If $a,b,c$ are integers, $a,b,c \ge 1$, $\gcd(ab,c) = 1$, then the equation $X^a + Y^b = Z^c$ has infinitely many solutions in integers.*

PROOF. It is trivial if $c = 1$. Let $c \ne 1$. We note that there exist integers $d,e$ such that $abd + 1 = ce$. We have $d \ne 0$, otherwise $ce = 1$, $c \ge 1$, hence $c = 1$, which has been excluded. So, there exists an integer $t$ such that $d+tc \ge 1$ and $ab(d+tc)+1 = c(e+abt)$ with $e + abt \ge 1$ since $c \ge 1$. Thus, there is no loss of generality to assume that $d \ge 1$, $e \ge 1$, and $abd + 1 = ce$.

Let $u \ge 1$ be arbitrary and let

$$\begin{cases} x = 2^{bd}u^{bc}, \\ y = 2^{ad}u^{ac}, \\ z = 2^{e}u^{ab}. \end{cases}$$

Then $x^a + y^b = 2^{abd}u^{abc} + 2^{abd}u^{abc} = 2^{abd+1}u^{abc} = 2^{ce}u^{abc} = z^c$. Since $u$ is arbitrary, the given equation has infinitely many solutions in integers.   $\square$

It follows at once:

**(2Q)**    *If $n,k \ge 1$ then $X^n + Y^n = Z^{n+1/k}$ has infinitely many solutions in integers.*

PROOF. First we note that $(x, y, z)$ is a solution of $X^n + Y^n = Z^{kn+1}$ if and only if $(x, y, z^k)$ is a solution of $X^n + Y^n = Z^{n+1/k}$. Taking $a = n$, $b = n$, $c = kn + 1$ in (2O), we see that the given equation has indeed infinitely many solutions in integers.  □

## L. Divisibility Condition for Pythagorean Triples. In 1913, Niewiadomski considered the polynomials in three indeterminates

$$D_0 = D_0(X, Y, Z) = -1,$$

and if $n \geq 1$,

$$D_n = D_n(X, Y, Z) = Z^n - X^n - Y^n.$$

He observed the identity, for $n \geq 1$,

$$(2.2) \quad D_{n+1} - (X + Y)D_n + XYD_{n-1} = Z^{n-1}(Z - X)(Z - Y),$$

which may be verified at once.

With this identity, Niewiadomski and Métrod proved (1913):

**(2R)**    *Let $x, y, z$ be positive integers such that $x^2 + y^2 = z^2$. Let $d_n = D_n(x, y, z)$ for all $n \geq 1$. Then $2d_n$ is divisible by $d_1^2$ when $n \geq 2$.*

PROOF. We may assume without loss of generality that $x, y, z$ are relatively prime. Indeed, let $e = \gcd(x, y, z)$, $x' = x/e$, $y' = y/e$, $z' = z/e$. So $x'^2 + y'^2 = z'^2$ and we may assume that $x'$ is even, while $y'$, $z'$ are odd. Let $d_n' = D_n(x', y', z')$, so $d_n = e^n d_n'$, in particular $d_1 = ed_1'$, and we note that $d_1 \neq 0$. Hence $2d_n/d_1^2 = e^{n-2}2d_n'/d_1'^2$. Since $n \geq 2$, it suffices to show that $2d_n'/d_1'^2$ is an integer.

With the assumption that $e = 1$, by Chapter I, (1A), there exist integers $a, b$ with $0 < b < a$, $\gcd(a, b) = 1$, such that

$$\begin{cases} x = 2ab, \\ y = a^2 - b^2, \\ z = a^2 + b^2. \end{cases}$$

If $n = 2$ then $d_2 = 0$ is divisible by $d_1^2$. Similarly by (2.2),

$$d_3 = (x + y)d_2 - xyd_1 + z(z - x)(z - y).$$

Hence noting that $d_1 = 2b(b - a)$, we have

$$2d_3 = -4ab(a^2 - b^2)d_1 + (a^2 + b^2)(a - b)^2 4b^2$$
$$= [2a(a + b) + a^2 + b^2]d_1^2 = [(a + b)^2 + 2a^2]d_1^2.$$

Assuming that $2d_{n-1}$ and $2d_n$ are divisible by $d_1^2$, it follows from (2.2) that

$$2d_{n+1} = (x + y)2d_n - xy \cdot 2d_{n-1} + 2z^{n-1}(z - x)(z - y)$$

is also divisible by $d_1^2$ because

$$2z^{n-1}(z - x)(z - y) = (a^2 + b^2)^{n-1}(a - b)^2 4b^2 = (a^2 + b^2)^{n-1}d_1^2. \quad \Box$$

## Bibliography

1657 Fermat, P. de, Lettre à Frénicle (Février 1657) et Second Défi aux Mathématiciens, *Oeuvres*, Vol. II, pp. 333–335, Gauthier-Villars, Paris, 1894.

1766 Lagrange, J.L., *Solution d'un problème d'arithmétique*, Miscellanea Taurinensia, **4** (1766–1769); reprinted in *Oeuvres*, Vol. I, pp. 671–731, Gauthier-Villars, Paris, 1867.

1895 Jonquières, E. de, *Sur une question d'algèbre qui a des liens avec le dernier théorème de Fermat*, C. R. Acad. Sci. Paris, **120** (1895), 1139–1143.

1907 Bini, U., *Sopra alcune congruenze*, Period. Mat., (3), **22** (1907), 180–183.

1907 Bôcher, M., *Introduction to Higher Algebra*, Macmillan, New York, 1907.

1907 Bottari, A., *Soluzioni intere in progressione aritmetica appartenenti a equazione indeterminate del tipo $\sum_{v=1}^{r} x_v^n = x_{r+i}^n$*, Period. Mat., (3), **22** (1907), 156–158.

1907 Mirimanoff, D., *Sur les congruences du troisième degré*, Enseign. Math., **9** (1907), 381–384.

1908 Cailler, R., *Sur les congruences du troisième degré*, Enseign. Math., **10** (1908), 474–487.

1908 Cattaneo, P., *Osservazioni sopra due articoli del Signor Amerigo Bottari*, Period. Mat., (3), **23** (1908), 218–220.

1909 Mirimanoff, D., *Sur le dernier théorème de Fermat*, Enseign. Math., **11** (1909), 49–51.

1913 Goldziher, H., *Hatványszamok Telbontása hatványszamok összegere*, Középiskolai Math. Lapok, **21** (1913), 177–184.

1913 Niewiadomski, R., *Question 4205*, L'Interm. Math., **20** (1913), 98–100.

1913 Métrod, G., *Sur la question 4205 de Niewiadomski*, L'Interm. Math., **20** (1913), 215–216.

1917 Pollaczek, F., *Über den großen Fermat'schen Satz*, Sitzungsber. Akad. Wiss., Wien, Abt. IIa, **126** (1917), 45–59.

1928 Pérez-Cacho, L., *Una proposición sobre el indicador*, Rev. Mat. Hisp.-Amer., (2), **3** (1928), 273–275

1932 Bussi, C., *Sull'ultimo teorema di Fermat*, Boll. Un. Mat. Ital., **11** (1932), 267–269.

1937 Skolem, T., *Zwei Sätze über kubische Kongruenzen*, Det Kongel. Norske Vidensk. Selskab Forhandlinger, Trondhejm, **10** (1937), no. 24, 89–92.

1941 Skolem, T., *Die Anzahl der Wurzeln der Kongruenz $x^3 + ax + b \equiv 0 \pmod{p}$ für die verschiedenen Paare $a, b$*, Det Kongel. Norske Vidensk. Selskab Forhandlinger, Trondhejm, **14** (1941), no. 43, 161–164.

1943 Bussi, C., *Osservazione sull'ultimo teorema di Fermat*, Boll. Un. Mat. Ital., (2), **5** (1943), 42–43.

1944 Pierre, C., *Remarques arithmétiques en connexion avec le dernier théorème de Fermat*, C. R. Acad. Sci. Paris, **218** (1944), 23–25.

1946 Inkeri, K., *Untersuchungen über die Fermatsche Vermutung*, Ann. Acad. Sci. Fenn., Ser. A1, Nr. 33, 1946, 60 pp.

1949 Kapferer, H., *Über ein Kriterium zur Fermatschen Vermutung*, Comment. Math. Helv., **23** (1949), 64–75.

1952 Mihaljinec, M., *Prilog Fermatovu problemu* (Une contribution au problème de Fermat), Hrvatsko Prirodoslovno Društvo, Glas. Mat. Fiz. Astr. Ser. II, **7** (1952), 12–18.

1958 Pérez-Cacho, L., *Sobre algunas cuestiones de la teoria de números*, Rev. Mat. Hisp.-Amer., (4), **18** (1958), 10–27 and 113–124.

1963 Pignataro, S., *Una osservazione sull'ultimo teorema di Fermat*, Rend. Accad. Sci. Fis. Mat. Napoli, (4), **30** (1963), 281–286.

1969 Rameswar Rao, D., *Some theorems on Fermat's last theorem*, Math. Student, **37** (1969), 208–210.

1969 Swistak, J.M., *A note on Fermat's last theorem*, Amer. Math. Monthly, **76** (1969), 173–174.

1972 Ribenboim, P., *Algebraic Numbers*, Wiley-Interscience, New York, 1972.

1973 Schaumberger, N., *Question 572*, Math. Mag., **46** (1973), 168.

1974 Klamkin, M.S., *Solution of the question 572*, Math. Mag., **47** (1974), 177–178.

1979 Kiss, P. and Phong, Bui Minh, *Divisibility properties in second order recurrences*, Publ. Math. Debrecen, **26** (1979), 187–197.

1979 Ribenboim, P., *13 Lectures on Fermat's Last Theorem*, Springer-Verlag, New York, 1979.

1980 Kiss, P., *Connection between second-order recurrences and Fermat's last theorem*, Period. Math. Hungar., **11** (1980), 151–157.

1999 Ribenboim, P., *The Classical Theory of Algebraic Numbers*, Springer-Verlag, New York, 1999.

# IX
# Interludes 9 and 10

We shall need the Gaussian periods and the Lagrange resolvents and Jacobi cyclotomic functions in the study of Fermat's congruence.

## IX.1. The Gaussian Periods

Let $q$ be an odd prime, $\rho$ a primitive $q$th root of 1, $h$ a primitive root modulo $q$, $L = \mathbb{Q}(\rho)$, $B = \mathbb{Z}[q]$, and let $\tau$ be the generator of the Galois group of $L \mid \mathbb{Q}$ defined by $\tau(\rho) = \rho^h$.

Every element $\alpha \in L$ may be indifferently written in a unique way as

$$\alpha = \sum_{i=0}^{q-2} a_i \rho^i \qquad \text{or as} \qquad \alpha = \sum_{j=0}^{q-2} a_j{}' \rho^{h^j}$$

(with $a_i, a_j{}' \in \mathbb{Q}$); moreover, $a \in B$ if and only if each $a_i, a_j{}' \in \mathbb{Z}$.

Comparing these two representations, and noting that

$$\rho^{h^{(q-1)/2}} = \rho^{q-1} = -(1 + \rho + \cdots + \rho^{q-2}),$$

it follows that $a_0 = -a'_{(q-1)/2}$ and $a_i = a_j{}' - a'_{(q-1)/2}$ where $i \equiv h^j$ (mod $q$) (for $i = 1, \ldots, q - 2$).

In the present situation, if $q - 1 = fr$, the $r$ periods with $f$ terms (relative to $\rho$ and $\tau$ or $h$) are:

(1.1)
$$
\begin{cases}
\mu_0 &= \rho + \rho^{h^r} + \rho^{h^{2r}} + \cdots + \rho^{h^{(f-1)r}}, \\
\mu_1 &= \rho^h + \rho^{h^{r+1}} + \rho^{h^{2n+1}} + \cdots + \rho^{h^{(f-1)r+1}}, \\
&\vdots \\
\mu_{r-1} &= \rho^{h^{r-1}} + \rho^{h^{2r-1}} + \rho^{h^{3r-1}} + \cdots + \rho^{h^{q-2}}.
\end{cases}
$$

We have $\sum_{j=0}^{r-1} \mu_j = -1$. For every $j$, we write $\mu_j = \mu_{j_0}$ if $0 \leq j_0 \leq r - 1$ and $j \equiv j_0 \pmod r$. The periods $\mu_j$ are conjugate to each other: $\tau^i(\mu_j) = \mu_{i+j}$ (for $i = 0, 1, \ldots, q - 2$, and any $j$). In particular, $\tau^r(\mu_j) = \mu_j$ for $j = 0, 1, \ldots, r - 1$.

Let $L'$ denote the subfield of $L$ which is fixed by $\tau^r$, so $[L : L'] = f$, $[L' : \mathbb{Q}] = f$; the Galois group of $L \mid L'$ is generated by $\tau^2$ and the Galois group of $L' \mid \mathbb{Q}$ is generated by the restriction $\tau'$ of $\tau$ to $L'$. Let $B'$ denote the ring of integers of $L'$.

**(1A)**

(1) $\{\mu_0, \mu_1, \ldots, \mu_{r-1}\}$ is a basis of the $\mathbb{Z}$-module $B'$.
(2) $L' = \mathbb{Q}(\mu_0, \ldots, \mu_{r-1})$, $B' = \mathbb{Z}[\mu_0, \ldots, \mu_{r-1}]$.
(3) $\{1, \rho, \rho^2, \ldots, \rho^{f-1}\}$ is a basis of the $B'$-module $B$.
(4) The polynomial of periods

(1.2)
$$
F_{\mu_0}(X) = \prod_{i=0}^{r-1} (X - \mu_i)
$$

has coefficients in $\mathbb{Z}$ and it is irreducible.

PROOF. (1) The elements $\mu_0, \mu_1, \ldots, \mu_{r-1}$ are linearly independent over $\mathbb{Z}$: if $\sum_{i=0}^{r-1} a_i \mu_i = 0$ (with $a_i \in \mathbb{Z}$), replacing each $u_i$ by its expression, we have a linear combination of $\rho, \rho^2, \ldots, \rho^q$ which is equal to 0, and with coefficients 0. $a_0, a_1, \ldots, a_{r-1} \in \mathbb{Z}$ so each $a_i = 0$.

On the other hand, if $\alpha \in B' \subseteq B$, we may write $\alpha = \sum_{i=0}^{q-2} a_i \rho^{h^i}$ with $a_i \in \mathbb{Z}$. Since $\tau^r(\alpha) = \alpha$ then

$$
\sum_{i=0}^{q-2} a_i \rho^{h^{i+r}} = \sum_{i=0}^{q-2} a_i \rho^{h^i}
$$

and from the uniqueness of the expression, we deduce that

$$\begin{cases}
a_0 = a_r = \cdots = a_{(f-1)r}, \\
a_1 = a_{r+1} = \cdots = a_{(f-1)r+1}, \\
\quad\vdots \\
a_{r-1} = a_{2r-1} = \cdots = a_{q-2}.
\end{cases}$$

Hence $a = \sum_{j=0}^{r-1} a_j \mu_j$.

(2) Clearly $L' \supseteq \mathbb{Q}(\mu_0, \dots, \mu_{r-1})$ and $B' \supseteq \mathbb{Z}[\mu_0, \dots, \mu_{r-1}]$. The converse follows from (1.1).

(3) Let $G(X) = \prod_{i=0}^{f-1}(X - \rho^{h^{ir}})$ be the polynomial whose roots are the summands of the period $\mu_0$. Then each coefficient of $G(X)$ is invariant by $\tau^r$, hence it belongs to $B \cap L' = B'$.

Thus $G(X) = X^f + \alpha_1 X^{f-1} + \cdots + \alpha_f$ and since $\rho$ is a root of $G(X)$, then $\rho^f = -(\alpha_1 \rho^{f-1} + \cdots + \alpha_f)$. So $\rho^f$ is a linear combination of $1, \rho, \dots, \rho^{f-1}$ with coefficients in $B'$. Multiplying the above relation successively by $\rho, \rho^2 \dots$, we deduce that $\rho^{f+1}, \rho^{f+2}, \dots, \rho^{q-1}$ are also linear combinations of $1, \rho, \rho^2, \dots, \rho^{f-1}$ with coefficients in $B'$. Thus every element of $B = \mathbb{Z}[\rho]$ is a linear combination of $1, \rho, \dots$ with coefficients in $B'$.

So $\{1, \rho, \dots, \rho^{f-1}\}$ is a system of generators of the $L'$-vector space $L$. Since $[L : L'] = f$ then $\{1, \rho, \dots, \rho^{f-1}\}$ are linearly independent over $L'$, hence over $B'$.

(4) The coefficients of $F_{\mu_0}(X)$ belong to $\mathbb{Q}$, since they are invariant by $\tau$; hence they are in $B' \cap \mathbb{Q} = \mathbb{Z}$.

Since $F_{\mu_0}(\mu_0) = 0$, the minimal polynomial of $\mu_0$ divides $F_{\mu_0}(X)$; its roots are all the conjugates of $\mu_0$, so it must coincide with $F_{\mu_0}(X)$, which is therefore irreducible.    $\square$

It is not true in general that

$$\mathbb{Z}[\mu_0, \dots, \mu_{r-1}] = \mathbb{Z}[\mu_0] = \cdots = \mathbb{Z}[\mu_{r-1}].$$

For example, let $q = 13$, $f = 3$, $r = 4$, and $k = 2$. The periods are:

$$\mu_0 = \rho + \rho^3 + \rho^{-4},$$
$$\mu_1 = \rho^2 + \rho^6 + \rho^5,$$
$$\mu_2 = \rho^4 + \rho^{-1} + \rho^{-3},$$
$$\mu_3 = \rho^{-5} + \rho^{-2} + \rho^{-6}.$$

We shall show that the unique expressions of $\mu_1, \mu_2, \mu_3$ as polynomials in $\mu_0$ with rational coefficients require some non-integral coefficients. Indeed:

$$\mu_0^2 = \mu_1 + 2\mu_2,$$
$$\mu_0\mu_1 = \mu_0 + \mu_1 + \mu_3,$$
$$\mu_0\mu_2 = 3 + \mu_1 + \mu_3,$$

and

$$\mu_0^3 = \mu_0\mu_1 + 2\mu_0\mu_2 = 6 + \mu_0 + 3\mu_1 + 3\mu_3$$
$$= 6 + \mu_0 + 3(-1 - \mu_0 - \mu_2),$$

hence

$$\mu_2 = \tfrac{1}{3}(-\mu_0^3 - 2\mu_0 + 3).$$

From this we obtain

$$\mu_1 = \mu_0^2 - 2\mu_2 = \tfrac{1}{3}(2\mu_0^3 + 3\mu_0^2 + 4\mu_0 - 6),$$
$$\mu_3 = -1 - \mu_0 - \mu_1 - \mu_2 = \tfrac{1}{3}(-\mu_0^3 - 3\mu_0^2 - 5\mu_0).$$

It follows from (1A) that given $i, j$, $0 \le i, j \le r - 1$, there exist integers $n_{ijk} \in \mathbb{Z}$ ($0 \le k \le r - 1$), which are unique such that $\mu_i\mu_j = \sum_{k=0}^{r-1} n_{ijk}\mu_k$. More precisely:

**(1B)**    *We have the relations*

$$\sum_{i=0}^{r-1} \mu_i\mu_{i+k} = n_k q - f \qquad (\textit{for } 0 \le k \le r - 1),$$

*where*

$$n_k = \begin{cases} 1 & \textit{when } f \textit{ is even and } k = 0, \\ 1 & \textit{when } f \textit{ is odd and } k = 0 \textit{ or } r/2, \\ 0 & \textit{otherwise.} \end{cases}$$

PROOF. First we evaluate the product

$$\mu_0\mu_k = \left( \sum_{l=0}^{f-1} \rho^{h^{lr}} \right) \left( \sum_{j=0}^{f-1} \rho^{h^{k+jr}} \right).$$

Writing $j \equiv i + l \pmod{q - 1}$ then the above product is equal to

$$\mu_0\mu_k = \sum_{l=0}^{f-1} \sum_{i=0}^{f-1} \rho^{h^{lr}(1 + h^{k+ir})}.$$

Let

$$\mu'_i = \sum_{l=0}^{f-1} \rho^{h^{lr}(1+h^{k+ir})}.$$

If $1 + h^{k+ir} \not\equiv 0 \pmod{q}$, there exists a unique $t$, $0 \leq t \leq q - 2$, such that $1 + h^{k+ir} \equiv h^t \pmod{q}$; hence $\mu'_i$ is equal to the period $\mu_t$. If $1 + h^{k+ir} \equiv 0 \pmod{q}$ then $\mu'_i = f$. Therefore, we may write

$$(1.3) \qquad \mu_0\mu_k = n_k f + m_{k,0}\mu_0 + m_{k,1}\mu_1 + \cdots + m_{k,r-1}\mu_{n-1},$$

with integers $n_k \geq 0$, $m_{k,0} \geq 0, \ldots, m_{k,r-1} \geq 0$. Now we determine $n_k$.

(I)    If $f$ is even and $k = 0$, let $f = 2f'$, then $1 + h^{f'r} \equiv 0$ (mod $q$) since $fr = q - 1$. So $\mu'_{f/2} = f$. On the other hand, if $0 \leq i < f$ and $\mu_i' = f$, then we have $1 + h^{ir} \equiv 0 \pmod{q}$, hence $2ir \equiv 0 \pmod{q-1}$, that is, $2ir = mrf$; but $mf = 2i < 2f$, $m = 0$ or 1. If $m = 0$ then $i = 0$, an absurdity because $q$ is odd. Thus $m = 1, i = f/2$. Therefore in this case $n_k = 1$.

(II)    If $f$ is odd (hence $r$ is even) and $k = r/2$, let $i = (f-1)/2$. Then $1 + h^{r/2+((f-1)/2)r} \equiv 0 \pmod{q}$, so $\mu'_{(f-1)/2} = f$. On the other hand, if $0 \leq i < f$ and $\mu_i' = f$, we have $1 + h^{r/2+ir} \equiv 0 \pmod{q}$, hence $r + 2ir = mrf$; thus $mf = 1 + 2i < 1 + 2f$; it follows that $m$ is odd, so $m = 1$ and $i = (f-1)/2$.

(III)    We consider the remaining cases. If $1 + h^{k+ir} \equiv 0 \pmod{q}$ then $2k + 2ir = mrf$ and $0 \leq r(mf - 2i) = 2k < 2r$, thus $mf - 2i = 0$ or 1.

If $mf = 2i < 2f$ then $m = 0$ or 1, and $k = 0$. If $m = 0$ then $i = 0$, $k = 0$, an absurdity, since $q$ is odd. Thus $m = 1$ and $f$ is even, which is a case already studied.

If $mf = 2i + 1 < 2f + 1$ then $m$ is odd, $m \leq 2$, so $m = 1$, $f$ is odd, $i = (f-1)/2$ and also $k = r/2$, which was Case (II) above. Therefore, in Case (III), $n_k = 0$.

Since $\mu_0\mu_k$ is the sum of $f^2$ terms of the form $\rho^i$ and since each period contains $f$ such terms, all appearing with different exponents $i$, $0 \leq i \leq q - 1$, it follows that

$$n_k + m_{k,0} + m_{k,1} + \cdots + m_{k,n-1} = f.$$

Applying the automorphisms $\tau^i$, we obtain from (1.3):

$$\mu_i\mu_{k+i} = n_k f + m_{k,0}\mu_i + m_{k,1}\mu_{i+1} + \cdots + m_{k,n-1}\mu_{r-1}.$$

Hence, from $\sum_{i=0}^{r-1} \mu_i = -1$ we conclude that $\sum_{i=0}^{r-1} \mu_i \mu_{i+k} = n_k(q - 1) - (m_{k,0} + m_{k,1} + \cdots + m_{k,r-1}) = n_k q - f.$  $\square$

## IX.2. Lagrange Resolvents and Jacobi Cyclotomic Function

We shall use the following notations:

- $p, q$ are prime numbers such that $q - 1 = 2kp$;
- $\zeta$ = primitive $k$th root of 1;
- $g$ = primitive root modulo $p$;
- $K = \mathbb{Q}(\zeta)$;
- $A = \mathbb{Z}[\zeta]$;
- $\sigma$ = generator of the Galois group of $K \mid \mathbb{Q}$, defined by $\sigma(\zeta) = \zeta^g$;
- $\rho$ = primitive $q$th root of 1 ;
- $L = \mathbb{Q}(\rho)$, $B = \mathbb{Z}[\rho]$;
- $\tau$ = generator of the Galois group of $L \mid \mathbb{Q}$, defined by $\tau(\rho) = \rho^h$;
- $\mu_0, \dots, \mu_{p-1}$: the $p$ periods with $2k$ terms (relative to $\rho$, $\tau$);
- $L' = \mathbb{Q}(\mu_0, \dots, \mu_{p-1}) = \mathbb{Q}(\mu_0) = \cdots = \mathbb{Q}(\mu_{p-1})$;
- $B' = \mathbb{Z}[\mu_0, \dots, \mu_{p-1}]$; and
- $\tau'$ = restriction of $\tau$ to $L'$.

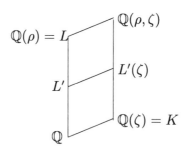

We note that $L \cap K = \mathbb{Q}$. Indeed, the prime $q$ is totally ramified in $L$ and unramified in $K$, hence it is both totally ramified and unramified in $L \cap K$, so $L \cap K = \mathbb{Q}$.

Thus $\mathbb{Q}(\rho, \zeta)$ is a Galois extension of $K$ with Galois group isomorphic to the one of $L \mid \mathbb{Q}$ and generated by the automorphism $\tilde{\tau}$, defined by

$$\begin{cases} \tilde{\tau}(\rho) = \rho^h, \\ \tilde{\tau}(\zeta) = \zeta. \end{cases}$$

Similarly, $\mathbb{Q}(\rho, \zeta)$ is a Galois extension of $L$ with Galois group isomorphic to the one of $K \mid \mathbb{Q}$ and generated by the automorphism $\tilde{\sigma}$, defined by

$$\begin{cases} \tilde{\sigma}(\rho) = \rho, \\ \tilde{\sigma}(\zeta) = \zeta^g. \end{cases}$$

It is convenient to work with indices, as we define now.

If $t$ is any integer, not a multiple of $q$, then there exists a unique integer $s$, $0 \leq s \leq q - 2$, such that $t \equiv h^s \pmod{q}$. $s$ is called the *index* of $t$ (with respect to $h, q$), and we write $s = \text{ind}_h(t)$, or simply $s = \text{ind}(t)$ if there is no ambiguity concerning the choice of the primitive root $h$.

For example, $\text{ind}(1) = 0$, $\text{ind}(-1) = (q-1)/2$. If $t \equiv t' \pmod{q}$ then $\text{ind}(t) = \text{ind}(t')$ and if $t, t'$ are not multiples of $q$ then $\text{ind}(tt') \equiv \text{ind}(t) + \text{ind}(t') \pmod{q-1}$. It is also clear that every integer $s$, $0 \leq s \leq q - 2$, is an index, namely $s = \text{ind}(h^s)$.

We define the *Lagrange resolvent* $\langle \alpha, \beta \rangle_\tau$, where $\alpha \in K$, $\beta \in L$, and $\tau$ is the automorphism defined above:

$$(2.1) \quad \langle \alpha, \beta \rangle_\tau = \beta + \alpha\tau(\beta) + \alpha^2\tau^2(\beta) + \cdots + \alpha^{p-1}\tau^{p-1}(\beta).$$

We write more simply $\langle \alpha, \beta \rangle$ in place of $\langle \alpha, \beta \rangle_\tau$. The element $\langle \alpha, \beta \rangle$ belongs to the field $\mathbb{Q}(\zeta, \rho)$. We note at once:

**(2A)**    *For every $n$ and $\beta \in \mathbb{Q}(\rho)$:*

$$\zeta^n \tilde{\tau}\left(\langle \zeta^n, \beta \rangle\right) = \langle \zeta^p, \beta \rangle \qquad and \qquad \langle \zeta^n, \beta \rangle^p \in \mathbb{Q}(\zeta).$$

PROOF.

$$\zeta^n \tilde{\tau}\left(\langle \zeta^n, \beta \rangle\right) = \zeta^n \tilde{\tau}(\beta + \zeta^n\tau(\beta) + \zeta^{2n}\tau^2(\beta) + \cdots + \zeta^{(p-1)n}\tau^{p-1}(\beta))$$
$$= \langle Z^n, \beta \rangle,$$

since $\zeta^p = 1$. So $\tilde{\tau}\left(\langle \zeta^n, \beta \rangle^p\right) = (\tilde{\tau}\langle \zeta^n, \beta \rangle)^p = \zeta^{-np} \cdot \langle \zeta^n, \beta \rangle^p = \langle \zeta^n, \beta \rangle^p$. Since $\langle \zeta^n, \beta \rangle$ is invariant by $\tilde{\tau}$ then $\langle \zeta^n, \beta \rangle^p \in \mathbb{Q}(\zeta)$. □

We consider the resolvents $\langle \zeta^n, \rho^m \rangle$. With the index notation introduced above, we may write

(2.2)        $$\langle \zeta^n, \rho^m \rangle_\tau = \sum_{u=0}^{q-2} \zeta^{nu} \rho^{mh^u} = \sum_{t=1}^{q-1} \zeta^{n\,\mathrm{ind}_h(t)} \rho^{mt}.$$

The complex conjugate of $\langle \zeta^n, \rho^m \rangle_\tau$ is $\overline{\langle \zeta^n, \rho^m \rangle_\tau} = \langle \zeta^{-n}, \rho^{-m} \rangle_\tau$. A first result to record is the following:

**(2B)**    *With the above notations, for every $n = 1, 2, \ldots, p-1$:*

$$\langle \zeta^n, \rho \rangle_\tau = \langle \zeta^n, \mu_0 \rangle_{\tau'} \neq 0$$

*and it belongs to $L'$.*

PROOF.

(2.3) $\langle \zeta^n, \rho \rangle_\tau = \rho + \zeta^n \rho^h + \zeta^{2n} \rho^{h^2} + \cdots + \zeta^{(q-2)n} \rho^{h^{q-2}}$

$\qquad = \rho + \zeta^n \rho^h + \cdots + \zeta^{(p-1)n} \rho^{h^{p-1}}$

$\qquad\quad + \rho^{h^p} + \zeta^n \rho^{h^{p+1}} + \cdots + \zeta^{(p-1)n} \rho^{h^{2p-1}}$

$\qquad\quad \cdots$

$\qquad\quad + \rho^{h^{(2k+1)p}} + \zeta^n \rho^{h^{(2k-1)p+1}} + \cdots + \zeta^{(p-1)n} \rho^{h^{q-1}}$

$\qquad = \mu_0 + \zeta^n \mu_1 + \cdots + \zeta^{(p-1)n} \mu_{p-1}$

$\qquad = \langle \zeta^n, \mu_0 \rangle_{\tau'} \in L'.$

Moreover, $\langle \zeta^n, \mu_0 \rangle_{\tau'} \neq 0$. Indeed, the periods $\mu_0, \mu_1, \ldots, \mu_{p-1}$, which are a basis of $L' \mid \mathbb{Q}$, are still a basis of $L'(\zeta) \mid \mathbb{Q}$ since this extension has still degree $p$.  $\square$

The theory of Gaussian periods, Lagrange resolvents and more general sums of roots of unity is very rich and important. We shall only indicate the results which will be needed in the sequel.

**(2C)**    *If $p \nmid n$, $q \nmid m$ then $\langle \zeta^n, \rho^m \rangle = \langle \zeta^n, \rho \rangle \zeta^{-n\,\mathrm{ind}(m)}$. In particular, $\langle \zeta^n, \rho^{n^i} \rangle = \langle \zeta^n, \rho \rangle \zeta^{-ni}$.*

PROOF. $\langle \zeta^n, \rho^m \rangle = \sum_{t=1}^{q-1} \zeta^{n \operatorname{ind}(t)} \rho^{mt}$. But $\operatorname{ind}(tm) \equiv \operatorname{ind}(t) + \operatorname{ind}(m)$ (mod $q - 1$), hence

$$\langle \zeta^n, \rho^m \rangle = \zeta^{-n \operatorname{ind}(m)} \sum_{t=1}^{q-1} \zeta^{n \operatorname{ind}(tm)} \rho^{tm}$$

$$= \zeta^{-n \operatorname{ind}(m)} \sum_{s=1}^{q-1} \zeta^{n \operatorname{ind}(s)} \rho^s$$

$$= \zeta^{-\operatorname{ind}(m)} \langle \zeta^n, \rho \rangle. \quad \square$$

**(2D)**    *If $p \nmid n$ then $\langle \zeta^n, \rho \rangle \langle \zeta^{-n}, \rho \rangle = q$.*

PROOF.

$$\langle \zeta^n, \rho \rangle \langle \zeta^{-n}, \rho \rangle = \left( \sum_{t=1}^{q-1} \zeta^{n \operatorname{ind}(t)} \rho^t \right) \left( \sum_{s=1}^{q-1} \zeta^{-n \operatorname{ind}(s)} \rho^s \right)$$

$$= \sum_{s=1}^{q-1} \sum_{t=1}^{q-1} \zeta^{n[\operatorname{ind}(t) - \operatorname{ind}(s)]} \rho^{t+s}.$$

For each $s$ let $r$ be defined by the congruence $t \equiv rs \pmod{q}$. Since $p \mid q - 1$ then $\zeta^{q-1} = 1$, so the above sum is equal to

$$\sum_{s=1}^{q-1} \sum_{r=1}^{q-1} \zeta^{n \operatorname{ind}(r)} \rho^{(r+1)s}.$$

But $q - 1 = 2kp$ so

$$\sum_{r=1}^{q-1} \zeta^{n \operatorname{ind}(r)} = \sum_{m=1}^{q-1} \zeta^m = 2k \sum_{m=1}^{p} \zeta^m = 0,$$

hence we may add $\sum_{r=1}^{q-1} \zeta^{n \operatorname{ind}(r)} = 0$ and write

$$\langle \zeta^n, \rho \rangle \langle \zeta^{-n}, \rho \rangle = \sum_{s=0}^{q-1} \sum_{r=1}^{q-1} \zeta^{n \operatorname{ind}(n)} \rho^{(r+1)s}$$

$$= \sum_{r=1}^{q-1} \zeta^{n \operatorname{ind}(r)} \left( \sum_{s=0}^{q-1} \rho^{(r+1)s} \right).$$

But

$$\sum_{s=0}^{q-1} \rho^{(r+1)s} = \begin{cases} q & \text{when } r = q - 1, \\ 0 & \text{when } 1 \leq r \leq q - 2, \end{cases}$$

hence $\langle \zeta^n, \rho \rangle \langle \zeta^{-n}, \rho \rangle = q\zeta^{n \, \text{ind}(q-1)} = q\zeta^{n\text{ind}(-1)} = q\zeta^{n(q-1)/2} = q\zeta^{nkp}$
$= q$. $\square$

**(2E)**    $|\langle \zeta^m, \rho \rangle| = q$ when $p \nmid m$.

PROOF. The absolute value of $\lambda_m = \langle \zeta^m, \rho \rangle$ is

$$
\begin{aligned}
|\lambda_m| = \lambda_m \overline{\lambda_m} &= \langle \zeta^m, \rho \rangle \langle \zeta^{-m}, \rho^{-1} \rangle \\
&= \langle \zeta^m, \rho \rangle \langle \zeta^{-m}, \rho \rangle \zeta^{-m \, \text{ind}(-1)} = q\zeta^{m(q-1)/2} \\
&= q\zeta^{mkp} = q,
\end{aligned}
$$

using (2D), (2C). $\square$

Now we give an expression of the Gaussian periods in terms of the Lagrange resolvents:

**(2F)**    *If $p \nmid n$ then*

$$
\mu_n = \frac{1}{p} \sum_{j=0}^{p-1} \zeta^{-jn} \langle \zeta^j, \rho \rangle.
$$

PROOF. We compute the right-hand side:

$$
\begin{aligned}
\sum_{j=0}^{p-1} \zeta^{-jn} \langle \zeta^j, \rho \rangle &= \sum_{j=0}^{p-1} \zeta^{-jn} \left( \sum_{u=0}^{q-2} \zeta^{ju} \rho^{h^u} \right) \\
&= \sum_{j=0}^{p-1} \sum_{u=0}^{q-2} \zeta^{j(u-n)} \rho^{h^u} \\
&= \sum_{j=0}^{p-1} \sum_{t=-n}^{q-2-n} \zeta^{j-t} \rho^{h^{t+n}} \\
&= \sum_{t=-n}^{q-2-n} \rho^{h^{t+n}} \left( \sum_{j=0}^{p-1} \zeta^{jt} \right).
\end{aligned}
$$

But $\sum_{j=0}^{p-1} \zeta^{jt} = p$ when $p \mid t$, and equal to 0 otherwise. Thus the above sum is equal to

$$
p \left( \sum_{i=-n, \, p|t}^{q-2-n} \rho^{h^{t+n}} \right) = p\mu_n. \quad \square
$$

# X
# The Local and Modular Fermat Problem

In this chapter we investigate some natural modifications of the original Fermat problem. In the first section, we search solutions in $q$-adic integers. In the following section we consider Fermat's congruence.

## X.1. The Local Fermat Problem

Our aim is to show that for every prime $q$, Fermat's equation has solutions in nonzero $q$-adic integers. Our tool will be Hensel's Lemma.

**(1A)** *For every prime $q$ and every prime $p$, the equation $X^p + Y^p = Z^p$ has nontrivial solution in $q$-adic integers.*

PROOF. If $p = 2$, this is trivial, by Chapter I, (1A). So we may assume $p \neq 2$ and for convenience, we consider the equation $X^p + Y^p + Z^p = 0$.

*First Case: $q \neq p$.*
   Let $F(X) = X^p + q^p - 1$; then $X^p + q^p - 1 \equiv X^p - 1 \equiv (X - 1)(X^{p-1} + X^{p-2} + \cdots + X + 1)$ (mod $q$). Since $1 \mod q$ is not a root of $X^{p-1} + X^{p-2} + \cdots + X + 1$ modulo $q$, by Hensel's Lemma (see

Chapter V, (1T) and (1U)), there exists a $q$-adic integer $\alpha$ such that $\alpha \equiv 1 \pmod{q}$ and $\alpha^q + p^q + (-1)^p = 0$.

*Second Case: $q = p$.*

Let $F(X) = X^p + p^p - 1$ and $G_0(X) = X - 1$, $H_0(X) = X^{p-1} + X^{p-2} + \cdots + X + 1$. By Chapter II, (4B), the resultant of $G_0$ and $H_0$ is equal to $R = H_0(1) = p$, so $v_p(R) = 1$. Since $G_0(X)H_0(X) = X^p - 1$ then $F(X) \equiv G_0(X)H_0(X) \pmod{p^p}$. But $p \geq 3 > 2v_p(R)$, so we may apply Hensel's Lemma of Chapter V, (1T) and (1U). Thus there exist monic polynomials $G(X), H(X)$ in $\hat{\mathbb{Z}}_p[X]$ such that $G(X) \equiv G_0(X) \pmod{p^{p-1}}$, $H(X) \equiv H_0(X) \pmod{p^{p-1}}$ and $F(X) = G(X) \times H(X)$. So $G(X) = X - \alpha$ where $\alpha$ is a $p$-adic integer, $\alpha \equiv 1 \pmod{p^{p-1}}$, so $\alpha \neq 0$ and $F(\alpha) = 0$, that is, $\alpha^p + p^p + (-1)^p = 0$. $\square$

In the proof we obtained solutions in which one of the numbers was not a unit in the $q$-adic field. So it is natural to ask if there is always a solution in $q$-adic units. In the sequel we give results which may be found in Klösgen's paper (1970).

**(1B)**   *Let $n \geq 1$ and let $p$ be an odd prime. The following conditions are equivalent:*

(a) *There exist integers $x, y, z$, not multiples of $p$, such that $x^{p^n} + y^{p^n} + z^{p^n} \equiv 0 \pmod{p^{n+1}}$.*

(b) *For every $m \geq 0$ there exist integers $x_m, y_m, z_m$, not multiples of $p$, such that $x_m^{p^n} + y_m^{p^n} + z_m^{p^n} \equiv 0 \pmod{p^{n+1+m}}$ and $x_{m+1} \equiv x_m \pmod{p^{m+1}}$, $y_{m+1} \equiv y_m \pmod{p^{m+1}}$, $z_{m+1} \equiv z_m \pmod{p^{m+1}}$.*

PROOF. It suffices to prove that (a) implies (b) and we proceed by induction on $m$. From $x_m^{p^n} + y_m^{p^n} + z_m^{p^n} \equiv 0 \pmod{p^{n+1+m}}$, with integers $x_m, y_m, z_m$ not multiples of $p$, we may write $x_m^{p^n} + y_m^{p^n} + z_m^{p^n} = r'p^{n+1+m}$, with $r' \in \mathbb{Z}$. Since $p \nmid z_m$, there exists an integer $r$ such that $rz_m^{p^n-1} \equiv r' \pmod{p}$. Let $x_{m+1} = x_m$, $y_{m+1} = y_m$, and $z_{m+1} = z_m - rp^{m+1}$. Then

$$z_{m+1}^{p^n} \equiv z_m^{p^n} - z_m^{p^n-1}rp^{n+1+m} + \binom{p^n}{2}z_m^{p^n-2}r^2p^{2(1+m)}$$
$$- \binom{p^n}{3}z_m^{p^n-3}r^3p^{3(1+m)} + \cdots$$

$$\equiv z_m^{p^n} - z_m^{p^n-1} r p^{n+1+m} \pmod{p^{n+2+m}},$$

because $m \geq 0$, $p \neq 2$, so $p^{n+2+m}$ divides all summands but the first two. Hence

$$x_{m+1}^{p^n} + y_{m+1}^{p^n} + z_{m+1}^{p^{n+1}} \equiv x_m^{p^n} + y_m^{p^n} + z_m^{p^n} - z_m^{p^n-1} r p^{n+1+m}$$
$$\equiv (r' - z_m^{p^n-1} r) p^{n+1+m}$$
$$\equiv 0 \pmod{p^{n+2+m}}. \quad \square$$

We note that we may in fact take $x_m = x_0$, $y_m = y_0$ for every $m \geq 0$. As a complement, we note the analogous result for $p = 2$:

**(1C)**    Let $n \geq 1$. The following conditions are equivalent:

(a) There exist odd integers $x, y, z$ such that
$$x^{2^n} + y^{2^n} + z^{2^n} \equiv 0 \pmod{2^{n+2}}.$$

(b) For every integer $m \geq 1$ there exist odd integers $x_m, y_m, z_m$ such that
$$x_m^{2^n} + y_m^{2^n} + z_m^{2^n} \equiv 0 \pmod{2^{n+2+m}}$$
and $x_{m+1} \equiv x_m \pmod{2^{m+1}}$, $y_{m+1} \equiv y_m \pmod{2^{m+1}}$, $z_{m+1} \equiv z_m \pmod{2^{m+1}}$.

PROOF. The proof is quite similar. It suffices to note that if $z_{m+1} = z_m - r2^{m+1}$, then

$$z_{m+1}^{2^n} = z_m^{2^n} - z_m^{2^n-1} r 2^{n+1+m} + \binom{2^n}{2} z_m^{2^n-2} r^2 2^{2(1+m)}$$

$$- \binom{2^n}{3} z_m^{2^n-3} r^3 2^{3(1+m)} + \cdots$$

$$\equiv z_m^{2n} - z_m^{2^n-1} r 2^{n+1+m} \pmod{2^{n+2+m}},$$

because $2^{n-1+2(1+m)}$ divides all but the first two summands and $n + 2 + m \leq n - 1 + 2(1 + m)$ since $m \geq 1$. $\quad \square$

Concerning the solutions of Fermat's equation by $p$-adic units, we have:

**(1D)**    *Let $p$ be an odd prime. The following conditions are equivalent:*

    (a) *There exist units $\alpha, \beta, \gamma$ of $\hat{\mathbb{Z}}_p$ such that $\alpha^p + \beta^p + \gamma^p = 0$.*

    (b) *There exist integers $x_0, y_0, z_0$, not multiples of $p$, such that $x_0^p + y_0^p + z_0^p \equiv 0 \pmod{p^2}$.*

    (c) *For every $n \geq 0$ there exist integers $x_n, y_n, z_n$, not multiples of $p$, such that $x_n^p + y_n^p + z_n^p \equiv 0 \pmod{p^{n+2}}$ and $x_{n+1} \equiv x_n \pmod{p^{n+1}}, y_{n+1} \equiv y_n \pmod{p^{n+1}}, z_{n+1} \equiv z_n \pmod{p^{n+1}}$.*

PROOF. (a) $\Rightarrow$ (b)  We write $\alpha = x_0 + \alpha' p$, $\beta = y_0 + \beta' p$, $\gamma = z_0 + \gamma' p$ where $x_0, y_0, z_0$ are integers, $0 \leq x_0, y_0, z_0 \leq p-1$, and $\alpha', \beta', \gamma' \in \hat{\mathbb{Z}}_p$. Since $\alpha, \beta, \gamma$ are units, then $p \nmid x_0 y_0 z_0$. From $\alpha^p + \beta^p + \gamma^p = 0$ it follows that $x_0^p + y_0^p + z_0^p \equiv 0 \pmod{p^2}$.

    (b) $\Rightarrow$ (c) This was proved in (1B).

    (c) $\Rightarrow$ (a) The sequences of integers $(x_n)_{n \geq 0}, (y_n)_{n \geq 0}, (z_n)_{n \geq 0}$ are $p$-adically convergent, since $x_{n+1} \equiv x_n \pmod{p^{n+1}}, y_{n+1} \equiv y_n \pmod{p^{n+1}}, z_{n+1} \equiv z_n \pmod{p^{n+1}}$ for every $n \geq 0$. Let $\alpha = \lim x_n$, $\beta = \lim y_n$, $\gamma = \lim z_n$. Since $x_n^p + y_n^p + z_n^p \equiv 0 \pmod{p^{n+2}}$ then, at the limit, $\alpha^p + \beta^p + \gamma^p = 0$.    $\square$

By the above proof, the conditions of (1D) are equivalent to:

    (a') There exist integers $x, y$, not multiples of $p$, and a unit $\gamma \in \hat{\mathbb{Z}}_p$, such that $x^p + y^p + \gamma^p = 0$.

Similarly, we have:

**(1E)**    *Let $q, p$ be distinct primes. The following conditions are equivalent:*

    (a) *There exist units $\alpha, \beta, \gamma$ of $\hat{\mathbb{Z}}_q$ such that $\alpha^p + \beta^p + \gamma^p = 0$.*

    (b) *There exist integers $x_0, y_0, z_0$, not multiples of $q$, such that $x_0^p + y_0^p + z_0^p \equiv 0 \pmod{q}$.*

    (c) *For every $n \geq 0$ there exist integers $x_n, y_n, z_n$, not multiples of $q$, such that $x_n^p + y_n^p + z_n^p \equiv 0 \pmod{q^{n+1}}$, and $x_{n+1} \equiv x_n \pmod{q^{n+1}}, y_{n+1} \equiv y_n \pmod{q^{n+1}}, z_{n+1} \equiv z_n \pmod{q^{n+1}}$.*

PROOF. (a) $\Rightarrow$ (b)  We write $\alpha = x_0 + \alpha' q$, $\beta = y_0 + \beta' q$, $\gamma = z_0 + \gamma' q$, where $x_0, y_0, z_0$ are integers, $0 \leq x_0, y_0, z_0 \leq q-1$, and $\alpha', \beta', \gamma' \in \hat{\mathbb{Z}}_q$. Since $\alpha, \beta, \gamma$ are units then $q \nmid x_0 y_0 z_0$. From $\alpha^p + \beta^p + \gamma^p = 0$ it follows that $x_0^p + y_0^p + z_0^p \equiv 0 \pmod{q}$.

(b) $\Rightarrow$ (c)   We prove the statement by induction on $n$. It is true for $n = 0$ and we assume it true for some $m \geq 0$. Let $x_m, y_m, z_m$ be integers, not multiples of $q$, such that $x_m^p + y_m^p + z_m^p \equiv 0 \pmod{q^{m+1}}$. Hence $x_m^p + y_m^p + z_m^p = r'q^{m+1}$ where $r'$ is an integer. Since $q \nmid z_m$, there is an integer $r$ satisfying the congruence $rpz_m^{p-1} \equiv r' \pmod{q}$. Let $x_{m+1} = x_m$, $y_{m+1} = y_m$ and $z_{m+1} = z_m - rq^{m+1}$. Then $z_{m+1}^p = (z_m - rq^{m+1})^p \equiv z_m^p - pz_m^{p-1}rq^{m+1} \pmod{q^{m+2}}$ and $x_{m+1}^p + y_{m+1}^p + z_{m+1}^p \equiv x_m^p + y_m^p + z_m^p - pz_m^{p-1}rq^{m+1} \equiv (r' - pz_m^{p-1}r)q^{m+1} \equiv 0 \pmod{q^{m+2}}$.

(c) $\Rightarrow$ (a)   The sequences of integers $(x_n)_{n \geq 0}$, $(y_n)_{n \geq 0}$, $(z_n)_{n \geq 0}$ are $q$-adically convergent, since $x_{n+1} \equiv x_n \pmod{q^{n+1}}$, $y_{n+1} \equiv y_n \pmod{q^{n+1}}$, $z_{n+1} \equiv z_n \pmod{q^{n+1}}$ for every $n \geq 0$. Let $\alpha = \lim x_n$, $\beta = \lim y_n$, $\gamma = \lim z_n$. Since $x_n^p + y_n^p + z_n^p \equiv 0 \pmod{p^{n+1}}$ then, at the limit, $\alpha^p + \beta^p + \gamma^p = 0$.   $\square$

As in (1D), the conditions of (1E) are also equivalent to:

(a'') There exist integers $x, y$, not multiples of $q$, and a unit $\gamma \in \hat{\mathbb{Z}}_q$ such that $x^p + y^p + \gamma^p = 0$.

We conclude by noting that Fermat equations (for $n \geq 3$) provide an interesting example where there is a nontrivial solution in every $q$-adic field (by (1A)) and only the trivial solution in integers, as proved by Wiles.

## Bibliography

1970 Klösgen, W., *Untersuchungen über Fermatsche Kongruenzen*, Gesellschaft Math. Datenverarbeitung, No. 36, 1970, 124 pp., Bonn.

## X.2. Fermat Congruence

We shall study the congruences

(2.1) $$X^n + Y^n + Z^n \equiv 0 \pmod{q},$$

(2.2) $$X^n + Y^n \equiv Z^n \pmod{q},$$

where $q$ is an odd prime number, $n \geq 0$, and $q$ does not divide $n$.

Let $N(n, q) = \#\{(x, y, z) \mid 1 \leq x, y, z < q, \, x^n + y^n + z^n \equiv 0 \pmod{q}\}$ and $N'(n, q) = \#\{(x, y, z) \mid 1 \leq x, y, z < q, \, x^n + y^n \equiv z^n$

(mod $q$)}. Clearly, if $n$ is odd then $N(n,q) = N'(n,q)$. In this connection, we consider the following problems:

    (1) To determine when $N(n,q)$, $N'(n,q)$ are greater than 0.
    (2) To find upper and lower bounds for $N(n,q)$, $N'(n,q)$.
    (3) If possible, to calculate the values of $N(n,q), N'(n,q)$.

    We recall from Chapter IV, (2A), that if $\gcd(n, q-1) = 1$ (hence $n$ is odd) then $N(n,q) = N'(n,q) > 0$. We have also shown in Chapter IV, (2D), that if $p$ and $q = 6kp + 1$ are primes then $N(p,q) > 0$.

    The following implication was proved by Libri (1832, p. 275) and again by Pepin (1880), Pellet (1887), and Matthews (1895).

**(2A)**    *Let $p$ be a prime number. If there exist infinitely many primes $q$ such that $N(p,q) = 0$ then Fermat's last theorem is true for the exponent $p$.*

PROOF. Assume that there exist nonzero integers $x, y, z$ such that $x^p + y^p + z^p = 0$. If $q$ is any prime number such that $q > \max\{|x|, |y|, |z|\}$ then $x^p + y^p + z^p \equiv 0 \pmod{q}$ and $N(p,q) > 0$. So $N(p,q) = 0$ for only finitely many primes $q$, proving the statement. $\square$

    This result shifts the proof of Fermat's last theorem to the proof that $N(p,q) = 0$ for infinitely many moduli $q$. The fact is that we shall actually prove the opposite, namely for every $p$ there exists a prime $q_0(p)$ such that if $q \geq q_0(p)$ then $N(p,q) > 0$.

    Before proving this theorem, we describe some of the numerous special results concerning these congruences.

    Legendre (1830) showed that $N(3,7) = N(3,13) = 0$ and $N(5,q) = 0$ for $q = 11, 41, 71, 101$.

    Libri wrote a series of papers (1824, 1832) in which he exposed a method to compute the number of solutions of very general congruences. Libri calculated $N(3,q)$ for many primes $q \equiv 1 \pmod{3}$ and showed that there exists $q_0(3)$ such that if $q \geq q_0(3)$ then $N(3,q) > 0$; it should be noted that some of his calculated values were incorrect. These results were published again by Pepin in 1880 (see also his paper of 1876).

    Since $q \equiv 1 \pmod{3}$ then $-3$ is not a square modulo $q$ and there exist integers $l, m$ of the same parity such that $4q = l^2 + 3m^2$. This may be explained by considering the decomposition of $q$ as a product of elements in the field $\mathbb{Q}(\sqrt{-3})$. From the fact that $-3$ is not a square modulo $q$ then $q = \alpha\alpha'$ where $\alpha = (l + m\sqrt{-3})/2$, $\alpha' = (l - m\sqrt{-3})/2$ with $l, m$ integers, both even, or both odd, so $4q = l^2 + 3m^2$. We choose a representation with minimal $|l|$. We note that $l$ is not a multiple of 3; by changing $l$ into $-l$ (if necessary) we may assume that $l \equiv 1 \pmod 3$ and this implies that $l$ is uniquely defined with the above property. Pepin showed:

$$(2.3) \qquad\qquad N(3, q) = (q - 1)(q - 8 + l).$$

Here are some numerical examples:

$$
\begin{aligned}
28 &= 1 + 3 \times 3^2 &\Rightarrow\quad N(3, 7) &= 6(7 - 8 + 1) = 0; \\
52 &= 5^2 + 3 \times 3^2 &\Rightarrow\quad N(3, 13) &= 12(13 - 8 - 5) = 0; \\
76 &= 1^2 + 3 \times 5^2 &\Rightarrow\quad N(3, 19) &= 18(19 - 8 + 1) = 216; \\
12\,4 &= 4^2 + 3 \times 6^2 &\Rightarrow\quad N(3, 31) &= 30(31 - 8 + 4) = 810.
\end{aligned}
$$

Pepin noted that since $l > -2\sqrt{q}$ then $N(3, q) > \sqrt{q}(\sqrt{q} - 2) - 8$. Hence if $q \geq 19$ then $N(3, q) > 0$.

Libri established that for every prime $p$ there exists $q_0(p)$ such that if $q \geq q_0(p)$ then $N(p, q) > 0$; however, he gave no bounds for $N(p, q)$ or a way of computing $q_0(p)$. Pellet used another method to show the same result in 1887; in a later note (1911), Pellet gave a bound for $N(p, q)$, but his value was erroneous.

In 1837, Lebesgue considered arbitrary polynomial congruences

$$(2.4) \qquad\qquad F(X_1, \dots, X_n) \equiv 0 \pmod q,$$

where $s \geq 2$, $q$ is an odd prime, and $F \in \mathbb{Z}[X_1, \dots, X_s]$. Let

$$
\begin{aligned}
N &= \#\{(x_1, \dots, x_s) \mid 1 \leq x_i \leq q - 1 \\
&\qquad \text{for all } i = 1, \dots, s, \text{ such that } F(x_1, \dots, x_s) \equiv 0 \pmod q\}, \\
N_0 &= \#\{(x_1, \dots, x_s) \mid 0 \leq x_1 \leq q - 1 \\
&\qquad \text{for all } i = 1, \dots, s, \text{ such that } F(x_1, \dots, x_s) \equiv 0 \pmod q\}.
\end{aligned}
$$

In the polynomial $F^{q-1}$ let $A$ (respectively, $A_0$) be the sum of the coefficients of all monomials $eX_1^{e_1} \cdots X_s^{e_s}$ such that $q - 1$ divides each $e_i$ (respectively, each $e_i$ is greater than 0 and divisible by $q - 1$).

Lebesgue showed that $N$, $N_0$ satisfy the congruences

$$
\begin{aligned}
N_0 &\equiv (-1)^{s+1} A_0 \pmod q, \\
N &\equiv (-1)^s (1 - A) \pmod q.
\end{aligned}
$$

Lebesgue applied his method to congruences like

(2.5)                    $A_1 X_1^n + \cdots + A_s X_s^n \equiv 0 \pmod{q}$,

with $n \geq 2$, $s \geq 2$, each $A_i$ is a nonzero integer, $q$ is an odd prime, $q \equiv 1 \pmod{n}$. He gave an expression for the number of solutions $N$ in terms of the periods of the cyclotomic equation. Lebesgue studied in detail the following special cases of the congruence (2.5): $s = 2$, $n = 2$; $s = 3$, $n = 3$; $s = 3$, $n = 4$. For (2.1) with $n = 3$, he derived once more some of the results of Libri.

Further results about the number of solutions of (2.5) appeared in Lebesgue's paper of 1838. In 1909 two papers by Dickson appeared, and one each by Cornacchia and Hurwitz, dealing with these congruences.

Cornacchia studied various special cases of (2.1) and (2.5) in detail and gave many explicit results, which had been in part previously indicated by Lebesgue, Pepin, and Pellet:

(a) If $n = 2$ and $q \equiv 1 \pmod{4}$ then

$$N'(2, q) = \begin{cases} \dfrac{q - 9}{8} & \text{when 2 is a square modulo } q, \\ \dfrac{q - 5}{8} & \text{otherwise.} \end{cases}$$

If $n = 2$ and $q \equiv -1 \pmod{4}$ then

$$N'(2, q) = \begin{cases} \dfrac{q - 7}{8} & \text{when 2 is a square modulo } q, \\ \dfrac{q - 3}{8} & \text{otherwise.} \end{cases}$$

(b) For $n = 3$, $q \equiv 1 \pmod{3}$, Cornacchia obtained once more Pepin's result and showed that if $N(3, q) = 0$ then $q = 7, 13$.

(c) If $n = 4$ and $q \equiv 1 \pmod{4}$, Cornacchia calculated $N(4, q)$. Moreover, he showed that $N'(4, q) = 0$ exactly when $q = 11, 17, 29, 41$.

(d) If $n = 6$ and $q \equiv 1 \pmod{6}$ then $N'(6, q)$ was also determined. Moreover, $N'(6, q) = 0$ exactly when $q = 7, 13, 19, 43, 61, 97, 157, 277$. On the other hand, for the congruence $X^6 + Y^6 + Z^6 \equiv 0 \pmod{q}$ we have $N(6, q) = 0$ exactly when $q = 7, 13, 31, 61, 67, 79, 97, 139, 157, 223, 277$.

(e) If $n = 8$ and $q \equiv 1$ (mod 8) then Cornacchia gave upper and lower estimates for $N'(8, q)$. Moreover, $N'(8, q) = 0$ exactly when $q = 17, 41, 113$. And for the congruence $X^8 + Y^8 + Z^8 \equiv 0$ (mod $q$), he established that $N(8, q) = 0$ exactly when $q = 17, 41, 113, 137, 233, 761$.

In his papers, Dickson dealt with the congruence (2.1). His method, involving the periods of cyclotomic equations, led to lower and upper bounds for $N(p, q)$, as well as an upper bound for $q_0(p)$, namely

$$q_0(p) \leq (p - 1)^2 (p - 2)^2 + 6p - 2.$$

As applications, Dickson showed that $N(5, q) = 0$ exactly when $q = 11, 41, 71, 101$ and that $N(7, q) = 0$ exactly when $q = 29, 71, 113, 491$. Using computations of Carey (1893) for squares and products of periods, Dickson applied his method to the congruence $X^4 + Y^4 \equiv Z^4$ (mod $q$).

Hurwitz's paper dealt with the more general congruence (2.5) with $n = p$ a prime. He considered the family of such congruences. For all possible values of the coefficients and indicated relations which must be satisfied by the numbers of solutions of these various congruences. From this information, Hurwitz deduced upper and lower bounds for the number of solutions of

$$(2.6) \qquad AX^p + BY^p + CZ^p \equiv 0 \pmod{q},$$

where $A, B, C$ are nonzero integers. He also determined a positive number $q_0(p)$ (depending on equation (2.6)) such that if $q \geq q_0(p)$ then (2.6) has a solution $(x, y, z)$, with $1 \leq x, y, z \leq q - 1$.

In 1917, Schur gave a proof that for every $n \geq 2$, if $q \geq (n!)e + 1$ then $X^n + Y^n + Z^n \equiv 0$ (mod $q$) has a solution $(x, y, z)$ with $1 \leq x, y, z \leq q - 1$. Schur's proof was based on the following interesting combinatorial lemma:

LEMMA 2.1. *Let $n \geq 1$ and $N \geq (n!)e + 1$. If the set of numbers $\{1, 2, \ldots, N\}$ is partitioned into $n$ disjoint subsets $L_1, \ldots, L_n$, there exists at least one subset $L_i$ such that if $m, m' \in L_i$ with $m < m'$ then $m' - m \in L_i$.*

Among further developments, we want to report that the equations

$$AX^e + BY^f + CZ^g = 0$$

and

$$A_1 X_1^{n_1} + A_2 X_2^{n_2} + \cdots + A_s X_s^{n_s} = 0,$$

with coefficients in a finite field with $q^d$ elements $(d \geq 1)$ and not necessarily equal exponents, have been the object of numerous papers. See Mitchell (1917), Vandiver (1944, 1945, 1946, 1947, 1948, 1949, 1954, 1955, 1956, 1959), Hua and Vandiver (1948, 1949), E. Lehmer and Vandiver (1957). For the vast, more recent, literature, the reader should consult the appropriate sections in *Mathematical Reviews*.

In his paper of 1949, Weil traveled a historical panorama of the evolution of Gauss' original method to deal, by means of Gaussian sums, with the congruence $AX^3 - BY^3 \equiv 1 \pmod{q}$, where $q$ is a prime, $q \equiv 1 \pmod 3$. These ideas were applied subsequently to wide classes of congruences. They were used by Hardy and Littlewood in connection with Waring's problem; Hasse expressed relations between Riemann's hypothesis for function fields and various kinds of exponential sums, and Weil published striking and definitive results on this question (1928).

Here we single out the following specific result (see Vandiver, 1946, pp. 47–52; Hua and Vandiver, 1948, pp. 258–263), which is more directly related with our subject matter.

Let $s \geq 1$, let $A_1, \ldots, A_s$ be nonzero integers, let $n_1, \ldots, n_s$ be integers, let $q$ be an odd prime, and let $d_i = \gcd(q - 1, |n_i|) > 1$ for $i = 1, \ldots, s$. Let $N$ denote the number of solutions in integers $(x_1, \ldots, x_s)$ with $1 \leq x_i \leq q - 1$ (for $i = 1, \ldots, s$), of the congruence

$$(2.7) \qquad A_1 X_1^{n_1} + \cdots + A_s X_s^{n_s} \equiv 0 \pmod{q}.$$

Then

$$\frac{(q-1)^s}{q} - d_1 \cdots d_s q^{s/2} < N < \frac{(q-1)^s}{q} + d_1 \cdots d_s q^{s/2}.$$

In particular, there exists a positive number $q_0$ such that if $q \geq q_0$ then $N > 0$. Another proof of this theorem, using the theory of group characters, was given by Feit (1967).

After this survey of results which are related to Fermat's congruence, we shall give the proof of Dickson's theorem. For this purpose, we recapitulate the following notation and facts from Chapter IX. Let $p \geq 3$, $q = 2kp + 1$ be prime numbers. Let

$g$ be a primitive root modulo $p$;

$\zeta$ be a primitive $p$th root of 1;

$h$ be a primitive root modulo $q$; and

$\rho$ be a primitive $q$th root of 1.

The $p$ periods of $2k$ terms in $\mathbb{Q}(\rho)$ are

(2.8)
$$
\begin{cases}
\mu_0 &= \rho + \rho^{h^p} + \rho^{h^{2p}} + \cdots + \rho^{h^{(2k-1)p}}, \\
\mu_1 &= \rho^h + \rho^{h^{p+1}} + \rho^{h^{2p+1}} + \cdots + \rho^{h^{(2k-1)p+1}}, \\
&\vdots \quad \vdots\ \vdots \\
\mu_i &= \rho^{h^i} + \rho^{h^{p+i}} + \rho^{h^{2p+i}} + \cdots + \rho^{h^{(2k-1)p+i}}, \\
&\vdots \quad \vdots\ \vdots \\
\mu_{p-1} &= \rho^{h^{p-1}} + \rho^{h^{2p-1}} + \rho^{h^{3p-1}} + \cdots + \rho^{h^{q-2}}.
\end{cases}
$$

In Chapter IX, we defined Lagrange resolvents for every $j = 0, 1, \ldots,$ $p - 1$:

(2.9)
$$
\lambda_j = \langle \zeta^j, \rho \rangle = \sum_{t=1}^{q-1} \zeta^{j \operatorname{ind}_h(t)} \rho^t,
$$

where $\operatorname{ind}_h(t) = s$, $0 \le s \le q-2$ when $t \equiv h^s \pmod{q}$. In particular, $\lambda_0 = \langle 1, \rho \rangle = \sum_{t=1}^{q-1} \rho^t = -1$.

For easy reference, we recall the following results from Chapter IX, (2D), (2E), (2F): For $j = 1, \ldots, p - 1$,

(2.10)    $\lambda_j \lambda_{p-j} = q,$

(2.11)    $|\lambda_j|^2 = \lambda_j \overline{\lambda_j} = q$

($\overline{\lambda_j}$ denotes the complex-conjugate of $\lambda_j$),

(2.12)
$$
\mu_i = \frac{1}{p} \sum_{j=0}^{p-1} \zeta^{-ji} \lambda_j.
$$

We shall give a simplified proof of Dickson's theorem, following Klösgen (1970). First, we give an expression of $N(p, q)$ in terms of the periods $\mu_i$:

(2B)

$$
N(p, q) = \frac{1}{q} \left[ (q-1)^3 + (q-1)p^2 \sum_{i=0}^{p-1} \mu_i^3 \right].
$$

PROOF. To begin, we note that if $x, y, z$ are integers such that $1 \leq x, y, z \leq q - 1$ then

$$\sum_{t=0}^{q-1} \rho^{t(x^p + y^p + z^p)} = \begin{cases} 0 & \text{when } x^p + y^p + z^p \not\equiv 0 \pmod{q}, \\ q & \text{when } x^p + y^p + z^p \equiv 0 \pmod{q}. \end{cases}$$

So

$$\begin{aligned} qN(p, q) &= \sum_{x,y,z=1}^{q-1} \left( \sum_{t=0}^{q-1} \rho^{t(x^p + y^p + z^p)} \right) \\ &= \sum_{t=0}^{q-1} \left( \sum_{x,y,z=1}^{q-1} \rho^{tx^p} \rho^{ty^p} \rho^{tz^p} \right) \\ &= \sum_{t=0}^{q-1} \left( \sum_{x=1}^{q-1} \rho^{tx^p} \right)^3 \\ &= (q-1)^3 + \sum_{t=1}^{q-1} \left( \sum_{x=1}^{q-1} \rho^{tx^p} \right)^3. \end{aligned}$$

If $t \equiv h^i \pmod{q}$ and $x \equiv h^j \pmod{q}$ (where $0 \leq i, j \leq q - 2$) then $\sum_{x=1}^{q-1} \rho^{tx^p} = \sum_{j=0}^{q-2} \rho^{h^{i+pj}} = p\mu_i$. Since $\mu_j = \mu_i$ when $j \equiv i \pmod{p}$, then

$$qN(p, q) = (q-1)^3 + 2k \sum_{i=0}^{p-1} p^3 \mu_i^3,$$

because $q = 2kp + 1$. Therefore

$$N(p, q) = \frac{1}{q} \left[ (q-1)^3 + (q-1)p^2 \sum_{i=0}^{p-1} \mu_i^3 \right]. \quad \square$$

And now, we prove Dickson's Theorem:

**(2C)**    *We have:*
  (1) $(q-1)[q + 1 - 3p - (p-1)(p-2)\sqrt{q}] < N(p, q)$
      $< (q-1)[q + 1 - 3p + (p-1)(p-2)\sqrt{q}].$
  (2) *If $q \geq (p-1)^2(p-2)^2 + 6p - 2$ then the congruence (2.1) has a nontrivial solution.*

PROOF. (1)  By (2B) and (2.11),

$$N(p,q) = \frac{1}{q}\left[(q-1)^3 + (q-1)p^2\frac{1}{p^3}\sum_{i=0}^{p-1}\left\{\sum_{j=0}^{p-1}\zeta^{-ji}\lambda_j\right\}^3\right]$$

$$= \frac{1}{q}\left[(q-1)^3 + \frac{q-1}{p}\sum_{i=0}^{p-1}\sum_{j_1,j_2,j_3=0}^{p-1}\zeta^{-i(j_1+j_2+j_3)}\lambda_{j_1}\lambda_{j_2}\lambda_{j_3}\right]$$

$$= \frac{1}{q}\left[(q-1)^3 + \frac{q-1}{p}\sum_{j_1,j_2,j_3=0}^{p-1}\lambda_{j_1}\lambda_{j_2}\lambda_{j_3}\left(\sum_{i=0}^{p-1}\zeta^{-i(j_1+j_2+j_3)}\right)\right].$$

But

$$\sum_{i=0}^{p-1}\zeta^{-i(j_1+j_2+j_3)} = \begin{cases} 0 & \text{when } j_1+j_2+j_3 \not\equiv 0 \pmod p, \\ p & \text{when } j_1+j_2+j_3 \equiv 0 \pmod p. \end{cases}$$

Thus

$$N(p,q) = \frac{1}{q}\left[(q-1)^3 + (q-1)\sum_{\substack{j_1,j_2,j_3=0 \\ j_1+j_2+j_3\equiv 0 \ (\text{mod } p)}}^{p-1}\lambda_{j_1}\lambda_{j_2}\lambda_{j_3}\right].$$

Since $\lambda_0 = -1$, it follows from (2.9) that the last sum of products $\lambda_{j_1}\lambda_{j_2}\lambda_{j_3}$ is equal to

$$\lambda_0^3 + 3\lambda_0\sum_{j=1}^{p-1}\lambda_j\lambda_{p-j} + S = -1 - 3q(p-1) + S,$$

where $S = \sum_{j_1,j_2,k_3}^{p-1}\lambda_{j_1}\lambda_{j_2}\lambda_{j_3}$. Hence

$$N(p,q) = \frac{q-1}{q}[(q-1)^2 - 1 - 3q(p-1) + S],$$

so

$$\frac{N(p,q)}{q-1} = \frac{1}{q}(q^2 - 3qp + q + S) = q - 3p + 1 + \frac{1}{q}S$$

and

$$\left|\frac{N(p,q)}{q-1} - (q+1-3p)\right| = \frac{1}{q}|S|.$$

By (2.10), $|\lambda_j| = \sqrt{q}$. We note also that for every $j_1$, $1 \le j_1 \le p-1$, there are $p-2$ pairs $(j_2,j_3)$, $1 \le j_2,j_3 \le p-1$, such that $j_1+j_2+j_3 = p$ or $2p$, namely $(1,p-j_1-1)$, $(2,p-j_1-2),\ldots,(p-$

$j_1-1, 1)$, and $(p-1, p-j_1+1)$, $(p-2, p-j_1+2), \dots, (p-j_1+1, p-1)$. Hence $|S| \le (p-1)(p-2)q^{3/2}$, so

$$\left| \frac{N(p,q)}{q-1} - (q+1-3p) \right| \le (p-1)(p-2)\sqrt{q}$$

and we conclude that

$$(q-1)[q+1-3p-(p-1)(p-2)\sqrt{q}] < N(p,q)$$

$$< (q-1)[q+1-3p+(p-1)(p-2)\sqrt{q}].$$

(2)    For later use, we prove more generally that if $\nu$ is an integer, $\nu \ge 0$, and if $q \ge (p-1)^2(p-2)^2 + 2(p\nu-1)$ then $q+1-(p-1)(p-2)\sqrt{q} - p\nu \ge 0$. Choosing $\nu = 3$ this gives the statement (2), in view of (1).

The inequality

$$\alpha^2 + 2\beta \ge \alpha\sqrt{\alpha^2 + 4\beta}$$

(for real numbers $\alpha, \beta$), which is easily verified, implies when $\alpha = (p-1)(p-2)$, $\beta = p\nu - 1$:

$$(p-1)^2(p-2)^2 + 2(p\nu - 1)$$
$$\ge (p-1)(p-2)\sqrt{(p-1)^2(p-2)^2 + 4(p\nu - 1)}.$$

Let $\delta = (p-1)^2(p-2)^2 + 4(p\nu-1) > 0$ and consider the polynomials

$$f(T) = T^2 - (p-1)(p-2)T - (p\nu - 1),$$

having discriminant $\delta$.

It suffices to show that

$$\sqrt{q} \ge \frac{(p-1)(p-2) + \sqrt{\delta}}{2};$$

indeed, this implies that $f(\sqrt{q}) \ge 0$, that is, $q+1-(p-1)(p-2)\sqrt{q} - p\nu \ge 0$. We have

$$4q \ge 4[(p-1)^2(p-2)^2 + 2(p\nu - 1)]$$
$$\ge 2[(p-1)^2(p-2)^2 + 2(p\nu - 1) + (p-1)(p-2)\sqrt{\delta}]$$
$$= (p-1)^2(p-2)^2 + \delta + 2(p-1)(p-2)\sqrt{\delta}$$
$$= [(p-1)(p-2) + \sqrt{\delta}]^2,$$

so $\sqrt{q} \ge ((p-1)(p-2) + \sqrt{\delta})/2$, as was required to show.

Taking $\nu = 3$, it follows from (1) that

$$N(p,q) > (q-1)[(q+1) - 3p - (p-1)(p-2)\sqrt{q}] \geq 0. \quad \square$$

The upper bounds for $q_0(p)$ given by Dickson are not sharp, as we see by explicit computation:

| Dickson bound | Actual value |
|---|---|
| $q_0(3) \leq 20$ | $q_0(3) = 13$ |
| $q_0(5) \leq 172$ | $q_0(5) = 101$ |
| $q_0(7) \leq 940$ | $q_0(7) = 491$ |

As a complement, Mantel showed in 1916 that if $N(p,q) = 0$ then $q$ must be of the form $q = 6mp/(p-3) - 1$ (for some integer $m$).

Taking into account (2C) and the result of Chapter IV, (2B), if $k \geq 1$, $q = 2kp + 1$ is a prime and $q \geq (p-1)^2(p-2)^2 + 6p - 2$ then $q$ divides the Wendt determinant $W_{2k}$. Therefore, for each prime $p$ there exist at most finitely many integers $k \geq 1$ such that $q = 2kp+1$ is a prime not dividing $W_{2k}$. What is not known is whether, for every prime $p$, there exists actually *one* prime $q$ with the above property (see Flye Sainte-Marie, 1890, and Landau, 1913).

## Bibliography

1824 Libri, G., *Mémoires sur divers points d'analyse (art. cinquième)*, Mém. Acad. Roy. Turin, **28** (1824), 152–280.

1830 Legendre, A.M., *Théorie des Nombres* (3$^e$ édition), Firmin Didot Frères, Paris, 1830; reprinted by A. Blanchard, Paris, 1955.

1832 Libri, G., *Mémoires sur la théorie des nombres*, J. Reine Angew. Math., **9** (1832), 54–80, 169–188, 261–276.

1832 Libri, G., *Mémoire sur la résolution de quelques équations indéterminées*, J. Reine Angew. Math., **9** (1832), 277–294.

1837 Lebesgue, V.A., *Recherches sur les nombres*, J. Math. Pures Appl., **2** (1837), 255–292; **3** (1838), 113–144.

1876 Pepin, T., *Etude sur la théorie des résidues cubiques*, J. Math. Pures Appl., (3), **2** (1876), 313–324.

1880 Pepin, T., *Sur diverses tentatives de démonstration du théorème de Fermat*, C. R. Acad. Sci. Paris, **91** (1880), 366–367.

1887 Pellet, A.E., *Mémoire sur la théorie algébrique des équations*, Bull. Soc. Math. France, **15** (1887), 61–102.

1890 Flye Sainte-Marie, C., *Question 1339*, L'Interm. Math., **5** (1890), 195.

1893 Carey, F.S., *Notes on the division of the circle*, Quart. J. Pure Appl. Math., **26** (1893), 322–371.

1895 Matthews, G.B., *Note in connexion with Fermat's last theorem*, Messenger Math., **24** (1895), 97–99.

1909 Cornacchia, G., *Sulla congruenza $x^n + y^n \equiv z^n$ (mod $p$)*, Giorn. Mat., **47** (1909), 219–268.

1909 Dickson, L.E., *On the congruence $x^n + y^n + z^n \equiv 0$ (mod $p$)*, J. Reine Angew. Math., **135** (1909), 134–141.

1909 Dickson, L.E., *Lower limit for the number of sets of solutions of $x^e + y^e + z^e \equiv 0$ (mod $p$)*, J. Reine Angew. Math., **135** (1909), 181–188.

1909 Hurwitz, A., *Über die Kongruenz $ax^e + by^e + cz^e \equiv 0$ (mod $p$)*, J. Reine Angew. Math., **136** (1909), 272–292.

1910 Dubouis, E., *Réponse à la question 1339 de C. Flye Sainte-Marie*, L'Interm. Math., **17** (1910), 103–104.

1910 Dubouis, E., *Problème 3771*, L'Interm. Math., **17** (1910), 241–242.

1911 Pellet, A.E., *Réponse à une question de M.E. Dubouis*, L'Interm. Math., **18** (1911), 81–82.

1913 Landau, E., *Réponse à la question 1339 de. C. Flye Sainte-Marie*, L'Interm. Math., **20** (1913), 154.

1916 Mantel, W., *Vraagstuk XCI (Problem 91)*, Wiskundige Opgaven, **12** (1916), 213–214.

1917 Schur, I., *Über die Kongruenz $x^m + y^m \equiv z^m$ (mod $p$)*, Jahresber. Deutsch. Math.-Verein., **25** (1917), 114–117; reprinted in Gesammelte Abhandlungen, Vol. II, Springer-Verlag, Berlin, 1973.

1917 Mitchell, H.H., *On the congruence $cx^\lambda + 1 \equiv dy^\lambda$ in a Galois field*, Ann. of Math., (2), **18** (1917), 120–131.

1919 Bachmann, P., *Das Fermatproblem in seiner bisherigen Entwicklung*, W. de Gruyter, Berlin, 1919; reprinted by Springer-Verlag, Berlin, 1976.

1928 Weil, A., *L'arithmétique sur les courbes algébriques*, Acta Math., **52** (1928), 281–315; reprinted in *Oeuvres Scientifiques*, Vol. I, pp. 11–35, Springer-Verlag, New York, 1980.

1944 Vandiver, H.S., *Some theorems in finite fields with applications to Fermat's last theorem*, Proc. Nat. Acad. Sci. U.S.A., **30** (1944), 362–367.

1944 Vandiver, H.S., *On trinomial congruences and Fermat's last theorem*, Proc. Nat. Acad. Sci. U.S.A., **30** (1944), 368–370.

1945 Vandiver, H.S., *On the number of solutions of certain non-homogeneous trinomial equations in a finite field*, Proc. Nat. Acad. Sci. U.S.A., **31** (1945), 170–175.

1946 Vandiver, H.S., *On the number of solutions of some general types of equations in a finite field*, Proc. Nat. Acad. Sci. U.S.A., **32** (1946), 47–52.

1946 Vandiver, H.S., *Cyclotomy and trinomial equations in a finite field*, Proc. Nat. Acad. Sci. U.S.A., **32** (1946), 317–319.

1946 Vandiver, H.S., *On some special trinomial equations in a finite field*, Proc. Nat. Acad. Sci. U.S.A., **32** (1946), 320–326.

1947 Vandiver, H.S., *Limits for the number of solutions of certain general types of equations in a finite field*, Proc. Nat. Acad. Sci. U.S.A., **33** (1947), 236–242.

1948 Hua, L.K. and Vandiver, H.S., *On the existence of solutions of certain equations in a finite field*, Proc. Nat. Acad. Sci. U.S.A., **34** (1948), 258–263.

1948 Vandiver, H.S., *Applications of cyclotomy to the theory of non-homogeneous equations in a finite field*, Proc. Nat. Acad. Sci. U.S.A., **34** (1948), 62–66.

1948 Vandiver, H.S., *Cyclotomic power characters and trinomial equations in a finite field*, Proc. Nat. Acad. Sci. U.S.A., **34** (1948), 196–203.

1949 Hua, L.K. and Vandiver, H.S., *Characters over certain types of rings with applications to the theory of equations in a finite field*, Proc. Nat. Acad. Sci. U.S.A., **35** (1949), 89–95.

1949 Hua, L.K. and Vandiver, H.S., *On the nature of the solutions of certain equations in a finite field*, Proc. Nat. Acad. Sci. U.S.A., **35** (1949), 481–487.

1949 Hua, L.K. and Vandiver, H.S., *On the number of solutions of some trinomial equations in a finite field*, Proc. Nat. Acad. Sci. U.S.A., **35** (1949), 477–481.

1949 Vandiver, H.S., *Quadratic relations involving the number of solutions of certain types of equations in a finite field*, Proc. Nat. Acad. Sci. U.S.A., **35** (1949), 681–685.

1949 Weil, A., *Numbers of solutions of equations in finite fields*, Bull. Amer. Math. Soc., **55** (1949), 497–508; reprinted in *Oeuvres Scientifiques*, Vol. I, pp. 399–410, Springer-Verlag, New York, 1980.

1954 Vandiver, H.S., *On trinomial equations in a finite field*, Proc. Nat. Acad. Sci. U.S.A., **40** (1954), 1008–1010.

1955 Vandiver, H.S., *On the properties of certain trinomial equations in a finite field*, Proc. Nat. Acad. Sci. U.S.A., **41** (1955), 651–653.

1955 Vandiver, H.S., *On cyclotomic relations and trinomial equations in a finite field*, Proc. Nat. Acad. Sci. U.S.A., **41** (1955), 775–780.

1956 Leveque, W.J., *Topics in Number Theory*, Vol. 2, Addison-Wesley, Reading, MA, 1956.

1956 Vandiver, H.S., *Diophantine equations in certain rings*, Proc. Nat. Acad. Sci. U.S.A., **42** (1956), 656–665.

1957 Lehmer, E. and Vandiver, H.S., *On the computation of the number of solutions of certain trinomial congruences*, J. Assoc. Comput. Mach., **4** (1957), 505–510.

1959 Vandiver, H.S., *On distribution problems involving the numbers of solutions of certain trinomial congruences*, Proc. Nat. Acad. Sci. U.S.A., **45** (1959), 1635–1641.

1967 Feit, W., *Characters of Finite Groups*, Benjamin, New York, 1967.

1970 Klösgen, W., *Untersuchungen über Fermatsche Kongruenzen*, Gesellschaft Math. Datenverarbeitung, No. 36, 1970, 124 pp., Bonn.

## X.3. Hurwitz Congruence

In this section we give the theorem of Hurwitz concerning the congruence

$$(3.1) \qquad A_1 X_1^p + \cdots + A_s X_s^p \equiv 0 \pmod{q},$$

where $p \geq 3$, $q = 2kp + 1$ are prime numbers and $A_1, \ldots, A_s$ are nonzero integers. Let $N = N(A_1, \ldots, A_s, p, q)$ be the number of nontrivial solutions of (3.1), i.e., of $(x_1, \ldots, x_s)$, with $1 \leq x_i \leq q - 1$, such that $\sum_{i=1}^{s} A_i x_i^p \equiv 0 \pmod{q}$. If $h$ is a primitive root modulo

$q$, let $a_i = \mathrm{ind}_h(A_i)$ (with $0 \le a_i \le q-2$), so $A_i \equiv h^{a_i} \pmod{q}$, for $i = 1, \ldots, s$.

Thus $N$ is equal to the number of $(t_1, \ldots, t_s)$, $0 \le t_i \le q-2$, such that $\sum_{i=1}^{s} h^{pt_i + a_i} \equiv 0 \pmod{q}$.

We consider the function $\chi : \mathbb{Z} \to \{0, 1\}$ defined by

$$\chi(z) = \begin{cases} 1 & \text{when } q \mid z, \\ 0 & \text{when } q \nmid z. \end{cases}$$

Then

(3.2) $$N = \sum_{t_1, \ldots, t_s = 0}^{q-2} \chi\left( \sum_{i=1}^{s} h^{pt_i + a_i} \right).$$

Noting that if $t_i \equiv t_i' \pmod{2k}$ then $h^{pt_i + a_i} \equiv h^{pt_i' + a_i} \pmod{q}$; so (3.2) may be rewritten as

(3.3) $$N = p^s \sum_{t_1, \ldots, t_s = 0}^{2k-1} \chi\left( \sum_{i=1}^{s} h^{pt_i + a_i} \right).$$

For convenience, we introduce the following "symbol":

(3.4) $$[a_1, \ldots, a_s] = \frac{1}{2k} \sum_{t_1, \ldots, t_s = 0}^{2k-1} \chi\left( \sum_{i=1}^{s} h^{pt_i + a_i} \right),$$

which is a nonnegative rational number. We may rewrite (3.3) as follows:

(3.5) $$N = 2kp^s[a_1, \ldots, a_s] = (q-1)p^{s-1}[a_1, \ldots, a_s],$$

and to determine $N$ we are led to study the symbol $[a_1, \ldots, a_s]$.

First we note that $[a_1] = 0$, since $q \nmid h^{pt_1 + a_1}$, for every $t_1$, $0 \le t_1 \le q-2$.

LEMMA 3.1.

$$[a_1, a_2] = \begin{cases} 1 & \text{when } a_1 \equiv a_2 \pmod{p}, \\ 0 & \text{when } a_1 \not\equiv a_2 \pmod{p}. \end{cases}$$

PROOF. $q$ divides $h^{pt_1 + a_1} + h^{pt_2 + a_2}$ if and only if $h^{pt_1 + a_1} \equiv -h^{pt_2 + a_2} = h^{(q-1)/2 + pt_2 + a_2} \pmod{q}$; this is equivalent to $pt_1 + a_1 \equiv (q-1)/2 + pt_2 + a_2 \pmod{q-1}$. Now if $a_1 \equiv a_2 \pmod{p}$, say $a_2 = mp + a_1$. For every $t_2$, $0 \le t_2 \le 2k-1$, let $t_1$, $0 \le t_1 \le 2k-1$, be the unique integer such that $t_1 \equiv k + m + t_2 \pmod{2k}$. Then $(q-1)/2 + pt_2 + a_2 = p(k + t_2 + m) + a_1 \equiv pt_1 + a_1 \pmod{2kp}$. So

$$[a_1, a_2] = \frac{1}{2k} \sum_{t_1, t_2 = 0}^{2k-1} \chi(h^{pt_1 + a_1} + h^{pt_2 + a_2}).$$

Conversely, if for some $t_2$, $0 \le t_2 \le 2k - 1$, there exists $t_1$, necessarily unique, such that $0 \le t_1 \le 2k - 1$ and $pt_1 + a_1 \equiv pt_2 + a_2 + kp \pmod{q-1}$ then $p(t_1 - t_2 - k) \equiv a_2 - a_1 \pmod{2k}$, hence $p \mid a_2 - a_1$. So if $a_1 \not\equiv a_2 \pmod{p}$ then $\chi(h^{pt_1 + a_1} + h^{pt_2 + a_2}) = 0$ for all $t_1, t_2$, so $[a_1, a_2] = 0$. $\square$

Now we collect some easy facts about $[a_1, \ldots, a_s]$:

LEMMA 3.2.    (1) *The value of* $[a_1, \ldots, a_s]$ *remains unchanged by any permutation of* $a_1, \ldots, a_s$.
  (2) $[a_1, \ldots, a_s] = [a_1', \ldots, a_s']$ *whenever* $a_1 \equiv a_1' \pmod{p}, \ldots,$ $a_s \equiv a_s' \pmod{p}$.
  (3) $[a_1 + u, \ldots, a_s + u] = [a_1, \ldots, a_s]$ *for any integer* $u$.

PROOF. (1)   This is obvious from the definition of $[a_1, \ldots, a_s]$.
  (2)   Indeed, let $a_i = pr_i + a_i'$ and let $t_i + r_i \equiv t_i' \pmod{2k}$ where $0 \le t_i' \le 2k - 1$. Then $h^{p(t_i + r_i)} \equiv h^{pt_i'} \pmod{q}$ and

$$\begin{aligned}
[a_1, \ldots, a_n] &= \frac{1}{2k} \sum_{t_1, \ldots, t_s = 0}^{2k-1} \chi\left(\sum_{i=1}^{s} h^{pt_i + a_i}\right) \\
&= \frac{1}{2k} \sum_{t_1, \ldots, t_s = 0}^{2k-1} \chi\left(\sum_{i=1}^{s} h^{p(t_i + r - i) + a_i'}\right) \\
&= \frac{1}{2k} \sum_{t_1', \ldots, t_s' = 0}^{2k-1} \chi\left(\sum_{i=1}^{s} h^{pt_i' + a_i'}\right) = [a_1', \ldots, a_s'].
\end{aligned}$$

  (3)   Since $h^u \not\equiv 0 \pmod{q}$ then for every $t_i = 0, 1, \ldots, q - 2$:

$$\chi\left(\sum_{i=1}^{s} h^{pt_i + a_i + u}\right) = \chi\left(\sum_{i=1}^{s} h^{pt_i + a_i}\right),$$

hence $[a_1 + u, \ldots, a_s + u] = [a_1, \ldots, a_s]$. $\square$

LEMMA 3.3. $[a_1, \ldots, a_s]$ *is a nonnegative integer and equal to*

$$[a_1, \ldots, a_s] = \sum_{t_1', \ldots, t_{s-1}' = 0}^{2k-1} \chi(h^{pt_1' + a_1} + \cdots + h^{pt_{s-1}' + a_{s-1}} + h^{a_s})$$

$$= \sum_{t_1', \ldots, t_{s-1}' = 0}^{2k-1} \chi(h^{pt_1' + a_1} + \cdots + h^{pt_{s-1}' + a_{s-1}} - h^{a_s}).$$

PROOF. Given $t_1, \ldots, t_{s-1}$, $0 \le t_i \le 2k - 1$, for every $t_s$ let $t_1', \ldots, t_{s-1}'$ be such that $0 \le t_i' \le 2k - 1$ and $t_i' \equiv t_1 - t_s \pmod{2k}$. Noting that $h^{pt_s} \not\equiv 0 \pmod{q}$, we have

$$[a_1, \ldots, a_s]$$

$$= \frac{1}{2k} \sum_{t_1, \ldots, t_{s-1} = 0}^{k-1} \sum_{t_s = 0}^{k-1} \chi\left(\sum_{i=1}^{s} h^{pt_i + a_i}\right)$$

$$= \frac{1}{2k} \sum_{t_1', \ldots, t_{s-1}' = 0}^{2k-1} \sum_{t_s = 0}^{2k-1} \chi(h^{pt_s}(h^{pt_1' + a_1} + \cdots + h^{pt_{s-1}' + a_{s-1}} + h^{a_s}))$$

$$= \frac{1}{2k} 2k \sum_{t_1', \ldots, t_{s-1}' = 0}^{2k-1} \chi(h^{pt_1' + a_1} + h^{pt_2' + a_2} + \cdots + h^{pt_{s-1}' + a_{s-1}} + h^{a_s}).$$

Hence $[a_1, \ldots, a_s]$ is a nonnegative integer. For the last equality, we note that $a_s \equiv a_s + (q-1)/2 \pmod{p}$ and $h^{(q-1)/2} \equiv -1 \pmod{q}$. Then, using Lemma 3.2 and what we have just proved,

$$[a_1, \ldots, a_s] = \left[a_1, \ldots, a_{s-1}, a_s + \frac{q-1}{2}\right]$$

$$= \sum_{t_1', \ldots, t_{s-1}' = 0}^{2k-1} \chi(h^{pt_1' + a_1} + \cdots + h^{pt_{s-1}' + a_{s-1}} - h^{a_s}). \quad \square$$

LEMMA 3.4. *If $r, s \ge 1$ then*

$$[a_1, \ldots, a_s, b_1, \ldots, b_r]$$

$$= 2k[a_1, \ldots, a_s][b_1, \ldots, b_r] + \sum_{c=0}^{p-1} [a_1, \ldots, a_s, c][b_1, \ldots, b_r, c].$$

PROOF. Let $A_i \equiv h^{a_1} \pmod{q}$, $B_j \equiv h^{b_j} \pmod{q}$. Then

$$2kp^{s+r}[a_1, \ldots, a_s, b_1, \ldots, b_r] = N,$$

the number of solutions $(x_1, \ldots, x_s, y_1, \ldots, y_r)$ with $1 \leq x_i, y_j \leq q-1$ of $\sum_{i=1}^{s} A_i X_i^p + \sum_{j=1}^{r} B_j Y_j^p \equiv 0 \pmod{q}$. Let $N'$ be the number of solutions $(x_1, \ldots, x_s)$, with $1 \leq x_i \leq q-1$ of $\sum_{i=1}^{s} A_i X_i^p \equiv 0 \pmod{q}$ and let $N''$ be the number of similar solutions of $\sum_{j=1}^{r} B_j Y_j^p \equiv 0 \pmod{q}$.

For every $d = 0, 1, \ldots, q-2$ let $N_d'$ be the number of solutions $(x_1, \ldots, x_s)$, $1 \leq x_i \leq q-1$, of

$$\sum_{i=1}^{s} A_i X_i^p + h^d \equiv 0 \pmod{q}$$

and let $N_d''$ be the number of solutions $(y_1, \ldots, y_r)$, $1 \leq y_j \leq q-1$, of

$$\sum_{j=1}^{r} B_j Y_j^p p - h^d \equiv 0 \pmod{q}.$$

Then $N' = 2kp^s[a_1, \ldots, a_s]$, $N'' = 2kp^r[b_1, \ldots, b_r]$.

$$N_d' = p^s \sum_{t_1', \ldots, t_s' = 0}^{2k-1} \chi\left(\sum_{i=1}^{s} h^{pt_i' + a_i} + h^d\right) = p^s[a_1, \ldots, a_s, d]$$

(as follows from Lemma 3.3),

$$N_d'' = p^r \sum_{t_1', \ldots, t_r' = 0}^{2k-1} \chi\left(\sum_{j=1}^{r} h^{pt_j' + b_j} - h^d\right) = p^r[a_1, \ldots, a_r, d]$$

(as follows from Lemma 3.3).

We have $N = N'N'' + \sum_{d=0}^{q-2} N_d' N_d''$. But

$$[a_1, \ldots, a_s, d] = [a_1, \ldots, a_s, c], \quad [b_1, \ldots, b_r, d] = [b_1, \ldots, b_r, c],$$

when $d \equiv c \pmod{p}$. Therefore

$$2kp^{s+r}[a_1, \ldots, a_s, b_1, \ldots, b_r]$$
$$= 4k^2 p^{s+r}[a_1, \ldots, a_s][b_1, \ldots, b_r]$$
$$+ 2kp^{s+r} \sum_{c=0}^{p-1} [a_1, \ldots, a_s, c][b_1, \ldots, b_r, c]$$

and dividing by $2kp^{s+r}$, we have the required relation.  $\square$

LEMMA 3.5. *For any integers $a_1, a_2, c$:*

$$\sum_{d=0}^{p-1} [a_1, a_2, c+d] = 2k - [a_1, a_2];$$

*in particular, the above sum is independent of c.*

PROOF.

$$\sum_{d=0}^{p-1} [a_1, a_2, c+d] = \sum_{d=0}^{p-1} [c+d, a_1, a_2]$$

$$= \sum_{d=0}^{p-1} \left( \sum_{t,u=0}^{2k-1} \chi(h^{pt+d+c} + h^{pu+a_1} + h^{a_2}) \right),$$

by Lemma 3.3. Let $U_1$ be the set of all $u$, $0 \le u \le 2k - 1$, such that $h^{pu+a_1} + h^{a_2} \equiv 0 \pmod{q}$. Then the number of elements of $U_1$ is $\#U_1 = \sum_{u=0}^{2k-1} \chi(h^{pu+a_1} + h^{a_2}) = [a_1, a_2]$, by Lemma 3.3. Let $U_2$ be the set of all $u$, $0 \le u \le 2k - 1$, not belonging to $U_1'$, so $\#U_2 = 2k - [a_1, a_2]$. We may write

$$\sum_{d=0}^{p-1} [a_1, a_2, c+d] = \sum_{u \in U_1} \sum_{d=0}^{p-1} \sum_{t=0}^{2k-1} \chi(h^{pt+d+c} + h^{pu+a_1} + h^{a_2})$$

$$+ \sum_{u \in U_2} \sum_{d=0}^{p-1} \sum_{t=0}^{2k-1} \chi(h^{pt+d+c} + h^{pu+a_1} + h^{a_2}).$$

If $u \in U_1$ then $\chi(h^{pt+d+c} + h^{pu+a_1} + h^{a_2}) = \chi(h^{pt+d+c}) = 0$. If $u \in U_2$, there exists exactly one couple $(d, t)$ such that $h^{pt+d+c} + h^{pu+a_1} + h^{a_2} \equiv 0 \pmod{q}$. So $\sum_{d=0}^{p-1} [a_1, a_2, c+d] = \#U_2 = 2k - [a_1, a_2]$. □

LEMMA 3.6. *For any integers $a_1, a_2, a_3, a_4$:*

$$\sum_{d=0}^{p-1} [a_1 + d, a_2 + d, a_3, a_4]$$

$$= (q-1)[a_1, a_2][a_3, a_4] + (2k - [a_1, a_2])(2k - [a_3, a_4]).$$

PROOF. By Lemma 3.4 we have

$$\sum_{d=0}^{p-1}[a_1 + d, a_2 + d, a_3, a_4] = 2k \left(\sum_{d=0}^{p-1}[a_1 + d, a_2 + d]\right)[a_3, a_4]$$
$$+ \sum_{d,c=0}^{p-1}[a_1 + d, a_2 + d, c][a_3, a_4, c]$$
$$= 2kp[a_1, a_2][a_3, a_4]$$
$$+ \left(\sum_{d'=0}^{p-1}[a_1, a_2, d']\right)\left(\sum_{c=0}^{p-1}[a_3, a_4, c]\right)$$

(where $d' \equiv c - d \pmod{p}$, $0 \leq d' \leq p - 1$), by virtue of Lemma 3.2. The above sum is therefore equal to $(q-1)[a_1, a_2][a_3, a_4] + (2k - [a_1, a_2])(2k - [a_3, a_4])$, by Lemma 3.5. □

For any integers $n, m$ let

(3.6)
$$\alpha_{n,m} = \sum_{d=0}^{p-1}[d, m + nd, 0].$$

We have:

LEMMA 3.7. $\alpha_{0,m} = 2k - [m, 0]$, $\alpha_{1,m} = 2k - [m, 0]$, $\alpha_{n,m} = \alpha_{n',m'}$ when $n \equiv n' \pmod{p}$, $m \equiv m' \pmod{p}$ and $\sum_{m=0}^{p-1}\alpha_{n,m} = q - 2$.

PROOF. By Lemmas 3.2 and 3.5,

$$\alpha_{0,m} = \sum_{d=0}^{p-1}[d, m, 0] = \sum_{d=0}^{p-1}[m, 0, d] = 2k - [m, 0].$$

Similarly,

$$\alpha_{1,m} = \sum_{d=0}^{p-1}[d, m + d, 0] = \sum_{d'=0}^{p-1}[0, m, d'] = \sum_{d'=0}^{p-1}[m, 0, d'] = 2k - [m, 0]$$

(where $d' \equiv -d \pmod{p}$, $0 \leq d' \leq p - 1$). From Lemma 3.2 it follows that $\alpha_{n,m}$ depends only on the classes of $n, m$ modulo $p$.

Finally, using Lemma 3.5,

$$\sum_{m=0}^{p-1} \alpha_{n,m} = \sum_{m=0}^{p-1}\sum_{d=0}^{p-1}[d, m+nd, 0]$$

$$= \sum_{d=0}^{p-1}\sum_{m=0}^{p-1}[d, 0, m+nd]$$

$$= \sum_{d=0}^{p-1}(2k - [d, 0])$$

$$= 2kp - 1 = q - 2. \quad \square$$

Next we consider the expressions (for any integers $n, m$)

(3.7)
$$\sigma_{n,m} = \sum_{d=0}^{p-1} \alpha_{n,d}\alpha_{n,d+m} = \sum_{d,c,e=0}^{p-1} [c, d+nc, 0][e, d+m+ne, 0].$$

We have:

LEMMA 3.8.

$$\sigma_{n,m} = \sum_{j=0}^{p-1}\{(q-1)[(n-1)j + m, 0][nj + m, 0]$$

$$+ (2k - [(n-1)j + m, 0])(2k - [nj + m, 0])\} - 2k,$$

and if $n \not\equiv 0, 1 \pmod{p}$ then

$$\sigma_{n,m} = \begin{cases} 2k(q-4) + q & \text{when } m \equiv 0 \pmod{p}, \\ 2k(q-4) & \text{when } m \not\equiv 0 \pmod{p}. \end{cases}$$

PROOF.

$$\sigma_{n,m} = \sum_{c,e=0}^{p-1}\left(\sum_{d=0}^{p-1}[c, d+nc, 0][e, d+m+ne, 0]\right)$$

$$= \sum_{c,e=0}^{p-1}\left(\sum_{d=0}^{p-1}[c - nc, -nc, d][e - ne - m, -ne - m, d]\right).$$

By Lemma 3.4 the above sum is equal to

$$\sum_{c,e=0}^{p-1} ([c - nc, -nc, e - ne - m, -ne - m]$$

$$-2k[c - nc, -nc][e - ne - m, -ne - m])$$

$$= \sum_{c,e=0}^{p-1} ([c - nc, -nc, e - ne - m, -ne - m] - 2k[c, 0][e, 0]).$$

By Lemmas 3.1 and 3.2, the above sum is equal to

$$\sum_{e=0}^{p-1} \left( \sum_{c=0}^{p-1} [c - nc + ne + m, -nc + ne + m, e, 0] \right) - 2k.$$

Let $j = e - c$ (for any fixed $e$), then $c - nc + ne + m = (n - 1)j + m + e$, $-nc + ne + m = nj + m$. Hence the sum is equal to

$$\sum_{e=0}^{p-1} \left\{ \sum_{j=0}^{p-1} [(n - 1)j + m + e, e, nj + m, 0] \right\} - 2k$$

$$= \sum_{j=0}^{p-1} \left\{ \sum_{e=0}^{p-1} [(n - 1)j + m + e, e, nj + m, 0] \right\} - 2k$$

and according to Lemma 3.6 the sum is equal to

$$\sum_{j=0}^{p-1} \{ (q - 1)[(n - 1)j + m, 0][nj + m, 0]$$

$$+ (2k - [(n - 1)j + m, 0])(2k - [nj + m, 0]) \} - 2k.$$

If $n \not\equiv 0, 1 \pmod{p}$ and if $m \equiv 0 \pmod{p}$ then

$$\sigma_{n,m} = \sum_{j=0}^{p-1} \{ (q - 1)[(n - 1)j, 0][nj, 0]$$

$$+ (2k - [(n - 1)j, 0])(2k - [nj, 0]) \} - 2k$$

$$= (q - 1) + 4pk^2 - 2k - 2k + 1 - 2k$$

$$= q + 2k(q - 1) - 6k$$

$$= q + 2k(q - 4).$$

If $n \not\equiv 0, 1 \pmod{p}$ and $m \not\equiv 0 \pmod{p}$ then a similar computation gives

$$\sigma_{n,m} = 4(p - 2) + 4k(2k - 1) - 2k = 2k(q - 4). \quad \square$$

LEMMA 3.9. *If $n \not\equiv 0, 1 \pmod{p}$, for every integer $m$ we have the inequalities*

$$q - 2 - (p-1)\sqrt{q} < p\alpha_{n,m} < q - 2 + (p-1)\sqrt{q}.$$

PROOF. We have seen that

$$\sigma_{n,0} = \alpha_{n,0}^2 + \alpha_{n,1}^2 + \cdots + \alpha_{n,p-1}^2 = 2k(q-4) + q$$

and if $r \not\equiv 0 \pmod{p}$ then

$$\sigma_{n,r} = \alpha_{n,0}\alpha_{n,r} + \alpha_{n,1}\alpha_{n,r+1} + \cdots + \alpha_{n,p-1}\alpha_{n,r+p-1} = 2k(q-4).$$

Then

$$(\alpha_{n,0} - \alpha_{n,r})^2 + (\alpha_{n,1} - \alpha_{n,r+1})^2 + \cdots + (\alpha_{n,p-1} - \alpha_{n,r+p-1})^2$$

$$= 2\left(\sum_{c=0}^{p-1} \alpha_{n,c}^2\right) - 2\left(\sum_{c=0}^{p-1} \alpha_{n,c}\alpha_{n,r+c}\right)$$

$$= 2[2k(q-4) + q] - 4k(q-4) = 2q.$$

But then

$$2q \geq (\alpha_{n,m} - \alpha_{n,m+r})^2 + (\alpha_{n,m-r} - \alpha_{n,m})^2$$

$$= 2\left[\left(\alpha_{n,m} - \frac{\alpha_{n,m+r} + \alpha_{n,m-r}}{2}\right)^2 + \left(\frac{\alpha_{n,m+r} - \alpha_{n,m-r}}{2}\right)^2\right]$$

$$\geq 2\left(\alpha_{n,m} - \frac{\alpha_{n,m+r} + \alpha_{n,m-r}}{2}\right)^2.$$

Since $\sqrt{q}$ is irrational, we have the strict inequalities

$$\sqrt{q} > \alpha_{n,m} - \frac{\alpha_{n,m+r} + \alpha_{n,m-r}}{2} > -\sqrt{q}.$$

The above inequalities hold for every $r = 1, 2, \ldots, p-1$. Adding them up, and noting that

$$\sum_{r=1}^{p-1} \frac{\alpha_{,m+r} + \alpha_{n,m-r}}{2} = -\alpha_{n,r} + \sum_{r=0}^{p-1} \frac{\alpha_{n,m+r} + \alpha_{n,m-r}}{2}$$

and that

$$\sum_{r=0}^{p-1} \alpha_{n,m+r} = \sum_{r=0}^{p-1} \alpha_{n,m-r} = \sum_{j=0}^{p-1} \alpha_{n,j} = q - 2$$

(by Lemma 3.7), then

$$(p-1)\sqrt{q} > (p-1)\alpha_{n,m} - (-\alpha_{n,m} + q - 2) > -(p-1)\sqrt{q},$$

hence

$$q - 2 - (p-1)\sqrt{q} < p\alpha_{n,m} < q - 2 + (p-1)\sqrt{q}. \quad \square$$

After these lemmas, we turn to the consideration of the congruence

(3.8)                     $AX^p + BY^p + CZ^p \equiv 0 \pmod{q}$

(with $A, B, C$ nonzero integers). Let $A \equiv h^a \pmod{q}$, $B \equiv h^b$ (mod $q$), $C \equiv h^c \pmod{q}$, with $0 \le a, b, c \le q - 2$. With these notations, we have:

**(3A)**

$$p[a, b, c] = 6k + 2 - q - \nu + \sum_{n=2}^{p-1} \alpha_{n, b-c-n(a-c)},$$

*where*

$\nu = [a, b] + [b, c] + [c, a]$

$= \begin{cases} 0 \text{ when } a, b, c \text{ are pairwise incongruent modulo } p, \\ 3 \text{ when } a, b, c \text{ are congruent to each other,} \\ 1 \text{ when two of } a, b, c \text{ are congruent, but not the other one.} \end{cases}$

PROOF. We shall compute the sum $S = \sum_{n=0}^{p-1} \alpha_{n, d-ne}$ where $d, e$ are integers. By definition,

$$S = \sum_{n=0}^{p-1} \sum_{j=0}^{p-1} [j, d - ne + nj, 0].$$

If $j \equiv e \pmod{p}$, then for every $n$ we have $d - ne + nj \equiv d \pmod{p}$. If $j \not\equiv e \pmod{p}$ then $\{d - ne + nj \mid n = 0, 1, \dots, p - 1\}$ is a set of pairwise incongruent integers modulo $p$. Therefore

$$S = p[e, d, 0] + \sum_{t=0}^{p-1}\sum_{j=0}^{p-1} [j, t, 0]$$

$$= p[e, d, 0] + \sum_{\substack{j=0 \\ j \not\equiv e \ (\mathrm{mod}\ p)}}^{p-1} (2k - [j, 0]),$$

by Lemma 3.5. So

$$S = p[e, d, 0] + 2k(p - 1) + [e, 0] - \sum_{j=0}^{p-1}[j, 0]$$
$$= p[e, d, 0] + 2kp - 2k + [e, 0] - 1$$
$$= p[e, d, 0] + [e, 0] + q - 2 - 2k.$$

But $\alpha_{0,d} = 2k - [d, 0]$ and $\alpha_{1,d-e} = 2k - [d - e, 0] = 2k - [d, e]$, hence

$$p[e, d, 0] = \sum_{n=2}^{p-1} \alpha_{n,d-ne} + 6k - (q - 2) - \{[d, 0] + [d, e] + [e, 0]\}.$$

Let $e = a - c$, $d = b - c$; then $p[a, b, c] = p[a - c, b - c, 0] = p[e, d, 0] = \sum_{n=2}^{p-1} \alpha_{n,b-c-n(a-c)} + 6k - (q - 2) - \nu.$  □

After these preliminaries, we may prove the theorem of Hurwitz (1909):

**(3B)**    *The number $N$ of solutions $(x, y, z)$, with $1 \leq x, y, z \leq q - 1$ of the congruence (3.8) satisfies the inequalities:*

$$(q - 1)[(q + 1) - (p - 1)(p - 2)\sqrt{q} - p\nu] < N$$
$$< (q - 1)[(q + 1) + (p - 1)(p - 2)\sqrt{q} - p\nu].$$

PROOF. By Lemma 3.9 we have

$$(p-2)[q-2-(p-1)\sqrt{q}] < p\sum_{n=2}^{p-1} \alpha_{n,b-c-n(a-c)} < (p-2)[q-2+(p-1)\sqrt{q}],$$

hence by (3A) we have

$$p(6k + 2 - q - \nu) + (p - 2)[(q - 2) - (p - 1)\sqrt{q}]$$
$$< p^2[a, b, c] < p(6k + 2 - q - \nu) + (p - 2)[(q - 2) + (p - 1)\sqrt{q}],$$

that is,

$$q + 1 - \nu p - (p - 2)(p - 1)\sqrt{q} < p^2[a, b, c]$$
$$< q + 1 - \nu p + (p - 2)(p - 1)\sqrt{q}.$$

But by (3.5), $N = (q-1)p^2[a, b, c]$, hence

$$(q-1)[(q+1) - (p-1)(p-2)\sqrt{q} - p\nu]$$
$$< N < (q-1)[(q+1) + (p-1)(p-2)\sqrt{q} - p\nu]. \quad \square$$

To guarantee that the congruence (3.8) has a solution, it suffices to show that $q + 1 - (p-1)(p-2)\sqrt{q} - p\nu \geq 0$.

**(3C)**    If $q \geq (p-1)^2(p-2)^2 + 2(p\nu - 1)$ then $N > 0$.

PROOF. It was shown in (2) of (2C) that if $q \geq (p-1)^2(p-2)^2 + 2(p\nu - 1)$ then $q \geq (p-1)(p-2)\sqrt{q} + p\nu - 1$. It follows from (3B) that $N > 0$.    $\square$

### Bibliography

1909 Hurwitz, A., *Über die Kongruenz* $ax^e + by^e + cz^e \equiv 0 \pmod{p}$, J. Reine Angew. Math., **136** (1909), 272–292.

## X.4. Fermat's Congruence Modulo a Prime-Power

We shall consider in this section the congruence

$$X^{p^m} + Y^{p^m} + Z^{p^m} \pmod{p^n},$$

where $p$ is an odd prime number and $n > m \geq 1$. According to (1B) we may restrict our attention to the congruence

(4.1)    $$X^{p^m} + Y^{p^m} + Z^{p^m} \equiv 0 \pmod{p^{m+1}}.$$

We have already begun the study of this congruence in Chapter VI, §1. We recall (Chapter VI, (1H)) that there exist integers $x, y, z$ not multiples of $p$, satisfying the congruence (4.1) if and only if there exists $a$, $1 \leq a \leq (p-3)/2$ such that $1 + a^{p^m} \equiv (1+a)^{p^m} \pmod{p^{m+1}}$.

More generally, let $k \geq 3$, $m \geq 1$ and $p$ be an odd prime. We wish to study the congruence

(4.2)    $$X_1^{p^m} + X_2^{p^m} + \cdots + X_k^{p^m} \equiv 0 \pmod{p^{m+1}}.$$

A (nontrivial) solution is a $k$-tuple $(x_1, x_2, \ldots, x_k)$ of integers, $1 \leq x_i \leq p^{m+1} - 1$, $p \nmid x_i$ (for all $i = 1, \ldots, k$), $x_1^{p^m} + x_2^{p^m} + \cdots + x_k^{p^m} \equiv 0$

$\pmod{p^{m+1}}$. Two solutions $(x_1, x_2, \ldots, x_k)$ and $(y_1, y_2, \ldots, y_k)$ are said to be *equivalent* if there exists some integer $a$, not a multiple of $p$, $1 \le a \le p^{m+1} - 1$, and a permutation $\pi$ of $\{1, 2, \ldots, k\}$, such that $y_i \equiv a x_{\pi(i)} \pmod{p^{m+1}}$ for $i = 1, \ldots, k$. This is clearly an equivalence relation in the set of solutions.

For each integer $a$ let $\bar{a} = a \pmod{p^{m+1}}$ and let $(\mathbb{Z}/p^{m+1})^\bullet$ be the multiplicative group of invertible residue classes modulo $p^{m+1}$. Let

$$U = U(p^{m+1})$$
$$= \{\bar{b} \mid \text{there exists } a, \text{ prime to } p \text{ and such that } \bar{b} = \bar{a}^{p^m}\}$$

and let

$$V = V(p^{m+1}) = \{\bar{b} \mid b \equiv 1 \pmod{p}\}.$$

As is well known, $U, V$ are subgroups of $(\mathbb{Z}/p^{m+1})$, $U$ has $p - 1$ elements, $V$ has $p^m$ elements, and

(4.3) $$(\mathbb{Z}/p^{m+1})^\bullet \cong U \times V$$

(see any standard text on elementary number theory, or Ribenboim's book on algebraic numbers, 1999).

Let $hU = \{\sum_{i=1}^h \bar{a}_i \mid \bar{a}_i \in U \text{ for } i = 1, \ldots, h\}$ for every $h \ge 1$. Then the congruence (4.2) has a solution if and only if $\bar{0} \in kU$.

For example, let $m = 1$. If $p = 3$ then $U = \{\bar{1}, \bar{8}\}$, and it is easy to verify that $\bar{0} \notin 3U, 5U, 7U$, but $\bar{0} \in hU$ for all $h \ne 1, 3, 5, 7$.

We deduce that a cube cannot be a sum of two, four, or six cubes, if these numbers are not divisible by 3. Otherwise, if, for example, $y^3 = \sum_{i=1}^6 x_i^3$ then $(-y)^3 + \sum_{i=1}^6 x_i^3 = 0$, hence also $-\bar{y}^3 + \sum_{i=1}^6 \bar{x}_i^3 = \bar{0}$ (with $-\bar{y}, \bar{x}_i$, belonging to $U$), which is a contradiction.

Similarly, if $p = 5$ then $U = \{\bar{1}, \bar{7}, \overline{18}, \overline{24}\}$. By computation we verify that $\bar{0} \notin U, 3U, 5U$, but $\bar{0} \in hU$ for all $h \ne 1, 3, 5$. Thus, a fifth power cannot be a sum of two or four fifth powers, if these numbers are not divisible by 5.

Let $g$ be a primitive root modulo $p$, $1 < g < p$, and let $r \equiv g^{p^m}$ $\pmod{p^{m+1}}$, $1 < r < p^{m+1}$. Then $1, r, r^2, r^3, \ldots, r^{p-2}$ are pairwise incongruent modulo $p^{m+1}$, and $U = \{\bar{1}, \bar{r}, \bar{r}^2, \bar{r}^3, \ldots, \bar{r}^{p-2}\}$. In other words, given $g$, every element $\bar{x}^{p^m} \in U$ is uniquely equal to some power $\bar{r}^i$, with $0 \le i \le p - 2$. So every solution of (4.2) corresponds bijectively to a representation of 0 as a sum of powers of $\bar{r}$ in $(\mathbb{Z}/p^{m+1})^\bullet$, that is, to a congruence

$$r^{i_1} + r^{i_2} + \cdots + r^{i_k} \equiv 0 \pmod{p^{m+1}},$$

with $0 \leq i_t \leq p - 2$ (for $t = 1, \ldots, k$).

Two such representations $(r^{i_1}, \ldots, r^{i_k})$ and $(r^{j_1}, \ldots, r^{j_k})$ of 0 are said to be *equivalent* when the corresponding solutions of (4.2) are equivalent. Explicitly, there is a permutation $\pi$ of $\{1, 2, \ldots, k\}$ and an integer $h$, $0 \leq h \leq p - 2$ such that $i_t = j_{\pi(t)} + h \pmod{p - 1}$ for $t = 1, \ldots, k$.

A representation $(r^{i_1}, \ldots, r^{i_k})$ of 0 is *normalized* when $i_1 = 0 \leq i_2 \leq \cdots \leq i_k \leq p - 2$. Every representation is equivalent to a normalized representation, as easily seen. However, as we shall see, an equivalence class of representations of 0 may contain more than one normalized representation of 0.

A *cyclic solution* of (4.2) is a solution $(x_1, x_2, \ldots, x_k)$ where $x_1 \equiv 1 \pmod{p^{m+1}}$, $x_j \equiv a^{j-1} \pmod{p^{m+1}}$ for $j = 2, \ldots, k$ for some integer $a$, not multiple of $p$. Each cyclic solution corresponds to a *cyclic representation* of 0, which is a representation (relative to a given primitive root modulo $p$) of the form

$$(4.4) \qquad 1 + r^i + r^{2i} + \cdots + r^{(k-1)i} \equiv 0 \pmod{p^{m+1}}$$

(for some $i$, $0 \leq i \leq p - 2$).

**(4A)**    If $p \equiv 1 \pmod{k}$ then there is a cyclic representation, namely taking $i = (p - 1)/k$.

PROOF.

$$(1 + r^i + \cdots + r^{(k-1)i})(1 - r^i) \equiv 1 - r^{ki} \equiv 1 - r^{p-1} \equiv 0 \pmod{p^{m+1}}.$$

But $r \equiv g^{p^m} \pmod{p^{m+1}}$, so $r \equiv g \pmod{p}$, hence $r$ is a primitive root modulo $p$. If $r^i \equiv 1 \pmod{p}$ then $p - 1$ divides $i = (p - 1)/k$ so $k = 1$, contrary to the hypothesis. This shows that $(1 - r^i) \pmod{p^{m+1}}$ is invertible, hence $1 + r^i + \cdots + r^{(k-1)i} \equiv 0 \pmod{p^{m+1}}$. $\square$

In the particular case where $k = 3$ and $m = 1$ we have the cyclic representation $1 + r^i + r^{2i} \equiv 0 \pmod{p^2}$ where $i = (p - 1)/3$, $r \equiv g^p \pmod{p^2}$. So $r^i \equiv (g^{(p-1)/3})^p \pmod{p^2}$ and $(r^i)^3 \equiv g^{(p-1)p} \equiv 1 \pmod{p^2}$, thus $r^i$ is a cubic root of 1 mod $p^2$.

The following criterion for Fermat's theorem relies on the existence of a representation which is not cyclic (see Klösgen, 1970):

**(4B)**   *If $m \geq 1$, $p$ is an odd prime and if there exist integers $x, y, z$, such that $p \nmid xyz$ and $x^{p^m} + y^{p^m} + z^{p^m} = 0$, then $0$ has a noncyclic representation modulo $p^{3m+1}$.*

PROOF. We have

$$(x^{p^{m-1}})^p + (y^{p^{m-1}})^p + (z^{p^{m-1}})^p = 0,$$

hence by the result of Pollaczek, quoted in Chapter VI, (2S), we have

$$x^{2p^{m-1}} + (xy)^{p^{m-1}} + y^{2p^{m-1}} \not\equiv 0 \pmod{p}$$

and therefore

$$x^2 + xy + y^2 \not\equiv 0 \pmod{p}.$$

Let $w$ be an integer such that $wx \equiv y \pmod{p}$. So

$$1 + w + w^2 \not\equiv 0 \pmod{p}.$$

From $x^{p^m} + y^{p^m} + z^{p^m} = 0$ we deduce that $x^{p^{3m}} + y^{p^{3m}} + z^{p^{3m}} \equiv 0$ $\pmod{p^{3m+1}}$ (see Chapter VI, (1M)). We have also $x + y + z \equiv 0$ $\pmod{p}$. So $z \equiv -(x+y) \equiv -x(1+w) \pmod{p}$ and therefore

$$z^{p^{3m}} \equiv -x^{p^{3m}}(1+w)^{p^{3m}} \pmod{p^{3m+1}}.$$

We deduce from the hypothesis that

$$x^{p^{3m}}(1 + w^{p^{3m}} - (1+w)^{p^{3m}}) \equiv 0 \pmod{p^{3m+1}},$$

so

$$1 + w^{p^{3m}} - (1+w)^{p^{3m}} \equiv 0 \pmod{p^{3m+1}}.$$

If $-(1+w)^{p^{3m}} \equiv w^{2p^{3m}} \pmod{p^{3m+1}}$ then $-(1+w) \equiv w^2 \pmod{p}$, hence $1 + w + w^2 \equiv 0 \pmod{p}$, which is a contradiction.

If $w^{p^{3m}} \equiv (1+w)^{2p^{3m}} \pmod{p^{3m+1}}$ then $w \equiv (1+w)^2 \pmod{p}$, so $1 + w + w^2 \equiv 0 \pmod{p}$, again a contradiction. Thus we have obtained a noncyclic representation of $0$ modulo $p^{3m+1}$.   □

Thus, for example, if $p \equiv 1 \pmod{3}$, if the only representation of $0$ modulo $p^4$ is the cyclic representation

$$1 + r^i + r^{2i} \equiv 0 \pmod{p^4},$$

where $g$ is a primitive root modulo $p$, and $r \equiv g^{p^3} \pmod{p^4}$, then the first case of Fermat's theorem holds for the exponent $p$.

Following Klösgen, and keeping the preceding notation, we show:

**(4C)**     (1)    *If* $1 + r^i + r^j \equiv 0 \pmod{p^{m+1}}$ *with* $1 \le i < j \le p - 2$,
*then the normalized representations equivalent to the above one are:*

$(R_1)$  $1 + r^i + r^j \equiv 0 \pmod{p^{m+1}}$;
$(R_2)$  $1 + r^{j-i} + r^{p-1-i} \equiv 0 \pmod{p^{m+1}}$; *and*
$(R_3)$  $1 + r^{p-1-j} + r^{p-1-j+i} \equiv 0 \pmod{p^{m+1}}$.

*If* $j = 2i$ *then the representations* $(R_1)$, $(R_2)$, $(R_3)$ *coincide. If* $j \ne 2i$, *the representations are distinct.*
    (2)    *If* $1 + 1 + r^j \equiv 0 \pmod{p^{m+1}}$, *then* $1 \le j$ *and the normalized representations equivalent to this one are:*

$(R_1')$  $1 + 1 + r^j \equiv 0 \pmod{p^{m+1}}$; *and*
$(R_2')$  $1 + r^{p-1-j} + r^{p-1-j} \equiv 0 \pmod{p^{m+1}}$.

*In this case the representations* $(R_1')$, $(R_2')$ *are distinct and* $2^{p^m} \equiv 2$ $\pmod{p^{m+1}}$.

PROOF. (1)   From $(R_1)$ we obtain $(R_2)$ by multiplying with $r^{p-1-i}$:

$$r^{p-1-i} + r^{p-1} + r^{p-1-i+j} \equiv 0 \pmod{p^{m+1}},$$

so $1 + r^{j-i} + r^{p-1-i} \equiv 0 \pmod{p^{m+1}}$, and this is a normalized representation (because $1 \le j - i < p - 1 - i \le p - 2$) which is equivalent to $(R_1)$.

In the same way, we see that $(R_3)$ is equivalent to $(R_2)$. If $r^h + r^{h+i} + r^{h+j} \equiv 0 \pmod{p^{m+1}}$ is an equivalent representation which is normalized then one of the three cases must happen:

(a) $h \equiv 0 \pmod{p - 1}$, which yields $(R_1)$;
(b) $h + i \equiv 0 \pmod{p - 1}$, which is equivalent to $(R_2)$; and
(c) $h + j \equiv 0 \pmod{p - 1}$, which is equivalent to $(R_3)$.

If $j = 2i$ then $1 + r^i + r^{2i} \equiv 0 \pmod{p^{m+1}}$, $1 + r^i + r^{p-1-i} \equiv 0 \pmod{p^{m+1}}$ and $1 + r^{p-1-2i} + r^{p-1-i} \equiv 0 \pmod{p^{m+1}}$. So $r^{2i} \equiv r^{p-1-i} \pmod{p^{m+1}}$. Therefore $2i \equiv p - 1 - i \pmod{p - 1}$. But $1 \le 2i$, $p - 1 - i \le p - 2$ hence $2i = p - 1 - i$, thus $i = (p-1)/3$. We conclude that the representations $(R_1)$, $(R_2)$, $(R_3)$ coincide with the cyclic representation $1 + r^{(p-1)/3} + r^{2(p-1)/3} \equiv 0 \pmod{p^{m+1}}$.

It remains to see that in all other cases these representations are distinct.

If $(R_1)$ and $(R_2)$ coincide then $i = j - i$, so $j = 2i$, contrary to the hypothesis. Similarly, if $(R_1)$ and $(R_3)$ coincide then $i = p - 1 - j$, $j = p - 1 - j + i$ and again $2j = p - 1 + i$, $2j = 2(p - 1) - 2i$, hence $i = (p - 1)/3$ and $j = 2(p - 1)/3 = 2i$, contrary to the hypothesis.

Finally, if $(R_2)$ and $(R_3)$ coincide, then $j - i = p - 1 - j$, $p - 1 - i = p - 1 - j + i$ and once more $2i = j$, contrary to the hypothesis.

(2)    If $1 + 1 + r^j \equiv 0 \pmod{p^{m+1}}$ then we have the normalized equivalent representation $(R_2')$ (obtained by multiplication with $r^{p-1-j}$). We have also $j \neq 0$, since $p^{m+1} > 3$. So $(R_2')$ is not the same representation as $(R_1')$. In this case $2 + r^j \equiv 0 \pmod{p^{m+1}}$. Recalling that $r \equiv g^{p^m} \pmod{p^{m+1}}$ and noting that $g^{p^m} \equiv g \pmod{p}$, then $2 + g^j \equiv 0 \pmod{p}$, so $g^j \equiv -2 \pmod{p}$ since $g^{jp^m} \equiv -2^{p^m} \pmod{p^{m+1}}$, and therefore $2^{p^m} \equiv 2 \pmod{p^{m+1}}$.  □

We may rephrase this result as follows:

**(4D)**    *The equivalence classes of solutions of*

$$1 + X_2^{p^m} + X_3^{p^m} \equiv 0 \pmod{p^{m+1}}$$

*consist of six distinct solutions, with the following exceptions:*

(a) $p \equiv 1 \pmod{6}$, $a \not\equiv 1 \pmod{p^2}$ *but* $a^3 \equiv 1 \pmod{p^2}$: *in this case* $(1, \bar{a}, \bar{a}^2)$ *and* $(1, \bar{a}^2, \bar{a})$ *form an equivalence class of solutions.*

(b) $2^{p^m} \equiv 2 \pmod{p^{m+1}}$: *in this case* $(\bar{1}, \bar{1}, -\bar{2})$, $(\bar{1}, -\bar{2}, \bar{1})$ *and* $(\bar{1}, (-\bar{2})^{p^m(p-2)}, (-\bar{2})^{p^m(p-2)})$ *form an equivalence class of solutions.*

PROOF. According to (4C), if $p \equiv 1 \pmod{6}$ the given congruence admits the cyclic solution; its equivalence class contains exactly two solutions (of which one is normalized). If $p \not\equiv 1 \pmod{6}$, there is no cyclic solution. If $1 + 1 + r^j \equiv 0 \pmod{p^{m+1}}$ then there are precisely three solutions in this equivalence class (of which two are normalized and if this happens then $2^{p^m} \equiv 2 \pmod{p^{m+1}}$. In all other cases, each equivalence class contains exactly six distinct solutions.  □

We show now that in certain cases, it is possible to obtain a new solution of (4.1) from a given one (Peschl, 1965):

**(4E)**    *If* $1 + r^i + r^j \equiv 0 \pmod{p^{m+1}}$, *where* $j \equiv 3i + (p - 1)/2$ $\pmod{p - 1}$, *then* $1 + r^{4i} + r^{(p-1)/2+5i} \equiv 0 \pmod{p^{m+1}}$ *and this representation is not equivalent to the given one.*

PROOF. Since $r^{(p-1)/2} \equiv -1 \pmod{p^{m+1}}$ then $1 + r^i - r^{3i} \equiv 0$ $\pmod{p^{m+1}}$. Hence $-r^i - r^{2i} + r^{4i} \equiv 0 \pmod{p^{m+1}}$ and $r^{2i} + r^{3i} - r^{5i} \equiv 0 \pmod{p^{m+1}}$. Adding these congruences, we obtain $1 + r^{4i} - r^{5i} \equiv 0 \pmod{p^{m+1}}$, that is, $1 + r^{4i} + r^{(p-1)/2+5i} \equiv 0$ $\pmod{p^{m+1}}$.

If this normalized representation is equivalent to the given one, by (4C) we must have one of the following three cases:

(a) $4i \equiv i \pmod{p-1}$ and $5i \equiv 3i \pmod{p-1}$, hence $i \equiv 0$ $\pmod{p-1}$, so $i = 0$ and $1 + 1 - 1 \equiv 0 \pmod{p^{m+1}}$, a contradiction.

(b) $4i \equiv 2i + (p-1)/2 \pmod{p-1}$ and $(p-1)/2 + 5i \equiv p-1-i$ $\pmod{p-1}$ and this leads, as before, to a contradiction.

(c) $4i \equiv (p-1)/2 - 3i \pmod{p-1}$ and $(p-1)/2 + 5i \equiv (p-1)/2 - 2i \pmod{p-1}$, leading again to a contradiction.    □

In a similar way, we have:

**(4F)**    *If* $p \equiv 1 \pmod 4$ *and*

$$1 \pm r^{(p-1)/4} + r^j \equiv 0 \pmod{p^{m+1}},$$

*then*

$$1 + 1 \pm r^{2j+p-1/4} \equiv 0 \pmod{p^{m+1}},$$

*and this representation is not equivalent to the given one.*

PROOF.

$$1 \pm r^{(p-1)/4} + r^j \equiv 0 \pmod{p^{m+1}},$$

hence

$$\mp r^{(p-1)/4} + 1 \mp r^{j+(p-1)/4} \equiv 0 \pmod{p^{m+1}}$$

and

$$\pm r^{j+(p-1)/4} - r^j \pm r^{2j+(p-1)/4} \equiv 0 \pmod{p^{m+1}}.$$

Adding these congruences, we have

$$1 + 1 \pm r^{2j+(p-1)/4} \equiv 0 \pmod{p^{m+1}}.$$

We note that $r^{(p-1)/4} \not\equiv \pm 2 \pmod{p^{m+1}}$, otherwise $-1 \equiv r^{(p-1)/2} \equiv 4$ $\pmod{p^{m+1}}$, which is impossible. So $j \not\equiv 0 \pmod{p-1}$ and $j \not\equiv (p-1)/4 \pmod{p-1}$, so the representation obtained is not equivalent to the given one.    □

We shall now concentrate more on the study of the number of solutions of (4.2). Our method will be similar to the one in §2, in connection with Dickson's theorem.

We introduce the following notations, where $p > 2$ is any prime number, $m \geq 0$, $k \geq 3$: Let $F(p, m, k)$ be the number of $(x_1, \ldots, x_k)$, such that $1 \leq x_i \leq p - 1$ (for $i = 1, \ldots, k$) and

$$x_1^{p^m} + x_2^{p^m} + \cdots + x_k^{p^m} \equiv 0 \pmod{p^{m+1}}.$$

Let $a$ be any integer, $1 \leq a \leq p - 1$ and let $F(p, m, k; a)$ be the number of $(x_1, \ldots, x_k)$ such that $1 \leq x_i \leq p - 1$ (for $i = 1, \ldots, k$) and

$$x_1^{p^m} + x_2^{p^m} + \cdots + x_k^{p^m} \equiv ap^m \pmod{p^{m+1}}.$$

Let $N(p, m, k)$ be the number of $(x_2, \ldots, x_k)$ such that $1 \leq x_i \leq p-1$ (for $i = 2, \ldots, k$) and

$$1 + x_2^{p^m} + \cdots + x_k^{p^m} \equiv 0 \pmod{p^{m+1}}.$$

If $k = 3, m = 1$, we shall simply write $F(p) = F(p, 1, 3)$, $F(p; a) = F(p, 1, 3; a)$, $N(p) = N(p, 1, 3)$.

First we indicate some relations between these various numbers. Then we shall derive inductive formulas in terms of certain periods of the cyclotomic field and, in turn, in terms of Jacobi cyclotomic sums.

For $m = 0$ it is easy to compute explicitly:

**(4G)**

$$F(p, 0, k) = (p - 1)\frac{(p - 1)^{k-1} + (-1)^k}{p},$$

$$N(p, 0, k) = \frac{(p - 1)^{k-1} + (-1)^k}{p}.$$

*In particular,*

$$F(p, 0, 3) = (p - 1)(p - 2),$$

$$N(p, 0, 3) = p - 2.$$

PROOF. If $1 \leq x_i \leq p-1$ for $i = 1, \ldots, k - 2$ and $x_1 + \cdots + x_{k-2} \not\equiv 0 \pmod{p}$ then we may choose $p - 2$ values for $x_{k-1}$, such that $1 \leq x_{k-1} \leq p-1$, $x_1 + \cdots + x_{k-2} + x_{k-1} \not\equiv 0 \pmod{p}$, and this determines a unique $x_k$, $1 \le x_k \le p-1$, such that $x_1 + \cdots + x_{k-1} + x_k \equiv 0 \pmod{p}$. So we have already $[(p-1)^{k-2} - F(p, 0, k-2)](p-2)$ solutions.

Now if $1 \le x_i \le p-1$ for $i = 1, \ldots, k-2$ and $x_1 + \cdots + x_{k-2} \equiv 0 \pmod{p}$ then we may choose $p-1$ values of $x_{k-1}$ (which determine $x_k$) and hence we have $F(p, 0, k-2)(p-1)$ solutions. Thus for $k \ge 3$:

$$
\begin{aligned}
F(p, 0, k) &= [(p-1)^{k-2} - F(p, 0, k-2)](p-2) + F(p, 0, k-2)(p-1) \\
&= (p-1)^{k-2}(p-2) + F(p, 0, k-2).
\end{aligned}
$$

In particular, $F(p, 0, 3) = (p-1)(p-2)$, $F(p, 0, 4) = (p-1)^2(p-2) + (p-1)$. From the above relations we obtain

$$
\begin{aligned}
F(p, 0, 2k+1) &= (p-1)^{2k-1}(p-2) + F(p, 0, 2k-1), \\
F(p, 0, 2k-1) &= (p-1)^{2k-3}(p-2) + F(p, 0, 2k-3),
\end{aligned}
$$

$$\vdots \quad \vdots \quad \vdots$$

$$F(p, 0, 3) = (p-1)(p-2).$$

Hence, adding up,

$$
\begin{aligned}
F(p, 0, 2k+1) &= (p-1)(p-2)\sum_{j=0}^{k-1}(p-1)^{2j} \\
&= (p-1)\left[\sum_{j=0}^{k-1}(p-1)^{2j+1} - \sum_{j=0}^{k-1}(p-1)^{2j}\right] \\
&= (p-1)\frac{(p-1)^{2k}-1}{(p-1)+1} = (p-1)\frac{(p-1)^{2k}-1}{p}.
\end{aligned}
$$

Similarly,

$$
\begin{aligned}
F(p, 0, 2k) &= (p-2)\sum_{j=0}^{k-1}(p-1)^{2j} + 1 \\
&= \sum_{j=0}^{k-1}(p-1)^{2j+1} - \sum_{j=0}^{k-1}(p-1)^{2j} + 1 \\
&= (p-1)\left[\sum_{j=0}^{k-1}(p-1)^{2j} - \sum_{j=1}^{k-1}(p-1)^{2j-1}\right] \\
&= (p-1)\frac{(p-1)^{2k-1}+1}{(p-1)+1} = (p-1)\frac{(p-1)^{2k-1}+1}{p}.
\end{aligned}
$$

Hence, whether $k$ be even or odd, we have

$$F(p, 0, k) = (p - 1)\frac{(p - 1)^{k-1} + (-1)^k}{p}.$$

The same argument gives

$$N(p, 0, k) = (p - 1)^{k-3}(p - 2) + N(p, 0, k - 2).$$

Thus

$$N(p, 0, 2k + 1) = (p - 1)^{2k-2}(p - 2) + N(p, 0, 2k - 1)$$

and

$$N(p, 0, 2k + 1) = (p - 2)\sum_{j=0}^{k-1}(p - 1)^{2j} = \frac{(p - 1)^{2k} - 1}{p}.$$

Similarly, $N(p, 0, 2k) = ((p - 1)^{2k-1} + 1)/p$, so for any values of $k$:

$$N(p, 0, k) = \frac{(p - 1)^{k-1} + (-1)^k}{p}. \quad \square$$

More generally:

**(4H)**    *With the above notations:*

   (1) $F(p, m, k) = (p - 1)N(p, m, k)$.
   (2) $F(p, m, k; 1) = F(p, m, k; 2) = \cdots = F(p, m, k; p - 1)$. *This number shall be denoted by $F^*(p, m, k)$.*
   (3) $N(p, m, k) = N(p, m - 1, k) - F^*(p, m, k)$ *for $m \geq 1$. In particular:* $N(p, 1, k) = ((p - 1)^{k-1} + (-1)^k)/p - F^*(p, 1, k)$.

PROOF. (1)    We consider the sets $\mathcal{F} = \{(x_1, \ldots, x_k) \mid 1 \leq x_i \leq p - 1$ for $i = 1, \ldots, k$ and $x_1^{p^m} + \cdots + x_k^{p^m} \equiv 0 \pmod{p^{m+1}}\}$ and $\mathcal{N} = \{(x_2, \ldots, x_k) \mid 1 \leq x_i \leq p - 1$ for $i = 2, \ldots, k$ and $1 + x_2^{p^m} + \cdots + x_k^{p^m} \equiv 0 \pmod{p^{m+1}}\}$. If $y$ is any integer, $1 \leq y \leq p - 1$ and $(x_2, \ldots, x_k) \in \mathcal{N}$, if $y_1 = y$, $y_i \equiv yx_i \pmod{p}$ for $i = 2, \ldots, k$, then $(y_1, y_2, \ldots, y_k) \in \mathcal{F}$. Different values of $y$ yield different solutions of $X_1^{p^m} + X_2^{p^m} + \cdots + X_k^{p^m} \equiv 0 \pmod{p^{m+1}}$.

   If $(x_2, \ldots, x_k) \in \mathcal{N}$ and $(x_2', \ldots, x_k') \in \mathcal{N}$, with $(x_2, \ldots, x_k) \neq (x_2', \ldots, x_k')$, if $y, y'$ are integers such that $1 \leq y, y' \leq p-1$, the above method leads to distinct solutions $(y_1, y_2, \ldots, y_k) \neq (y_1', y_2', \ldots, y_k')$ because if $y \neq y'$ then $y_1 \neq y_1'$, and if $y = y'$ and say $x_1 \neq x_1'$, then $y_i \neq y_i'$.

It is also clear that every $(y_1, y_2, \ldots, y_k) \in \mathcal{F}$ may be obtained in this manner, namely taking $x_i$, $1 \leq x_i \leq p - 1$, such that $y_i \equiv y_1 x_i$ (mod $p$) for $i = 2, \ldots, k$. Thus, $F(p, m, k) = (p - 1)N(p, m, k)$.

(2)    Let $1 \leq a, b \leq p - 1$, and let $c$ be such that $b \equiv ca$ (mod $p$), $1 \leq c \leq p - 1$. If $(x_1, x_2, \ldots, x_k)$ is such that $1 \leq x_i \leq p - 1$ and

$$x_1^{p^m} + x_2^{p^m} + \cdots + x_k^{p^m} \equiv ap^m \pmod{p^{m+1}},$$

then letting $y_i$ be such that $1 \leq y_i \leq p - 1$, $y_i \equiv cx_i$ (mod $p$) then $y_i^{p^m} \equiv c^{p^m} x_i^{p^m}$ (mod $p^{m+1}$). From $c^{p^m} \equiv c$ (mod $p$) we conclude that

$$y_1^{p^m} + y_2^{p^m} + \cdots + y_k^{p^m} \equiv c^{p^m} ap^m \equiv bp^m \pmod{p^{m+1}}.$$

In this way we establish a bijection between the sets of solutions of

$$X_1^{p^m} + X_2^{p^m} + \cdots + X_k^{p^m} \equiv ap^m \pmod{p^{m+1}}$$

and of

$$X_1^{p^m} + X_2^{p^m} + \cdots + X_k^{p^m} \equiv bp^m \pmod{p^{m+1}}.$$

Therefore $F(p, m, k; a) = F(p, m, k; b)$.

(3)    Let $x_1^{p^{m-1}} + \cdots + x_k^{p^{m-1}} \equiv 0$ (mod $p^m$) with $1 \leq x_i \leq p - 1$ (for $i = 1, \ldots, k$). Then $x_1^{p^m} + \cdots + x_k^{p^m} \equiv 0$ (mod $p^m$) since $x_i^{p^m} \equiv x_i^{p^{m-1}}$ (mod $p^m$). Hence there exists $a$, $0 \leq a \leq p - 1$, such that

$$x_1^{p^m} + \cdots + x_k^{p^m} \equiv ap^m \pmod{p^{m+1}}.$$

Thus $(x_1, \ldots, x_k)$ is a solution of

$$X_1^{p^m} + \cdots + X_k^{p^m} \equiv 0 \pmod{p^{m+1}}$$

or a solution of

$$X_1^{p^m} + \cdots + X_k^{p^m} \equiv ap^m \pmod{p^{m+1}}$$

for some $a$, $1 \leq a \leq p - 1$; and conversely.

Hence, by Part (2),

$$F(p, m - 1, k) = F(p, m, k) + (p - 1)F^*(p, m, k).$$

By Part (1),

$$N(p, m - 1, k) = N(p, m, k) + F^*(p, m, k).$$

In the special case where $m = 1$, by (4F) we have

$$N(p, 1, k) = \frac{(p - 1)^{k-1} + (-1)^k}{p} - F^*(p, 1, k). \quad \square$$

If $p > 3$, let

$$\delta(p) = \begin{cases} 0 & \text{when } p \equiv -1 \pmod 6, \\ 1 & \text{when } p \equiv 1 \pmod 6, \end{cases}$$

and

$$\gamma(p, m) = \begin{cases} 1 & \text{when } 2^{p^m} \equiv 2 \pmod{p^{m+1}}, \\ 0 & \text{when } 2^{p^m} \not\equiv 2 \pmod{p^{m+1}}. \end{cases}$$

Then we have

**(4I)**

$$F^*(p, m, 3) \equiv 3\gamma(p, m) + 3\gamma(p, m - 1) \pmod 6,$$
$$N(p, m, 3) \equiv 3\gamma(p, m) + 2\delta(p) \pmod 6.$$

PROOF. The solutions of $1 + X_2^{p^m} + X_3^{p^m} \equiv 0 \pmod{p^{m+1}}$ are organized into equivalence classes. By (4D), these classes consist of six elements, except when $p \equiv 1 \pmod 6$, where there is a class of only two elements, and when $2^{p^m} \equiv 2 \pmod{p^{m+1}}$, where there is a class of three elements. Thus

$$N(p, m, 3) \equiv 3\gamma(p, m) + 2\delta(p) \pmod 6.$$

By (4H),

$$\begin{aligned} F^*(p, m, 3) &= N(p, m - 1, 3) - N(p, m, 3) \\ &\equiv 3\gamma(p, m - 1) - 3\gamma(p, m) \\ &\equiv 3\gamma(p, m - 1) + 3\gamma(p, m) \pmod 6. \quad \square \end{aligned}$$

For the special case where $m = 1$, $k = 3$, we have

$$F^*(p) = F^*(p, 1, 3) \equiv 3\gamma(p, 1) + 3 \pmod 6,$$
$$N(p) \equiv 3\gamma(p, 1) + 2\delta(p) \pmod 6.$$

In order to indicate an upper bound for $N(p)$ (with $p > 3$) we need to study in more detail the Cauchy polynomials modulo $p$. We recall from Chapter VII, §2, that

$$(X + 1)^p - X^p - 1 = pX(X + 1)(X^2 + X + 1)^\varepsilon C_p(X),$$

where $C_p(X) \in \mathbb{Z}[X]$,

$$\varepsilon = \begin{cases} 1 & \text{when } p \equiv -1 \pmod 6, \\ 2 & \text{when } p \equiv 1 \pmod 6. \end{cases}$$

$X^2 + X + 1$ does not divide $C_p(X)$, $C_p(X)$ is a symmetric monic polynomial, $C_p(-1 - X) = C_p(X)$, hence $C_p(0) = C_p(-1) = 1$. Let

$$q(X) = \frac{(X+1)^p - X^p - 1}{p} \in \mathbb{Z}[X]$$

and $\overline{q}(X) \equiv q(X) \pmod{p}$, $\overline{C}_p(X) \equiv C_p(X) \pmod{p}$.

**(4J)**    *For $p > 3$, we have:*

> (1) *All the roots of $\overline{q}(X)$ (different from $\overline{0}$ and $\overline{-1}$) in $\mathbb{F}_p$, and all roots of $\overline{C}_p(X)$ in $\mathbb{F}_p$ are double roots.*
> (2) *If $\alpha$ is a root of $\overline{C}_p(X)$ then each element in the set*
>
> $$M_\alpha = \left\{ \alpha, \frac{1}{\alpha}, -(1+\alpha), -\frac{1}{1+\alpha}, -\frac{\alpha}{1+\alpha}, -\frac{1+\alpha}{\alpha} \right\}$$
>
> *is also a root of $\overline{C}_p(X)$. If $M_\alpha$ has less than six distinct elements, then $M_\alpha = \{\overline{1}, -\overline{2}, (p-1)/2 \pmod{p}\}$ (in this case $2^p \equiv 2 \pmod{p^2}$) or $p \equiv -1 \pmod 6$ and $\alpha^2 + \alpha + \overline{1} = \overline{0}$, $\alpha \notin \mathbb{F}_p$.*
> (3) *$X^2 + X + \overline{1} \in \mathbb{F}_p[X]$ does not divide $\overline{C}_p(X)$.*

PROOF. (1)    Let $\alpha \in \mathbb{F}_p$ be such that $\overline{q}(\alpha) = \overline{0}$. We have $\overline{q}'(X) = (X + \overline{1})^{p-1} - X^{p-1}$. If $\alpha \neq \overline{0}, \overline{-1}$ then $(\alpha + \overline{1})^{p-1} = \alpha^{p-1} = \overline{1}$, so $\overline{q}'(\alpha) = \overline{0}$. Since $\overline{q}''(X) = (p - \overline{1})[(X + \overline{1})^{p-2} - X^{p-2}]$ then

$$\overline{q}''(\alpha) = -[(\alpha + \overline{1})^{p-2} - \alpha^{p-2}] = \frac{\overline{1}}{\alpha + \overline{1}} - \frac{\overline{1}}{\alpha} = -\frac{\overline{1}}{\alpha(\alpha + \overline{1})} \neq \overline{0}.$$

This shows that $\alpha \neq \overline{0}, -\overline{1}$ is a double root of $\overline{q}(X)$.

From $\overline{q}(X) = X(X + \overline{1})(X^2 + X + \overline{1})^\varepsilon \overline{C}_p(X)$ if $\overline{C}_p(\alpha) = \overline{0}$ then $\overline{q}(\alpha) = 0$, so taking derivatives: $\overline{0} = (\alpha^2 + \alpha + \overline{1})^\varepsilon \overline{C}'_p(\alpha)$. We have $\alpha^2 + \alpha + \overline{1} \neq \overline{0}$, otherwise since $\alpha \in \mathbb{F}_p$ then necessarily $p \equiv 1 \pmod 6$ (Chapter I, Lemma 4.1). Hence $\varepsilon = 2$ and $\overline{C}_p(\alpha) = \overline{0}$, since $\alpha$ is a double root of $\overline{q}(X)$. Thus $\overline{C}'_p(\alpha) = \overline{0}$, and $\alpha$ is necessarily a double root of $\overline{C}_p(X)$.

(2)    Since $C_p(X)$ is a symmetric polynomial such that $C_p(-1 - X) = C_p(X)$ and $C_p(0) = C_p(-1) = 1$, then if $\overline{C}_p(\alpha) = \overline{0}$ then each $\beta \in M_\alpha$ is also a root of $\overline{C}_p(X)$. We suppose that $M_\alpha$ has less than six elements. Then one of the following cases happens:

> (i) $\alpha = 1/\alpha$: then $\alpha = \pm\overline{1}$.

(ii) $\alpha = -(1 + \alpha)$: then $\alpha = (p - 1)/2 \bmod p$.
(iii) $\alpha = -1/(1 + \alpha)$: then $\alpha^2 + \alpha + \bar{1} = 0$.
(iv) $\alpha = -\alpha/(1 + \alpha)$: then $\alpha = 0$ or $\alpha = -\bar{2}$.
(v) $\alpha = -(1 + \alpha)/\alpha$: then $\alpha^2 + \alpha + \bar{1} = 0$.

But $\bar{0}, -\bar{1}$ are not roots of $\overline{C}_p(X)$, $\alpha = \bar{1}, -\bar{2}$ or $(p - 1)/2 \bmod p$. $M_\alpha = \{\bar{1}, -\bar{2}, (p - 1)/2 \bmod p\}$ and from $((1 + 1)^p - 1^p - 1)/p \equiv 0 \pmod{p}$ we have $2^p \equiv 2 \pmod{p^2}$. If $\alpha^2 + \alpha + \bar{1} = \bar{0}$ with $\alpha \in \mathbb{F}_p$ then $p \equiv 1 \pmod 6$ so $\varepsilon = 2$. Since $\alpha$ is a double root of $\bar{q}(X)$ it cannot be a root of $\overline{C}_p(X)$. Thus $\alpha \notin \mathbb{F}_p$ and $p \equiv -1 \pmod 6$.

(3)   If $p \equiv 1 \pmod 6$ then $\varepsilon = 2$, $X^2 + X + \bar{1}$ has root $\alpha \in \mathbb{F}_p$ which is a double root of $\bar{q}(X)$, hence not a root of $\overline{C}_p(X)$. Hence $X^2 + X + \bar{1}$ does not divide $\overline{C}_p(X)$.

Let $p \equiv -1 \pmod 6$, so $\varepsilon = 1$. The roots of $\overline{C}_p(X)$ appear in groups of six distinct roots, with the following exceptions:

(i) the group of three double roots $\{\bar{1}, -\bar{2}, (p - 1)/2 \bmod p\}$; and
(ii) the two roots of $X^2 + X + \bar{1}$ (which are necessarily outside $\mathbb{F}_p$).

If $\overline{C}_p(X) = (X^2 + X + \bar{1})^r \overline{H}(X)$ with $r \geq 1$ and $X^2 + X + \bar{1}$ not dividing $\overline{H}(X) \in \mathbb{F}_p[X]$, then

$$\deg \overline{C}_p(X) = 2r + \deg \bar{h}(X) \equiv 2r \pmod 6,$$

in view of the grouping of roots of $\overline{H}(X)$. But if $p = 6n - 1$ then $\deg \overline{C}_p(X) = 6(n - 1)$, so $2r \equiv 0 \pmod 6$, hence $r \equiv 0 \pmod 3$ and therefore $r \geq 3$. Thus if $\alpha^2 + \alpha + \bar{1} = \bar{0}$ then $\alpha$ is a triple root of $\overline{C}_p(X)$, hence also of $\bar{q}(X)$. Therefore

$$\bar{q}'(X) = (X + \bar{1})^{p-1} - X^{p-1}$$

and

$$\bar{q}''(X) = (p - 1)[(X + \bar{1})^{p-2} - X^{p-2}]$$

vanish at $\alpha$:

$$(\alpha + \bar{1})^{p-2} = \alpha^{p-1}, \qquad (\alpha + \bar{1})^{p-2} = \alpha^{p-2}.$$

Comparing, we have

$$\alpha^{p-1} = (\alpha + \bar{1})(\alpha + \bar{1})^{p-2} = (\alpha + \bar{1})\alpha^{p-2} = \alpha^{p-1} + \alpha^{p-2},$$

hence $\alpha^{p-2} = \bar{0}$, so $\alpha = \bar{0}$, a contradiction.   □

We may now determine an upper bound for $N(p)$:

**(4K)**

(1) *If $2^p \not\equiv 2 \pmod{p^2}$ then*

$$N(p) \leq \begin{cases} \dfrac{p-9}{2} & \text{when } p \equiv 1 \pmod{12}, \\[2mm] \dfrac{p-5}{2} & \text{when } p \equiv 5 \pmod{12}, \\[2mm] \dfrac{p-3}{2} & \text{when } p \equiv 7 \pmod{12}, \\[2mm] \dfrac{p-11}{2} & \text{when } p \equiv 11 \pmod{12}. \end{cases}$$

(2) *If $2^p \equiv 2 \pmod{p^2}$ then*

$$N(p) \leq \begin{cases} \dfrac{p-3}{2} & \text{when } p \equiv 1 \pmod{12}, \\[2mm] \dfrac{p-11}{2} & \text{when } p \equiv 5 \pmod{12}, \\[2mm] \dfrac{p-9}{2} & \text{when } p \equiv 7 \pmod{12}, \\[2mm] \dfrac{p-5}{2} & \text{when } p \equiv 11 \pmod{12}. \end{cases}$$

PROOF. We recall that the nontrivial solutions of $1 + X_1^p + X_2^p \equiv 0$ $\pmod{p^2}$ correspond to the nontrivial solutions of $1 + X^p \equiv (1+X)^p$ $\pmod{p^2}$, that is, to the zeros in $\mathbb{F}_p$ distinct from $\bar{0}$, $-\bar{1}$ of

$$\bar{q}(X) = \frac{(X+1)^p - X^p - 1}{p} \pmod{p} \in \mathbb{F}_p[X].$$

(1)   If $p = 12n + 1$ then $q(X) = X(X+1)(X^2 + X + 1)^2 C_p(X)$, $\deg q(X) = p - 1$, $\deg C_p(X) = p - 7 = 12n - 6$. Since every root of $\overline{C_p}(X)$ in $\mathbb{F}_p$ is a double root and the roots appear in groups of six, then

$$N(p) \leq 2 + 6\left[\frac{6n-1}{6}\right] = 2 + 6n - 6 = 6n - 4$$

$$= \frac{p-1}{2} - 4 = \frac{p-9}{2}.$$

If $p = 12n + 5$ then $q(X) = X(X+1)(X^2 + X + 1)C_p(X)$. The same argument gives (noting that $X^2 + X + \bar{1}$ has no root in $\mathbb{F}_p$):

$\deg C_p(X) = 12n$, $N(p) \le 6n = (p-5)/2$. If $p = 12n + 7$ and $p = 12n + 11$, we proceed similarly.

(2) If $2^p \equiv 2 \pmod{p^2}$ then $\overline{q}(X) = X(X+\overline{1})(X^2+X+\overline{1})^\varepsilon(X-\overline{1})^2(X+\overline{2})^2(X-(p-1)/2 \bmod p)^2\overline{A}(X)$ where the roots of $\overline{A}(X) \in \mathbb{F}_p[X]$ appear in groups of six distinct double roots.

If $p = 12n + 1$ then $\gcd \overline{A}(X) = 12(n-1)$ and $N(p) \le 2+3+6(n-1) = 6n-1 = (p-3)/2$.

In the same way we derive the other upper bounds. $\square$

All the preceding considerations do not yet provide any explicit formula for the numbers of solutions of the congruences in question. As we shall now see, such formulas may be obtained using Gaussian periods and Jacobi cyclotomic sums, as in §2.

Let $p$ be an odd prime, $m \ge 1$, let $h$ be a primitive root modulo $p^{m+1}$, let $\zeta$ be a primitive root of 1 of order $p^m$, and let $\rho$ be a primitive root of 1 of order $p^{m+1}$, $\rho^p = \zeta$. The Gaussian periods $\eta_i = \eta_i(p, m, h)$ are defined as follows:

(4.5)
$$
\begin{cases}
\eta_0 &= \rho + \rho^{h^{p^m}} + \rho^{h^{2p^m}} + \cdots + \rho^{h^{(p-2)p^m}}, \\
\vdots & \quad \vdots \quad \vdots \\
\eta_i &= \rho^{h^i} + \rho^{h^{p^m+i}} + \rho^{h^{2p^m+i}} + \cdots + \rho^{h^{(p-2)p^m+i}}, \\
\vdots & \quad \vdots \quad \vdots \\
\eta_{p^m-1} &= \rho^{h^{p^m-1}} + \rho^{h^{2p^m-1}} + \rho^{h^{3p^m-1}} + \cdots + \rho^{(p-1)p^m-1}.
\end{cases}
$$

It is also convenient to agree that $\eta_j$ is defined for any index $j$, by letting $\eta_j = \eta_i$ when $j \equiv i \pmod{p^m}$, $0 \le i \le p^m - 1$.

If $g$ is another primitive root modulo $p^{m+1}$, then $g \equiv h^r \pmod{p^{m+1}}$ where $\gcd(r, p^m(p-1)) = 1$. If $\eta_i' = \eta_i(p, m, g)$ then $\eta_i' = \eta_{ri}$ (for $i = 0, 1, \ldots, p^m - 1$). Indeed,

$$
\eta_i' = \sum_{j=0}^{p-2} \rho^{g^{jp^m+i}} = \sum_{j=0}^{p-2} \rho^{h^{jrp^m+ri}} = \sum_{t=0}^{p-2} \rho^{h^{tp^m+ri}} = \eta_{ri},
$$

because if $jr \equiv t \pmod{p-1}$, $0 \le t \le p-2$ then $\{(jrp^m + ri) \bmod p^m(p-1) \mid j = 0, 1, \ldots, p-2\} = \{(tp^m + ri) \bmod p^m(p-1) \mid t = 0, 1, \ldots, p-2\}$ as we may easily verify: $j \not\equiv j' \pmod{p^m}$ if and only if $jr \not\equiv j'r \pmod{p^m}$. Hence up to a change of numbering, the

Gaussian periods are independent of the choice of the primitive root modulo $p^{m+1}$.

Let us observe now that each period $\eta_i$ is a real number. Indeed, $\eta_i$ is the sum of $(p-1)/2$ pairs of complex-conjugate numbers:

$$\rho^{h^{j p^m + i}} + \rho^{h^{((p-1)/2+j)p^m + i}} = \rho^{h^{j p^m + i}} + \rho^{-h^{j p^m + i}} = \rho^{h^{j p^m + i}} + \overline{\rho^{h^{j p^m + i}}} \in \mathbb{R}.$$

And we also have the gross estimations

$$(4.6) \qquad |\eta_i| \le \sum_{j=0}^{p-2} \left| \rho^{h^{j p^m + i}} \right| < p - 1$$

(the equality would hold only if the numbers $\rho^{h^{j p^m + i}}$ would be all multiples of one of them, which is not the case).

For every $t$ not a multiple of $p$, let $\mathrm{ind}_h(t) = s$, where $0 \le s \le (p-1)p^m - 1$ and $t \equiv h^s \pmod{p^{m+1}}$.

The Jacobi sums $\tau_j = \tau_j(p, m, h)$ (for $j = 0, 1, \dots, p^m - 1$) are defined by

$$(4.7) \qquad \tau_j = \langle \zeta^j, \rho \rangle = \sum_{\substack{t=1 \\ p \nmid t}}^{p^{m+1} - 1} \zeta^{j \, \mathrm{ind}_h(t)} \rho^t.$$

In particular,

$$(4.8) \qquad \tau_0 = 0.$$

Indeed,

$$\tau_0 = \langle 1, \rho \rangle = \sum_{\substack{t=1 \\ p \nmid t}}^{p^{m+1} - 1} \rho^t,$$

so $\tau_0$ is the sum of the primitive $(p^{m+1})$th roots of 1. Hence $\tau_0$ is the coefficient of the term of degree $\varphi(p^{m+1}) - 1$ of the cyclotomic polynomial

$$\Phi_{p^{m+1}}(X) = \frac{X^{p^{m+1}} - 1}{X^{p^m} - 1} = X^{(p-1)p^m} + X^{(p-2)p^m} + \cdots + X^{p^m} + 1,$$

thus $\tau_0 = 0$.

We shall see that $\tau_j \ne 0$ if and only if $p \nmid j$. For this purpose we require the following lemma about sums of roots of unity:

LEMMA 4.1.    (1) *Let $n > 1$ and let $\xi$ be a primitive $n$th root of $1$. For every integer $a$:*

$$\sum_{x=1}^{n} \xi^{ax} = \begin{cases} n & \text{when } n \mid a, \\ 0 & \text{when } n \nmid a. \end{cases}$$

(2) *Let $p$ be an odd prime, $m \geq 0$, and let $\rho$ be a primitive root of $1$ of order $p^{m+1}$. For every integer $a$:*

$$\sum_{\substack{x=1 \\ p \nmid x}}^{p^{m+1}} \rho^{ax} = \begin{cases} \varphi(p^{m+1}) & \text{when } p^{m+1} \mid a, \\ -p^m & \text{when } p^m \mid a,\ p^{m+1} \nmid a, \\ 0 & \text{when } p^m \nmid a. \end{cases}$$

PROOF. (1)    Let $d = \gcd(n, a)$, $n = dn'$, $a = da'$, so $\gcd(n', a') = 1$. Since $\xi^d$ is a primitive root of $1$ of order $n$ then $\xi^{da'}$ is also a primitive root of $1$ of order $n'$. Hence

$$\sum_{x=1}^{n} \xi^{ax} = \sum_{x=1}^{n} (\xi^{da'})^x.$$

But each $x$, $1 \leq x \leq n$, may be written as $x = hn' + y$, $1 \leq y \leq n'$, $0 \leq h \leq d - j$. So

$$\begin{aligned}
\sum_{x=1}^{n} \xi^{ax} &= \sum_{x=1}^{n} \left(\xi^{da'}\right)^x \\
&= \sum_{h=0}^{d-1} \sum_{y=1}^{n'} \left(\xi^{da'}\right)^{hn'+y} \\
&= \sum_{h=0}^{d-1} \sum_{y=1}^{n'} \left(\xi^{da'}\right)^y \\
&= \begin{cases} n'd = n & \text{when } n' = 1, \text{ i.e., } n \mid a, \\ 0 & \text{when } n' > 1, \text{ i.e., } n \nmid a. \end{cases}
\end{aligned}$$

(2)    If $p^{m+1} \mid a$ then the sum

$$S = \sum_{\substack{x=1 \\ p \nmid x}}^{p^{m+1}} \rho^{ax}$$

is obviously equal to $\Phi(p^{m+1})$. If $p^{m+1} \nmid a$, $p^m \mid a$, let $a = bp^m$, with $p \nmid b$. Then $\zeta = \rho^{bp^m}$ is a primitive $p$th root of $1$. Every $x$,

$1 \leq x \leq p^{m+1}$ is of the form $x = hp + y$, $0 \leq h \leq p^m - 1$, $1 \leq y \leq p$; moreover, $p \nmid x$ when $y \neq p$. So

$$S = \sum_{\substack{x=1 \\ p \nmid x}}^{p^{m+1}} \zeta^x = \sum_{h=0}^{p^m-1} \sum_{y=1}^{p-1} \zeta^y = \sum_{h=0}^{p^m-1} (-1) = -p^m.$$

If $p^m \nmid a$ then

$$S + \sum_{\substack{p \mid x \\ 1 \leq x \leq p^{m+1}}} \rho^{ax} = \sum_{x=1}^{p^{m+1}} \rho^{ax} = 0$$

by (1), since $p^{m+1} \nmid a$.

Next we note that $\rho^p = \zeta$ is a primitive root of 1 of order $p^m$. Each $x$, multiple of $p$, $1 \leq x \leq p^{m+1}$, is written as $x = py$, $1 \leq y \leq p^m$. So

$$\sum_{\substack{x=1 \\ p \mid x}}^{p^{m+1}} \rho^{ax} = \sum_{y=1}^{p^m} \zeta^{ay} = 0$$

by (1), since $p^m \nmid a$. We conclude that $S = 0$.    □

The following lemmas concern indices:

LEMMA 4.2. *Let $p$ be an odd prime, $m \geq 0$, $h$ a primitive root modulo $p^{m+1}$, and $i$ an integer, $1 \leq i \leq p - 1$. Then*

$$\mathrm{ind}_h(1 + ip^m) \equiv iap^{m-1} \pmod{p^m}$$

*for some integer $a$, not a multiple of $p$.*

PROOF. Let $s = \mathrm{ind}_h(1 + ip^m)$, so $h^s \equiv 1 + ip^m \pmod{p^{m+1}}$. Then

$$h^{sp} \equiv (1 + ip^m)^p \equiv 1 + ip^{m+1} \equiv 1 \pmod{p^{m+1}}.$$

Hence $sp \equiv 0 \pmod{p^m(p-1)}$, that is, $sp = -bp^m(p-1) = bp^m - bp^{m+1}$, then $s = bp^{m-1} - bp^m$, so $s \equiv bp^{m-1} \pmod{p^m}$. Since $p \nmid i$ there exists $a$ such that $b \equiv ia \pmod{p}$, hence $s \equiv iap^{m-1} \pmod{p^m}$.

It remains to show that $p \nmid a$. Otherwise $s \equiv 0 \pmod{p^m}$, that is, $1 + ip^m \equiv h^{p^m c} \pmod{p^{m+1}}$ (for some integer $c$). By (3.3), $(1 + ip^m)$ mod $p^{m+1} = h^{p^m c}$ mod $p^{m+1} \in U \cap V = \{\bar{1}\}$, and therefore $i \equiv 0 \pmod{p}$, which is a contradiction.    □

For every $j = 0, 1, \ldots, p^m(p-1) - 1$ let $U_j = \{\bar{a} \in (\mathbb{Z}/p^{m+1})^\bullet \mid \mathrm{ind}_h(\bar{a}) \equiv j \pmod{p^m}\}$. Since $U = \{\bar{a}^{p^m} \mid \bar{a} \in (\mathbb{Z}/p^{m+1})^\bullet\}$ then $U_j$ is a coset of $(\mathbb{Z}/p^{m+1})^\bullet$ modulo $U$. In view of a later result concerning sums of squares of periods, we need the following description of these cosets.

Let $1 \le k \le m$, let $i = 0, 1, \ldots, p^{m-k} - 1$, and let

$$\mathcal{S}_i = \{U_i, U_{i+p^{m-k}}, U_{i+2p^{m-k}}, \ldots, U_{i+(p^k-1)p^{m-k}}\}.$$

Similarly, let

$$\mathcal{S}_i' = \{\overline{(1+ip)} \cdot U, \overline{[(1+i+p^{m-k})p]} \cdot U, \overline{[1+(i+2p^{m-k})p]} \cdot U, \\ \ldots, \overline{[1+(i+(p^k-1)p^{m-k})p]} \cdot U\}.$$

First we note that for each $i$, the cosets $U_{i+lp^{m-k}}$, $U_{i+l'p^{m-k}}$ (with $l \ne l'$) are distinct. Otherwise, $i + lp^{m-k} \equiv i + l'p^{m-k} \pmod{p^m}$, hence $l \equiv l' \pmod{p^k}$, contrary to the hypothesis. So $\#\mathcal{S}_i = p^k$.

If $i \ne i'$ then $\mathcal{S}_i \cap \mathcal{S}_{i'} = \emptyset$: if $i + lp^{m-k} \equiv i' + l'p^{m-k} \pmod{p^m}$ then necessarily $i \equiv i' \pmod{p^{m-k}}$, so $i = i'$.

So $\bigcup_{i=0}^{p^{m-k}-1} \mathcal{S}_i$ has $p^m$ cosets, that is, it consists of all the cosets modulo $U$. Similarly, the cosets

$$\overline{[1+(i+lp^{m-k})p]} \cdot U, \qquad \overline{[1+(i+l'p^{m-k})p]} \cdot U$$

(with $l \ne l'$) are distinct. Otherwise

$$\overline{1 + (i+lp^{m-k})p} \cdot \left(\overline{[1+(i+l'p^{m-k})p]}\right)^{-1} \in V \cap U = \{\bar{1}\},$$

so

$$1 + (i+lp^{m-k})p \equiv 1 + (i+l'p^{m-k})p \pmod{p^{m+1}},$$

hence $l \equiv l' \pmod{p^k}$. So $\#\mathcal{S}_i' = p^k$.

If $i \ne i'$ then $\mathcal{S}_i' \cap \mathcal{S}_{i'}' = \emptyset$: if

$$\overline{[1+(i+lp^{m-k})p]} \cdot U = \overline{[1+(i'+l'p^{m-k})p]} \cdot U$$

then as before

$$1 + (i+lp^{m-k})p \equiv (1+i'+l'p^{m-k})p \pmod{p^{m+1}},$$

hence necessarily $i \equiv i' \pmod{p^{m-k}}$, so $i = i'$. Thus $\bigcup_{i=0}^{p^{m-k}-1} \mathcal{S}_i'$ has $p^m$ cosets, that is, it consists of all cosets modulo $U$.

Now we prove:

LEMMA 4.3. *There exists a permutation $\pi$ of $\{0, 1, \ldots, p^{m-k} - 1\}$ such that $\mathcal{S}_i' = \mathcal{S}_{\pi(i)}$.*

PROOF. Let $\overline{\mathrm{ind}(1+ip)} = g$, that is, $\overline{(1+ip)} \cdot U = U_j$. We show first that for every $l = 0, 1, \ldots, p^k - 1$,

$$\mathrm{ind}\overline{[1 + (i + lp^{m-k})p]} = j + sp^{m-k}$$

for some $s$, $0 \le s \le p^k - 1$; this is equivalent to showing that

$$\overline{[1 + (i + lp^{m-k})p]} \cdot U = U_{j+sp^{m-k}}$$

and allows us to define $\pi(i) = j$, with $\mathcal{S}'_i = \mathcal{S}_{\pi(i)}$. Indeed, let $i'$ be such that $(1 + i'p)(1 + ip) \equiv 1 \pmod{p^{m+1}}$ and let $l = l'p^r$ with $p \nmid l'$, $0 \le r \le k$. We define $b$, $1 \le b \le p^{k-r} - 1$ by the congruence $l'(1 + i'p) \equiv b \pmod{p^{k-r}}$. Then

$$\begin{aligned}
[1 + (i + lp^{m-k})p][1 + i'p] &\equiv 1 + l(1 + i'p)p^{m-k+1} \\
&\equiv 1 + l'(1 + i'p)p^{m-k+r+1} \\
&\equiv 1 + bp^{m-k+r+1} \pmod{p^{m+1}}.
\end{aligned}$$

But

$$(1 + bp^{m-k+r+1})^{p^{k-r}} \equiv 1 \pmod{p^{m+1}},$$

so

$$1 + bp^{m-k+r+1} \equiv h^{t(p-1)p^{m-k+r}} \pmod{p^{m+1}}.$$

Taking $s = t(p-1)p^r$ then

$$1 + (i + lp^{m-k})p \equiv (1 + ip)h^{sp^{m-k}} \equiv h^{j+sp^{m-k}} \pmod{p^{m+1}},$$

as we needed to prove.

We show that if $i \ne i'$ then $\pi(i) \ne \pi(i')$. If $h^j \equiv 1 + ip \pmod{p^{m+1}}$, $h^{j'} \equiv 1 + i'p \pmod{p^{m+1}}$ and $j \equiv j' \pmod{p^{m-k}}$, let $j' = j + sp^{m-k}$. So $1 + i'p \equiv (1 + ip)h^{sp^{m-k}} \pmod{p^{m-k}}$, hence $h^{sp^{m-k}} \in V$. So $sp^{m-k}$ is a multiple of $(p-1)$, hence also of $(p-1)p^{m-k}$, that is, $sp^{m-k} = t(p-1)p^{m-k}$. Let $h^{p-1} \equiv 1 + ap \pmod{p^{m+1}}$. Then $h^{sp^{m-k}} \equiv (1 + ap)^{tp^{m-k}} \equiv 1 + atp^{m-k+1} \pmod{p^{m-k+1}}$. Therefore $1 + i'p \equiv (1 + ip)(1 + atp^{m-k+1}) \equiv 1 + (i + atp^{m-k})p \pmod{p^{m-k+2}}$. We conclude that $i' \equiv i \pmod{p^{m-k}}$, so $i' = i$.

We conclude therefore that $\pi$ is a permutation with the required property. $\square$

We now derive some formulas for Jacobi sums and Gauss periods, which are analogues to the ones already proved in Chapter IX, §2.

LEMMA 4.4.    (1) *If $j = 1, \ldots, p^m - 1$ then $\overline{\tau}_j = \tau_{p^m - j}$ ($\overline{\tau}_j$ denotes the complex conjugate of $\tau_j$);*

(2) $\tau_j \overline{\tau}_j = \begin{cases} p^{m+1} & \text{when } p \nmid j; \\ 0 & \text{when } p \mid j; \text{ and} \end{cases}$

(3) *$\tau_j \neq 0$ if and only if $p$ does not divide $j$.*

PROOF. (1)

$$\overline{\tau}_j = \sum_{\substack{x=1 \\ p \nmid x}}^{p^{m+1}} \zeta^{-j \, \text{ind}_h(x)} \rho^{-x}.$$

But $\text{ind}_h(-x) \equiv \text{ind}_h(-1) + \text{ind}_h(x) \pmod{p^m(n-1)}$ and $\text{ind}_h(-1) = \frac{1}{2} p^m(p-1)$, so $\zeta^{\text{ind}_h(-1)} = 1$. Hence

$$\overline{\tau}_j = \sum_{\substack{x=1 \\ p \nmid x}}^{p^{m+1}} \zeta^{-j \, \text{ind}_h(-x)} \rho^{-x} = \tau_{p^m - j}.$$

(2)    We have

$$\tau_j \overline{\tau}_j = \left( \sum_{\substack{x=1 \\ p \nmid x}}^{p^{m+1}-1} \zeta^{j \, \text{ind}_h(x)} \rho^x \right) \left( \sum_{\substack{y=1 \\ p \nmid y}}^{p^{m+1}-1} \zeta^{-j \, \text{ind}_h(y)} \rho^{-y} \right)$$

$$= \sum_{\substack{x=1 \\ p \nmid x}}^{p^{m+1}-1} \sum_{\substack{y=1 \\ p \nmid y}}^{p^{m+1}-1} \zeta^{j(\text{ind}_h(x) - \text{ind}_h(y))} \rho^{x-y}.$$

For every $x, y$ as above, let $t$, $1 \leq t \leq p^{m+1} - 1$, be such that $y \equiv xt \pmod{p^{m+1}}$, hence $p \nmid t$ and $\text{ind}_h(y) \equiv \text{ind}_h(x) + \text{ind}_h(t)$ $\pmod{p^m(p-1)}$. Then

$$\tau_j \overline{\tau}_j = \sum_{\substack{x=1 \\ p \nmid x}}^{p^{m+1}-1} \sum_{\substack{t=1 \\ p \nmid x}}^{p^{m+1}-1} \zeta^{-j \, \text{ind}_h(t)} \rho^{x(1-t)}$$

$$= \sum_{\substack{t=1 \\ p \nmid t}}^{p^{m+1}-1} \zeta^{j \, \text{ind}_h(t)} \left( \sum_{\substack{x=1 \\ p \nmid x}}^{p^{m+1}-1} \rho^{x(1-t)} \right).$$

Using Lemma 4.1, we have

$$\tau_j \overline{\tau}_j = p^m(p-1) + (-p^m) \sum_{i=1}^{p-1} \zeta^{-j \, \text{ind}_h(1+ip^m)},$$

where the first summand corresponds to $t = 1$ and the second summation to the terms $t = 1 + ip^m$; the other values of $t$ give the sum 0. By Lemma 4.2

$$\text{ind}_h(1 + ip^m) \equiv iap^{m-1} \pmod{p^m}$$

(where $p \nmid a$), so

$$\tau_j \overline{\tau}_j = p^m(p-1) + (-p^m) \sum_{i=1}^{p-1} \zeta^{-jiap^{m-1}}$$

$$= \begin{cases} p^m(p-1) + p^m = p^{m+1} & \text{if } p \nmid j, \\ p^m(p-1) - p^m(p-1) = 0 & \text{if } p \mid j. \end{cases}$$

(3)   This is obvious from (2).   □

We note the following connection between the periods and the Jacobi sums:

LEMMA 4.5. *Consider the matrix* $Z = (\zeta^{ij})_{i,j=0,1,\ldots,p^m-1}$ *and the vectors*

$$\tau = \begin{pmatrix} \tau_0 \\ \tau_1 \\ \vdots \\ \tau_{p^m-1} \end{pmatrix}, \qquad \eta = \begin{pmatrix} \eta_0 \\ \eta_1 \\ \vdots \\ \eta_{p^m-1} \end{pmatrix}.$$

*Then:*

(1) $Z\overline{Z} = p^m I$ (*I identity matrix*)*, that is,*

$$\sum_{k=1}^{p^n} \zeta^{ik} \zeta^{-kj} = p^m \delta_{ik}$$

*(for* $i, j = 0, 1, \ldots, p^m - 1$*).*

(2) $\overline{Z}\tau = p^m \eta$*, that is,*

$$\sum_{j=0}^{p^m-1} \zeta^{-ij} \tau_j = p^m \eta_i$$

*(for* $i = 0, 1, \ldots, p^m - 1$*).*

(3) $\tau = Z\eta$*, that is,*

$$\sum_{j=0}^{p^m-1} \zeta^{ij} \eta_j = \tau_i$$

*(for* $i = 0, 1, \ldots, p^m - 1$*).*

PROOF. (1)  By Lemma 4.1(1),

$$\sum_{k=1}^{p^m} \zeta^{ik}\zeta^{-kj} = \sum_{k=1}^{p^m-1} \zeta^{(i-j)k} = \begin{cases} p^m & \text{when } p^m \mid i - j, \text{ that is, } j = i, \\ 0 & \text{when } j \neq i. \end{cases}$$

(2)

$$\sum_{j=0}^{p^m-1} \zeta^{ij}\tau_j = \sum_{j=0}^{p^m-1} \zeta^{-ij} \sum_{\substack{x=1 \\ p \nmid x}}^{p^{m+1}} \zeta^{j\,\mathrm{ind}_h(x)}\rho^x$$

$$= \sum_{\substack{x=1 \\ p \nmid x}}^{p^{m+1}} \rho^x \left( \sum_{j=0}^{p^m-1} \zeta^{j(-i+\mathrm{ind}_h(x))} \right).$$

But by Lemma 4.1(1)

$$\sum_{j=0}^{p^m-1} \zeta^{j(-i+\mathrm{ind}_h(x))} = \begin{cases} p^m & \text{when } \mathrm{ind}_h(x) \equiv i \pmod{p^m}, \\ 0 & \text{otherwise,} \end{cases}$$

hence the sum to be evaluated is equal to

$$\sum_{\mathrm{ind}_h(x) \equiv i \pmod{p^m}} \rho^x = \sum_{a=0}^{p-2} \rho^{h^{ap^m+i}} = \eta_i,$$

since $\mathrm{ind}_h(x) \equiv i \pmod{p^m}$, $1 \leq x \leq p^m - 1$ is equivalent to $x \equiv h^{ap^m+i} \pmod{p^{m+1}}$, with $0 \leq a \leq p - 1$.

(3)  From $\overline{Z}\tau \equiv p^m\eta$, by multiplication with $Z$ we obtain $p^m\tau = Z\overline{Z}\tau = p^m Z\eta$ hence $\tau = Z\eta$.  $\square$

As a corollary we obtain the vanishing of special sums of periods:

LEMMA 4.6. *Let $p$ be an odd prime, $m \geq 1$, let $1 \leq k \leq m$, $i = 0, 1, \ldots, p^{m-k} - 1$. Then*

$$\sum_{x=0}^{p^k-1} \eta_{i+xp^{m-k}} = 0.$$

PROOF. By Lemma 4.5(2):

$$p^m \left( \sum_{x=0}^{p^k-1} \eta_{i+xp^{m-k}} \right) = \sum_{x=0}^{p^k-1} \sum_{j=0}^{p^m-1} \zeta^{-(i+xp^{m-k})j} \tau_j$$

$$= \sum_{j=0}^{p^m-1} \zeta^{-ij} \tau_j \left( \sum_{x=0}^{p^k-1} \zeta^{-xp^{m-k}j} \right)$$

$$= \sum_{j=0}^{p^m-1} \zeta^{-ij} \tau_j \left( \sum_{x=0}^{p^k-1} \xi^{-xj} \right),$$

where $\xi = \zeta^{p^{m-k}}$. Since $\xi$ is a primitive root of 1 of order $p^k$, by Lemma 4.1(1),

$$\sum_{x=0}^{p^k-1} \xi^{-xj} = \begin{cases} p^k & \text{if } p^k \mid j, \\ 0 & \text{otherwise.} \end{cases}$$

Hence

$$p^m \left( \sum_{x=0}^{p^k-1} \eta_{i+xp^{m-k}} \right) = p^k \left( \sum_{\substack{j=0 \\ p^k\mid j}}^{p^m-1} \zeta^{-ij} \tau_j \right) = 0,$$

by Lemma 4.4(3). $\quad\square$

For example, if $m = 1$ then $k = 1$ and $\sum_{j=0}^{p-1} \eta_j = 0$, which was already known. If $m = 2$, $k = 1$ then

(4.9) $$\sum_{j=0}^{p-1} \eta_{i+jp} = 0 \qquad \text{for} \quad i = 0, 1, \ldots, p-1.$$

If $m = 2$, $k = 2$ then $\sum_{j=0}^{p^2-1} \eta_d = 0$, as already known. If $m = 3, k = 1$ then

(4.10) $$\sum_{j=0}^{p-1} \eta_{i+jp^2} = 0 \qquad \text{for} \quad i = 0, 1, \ldots, p^2-1.$$

If $m = 3$, $k = 2$ then

(4.11) $$\sum_{j=0}^{p^2-1} \eta_{i+jp} = 0 \qquad \text{for} \quad i = 0, 1, \ldots, p-1,$$

and finally, if $m = 3, k = 3$ then $\sum_{j=0}^{p^3-1} \eta_j = 0$, as already known.

We may also evaluate the following sums of squares of periods:

LEMMA 4.7. *Let $p$ be an odd prime, $m \geq 1$, let $1 \leq k \leq m$, $i = 0, 1, \ldots, p^{m-k} - 1$. Then*

$$\sum_{x=0}^{p^k-1} \eta_{i+xp^{m-k}}^2 = p^k(p-1).$$

PROOF. Let $U_j = \{\bar{a} \in (\mathbb{Z}/p^{m+1})^{\bullet} \mid \text{ind}_h(a) \equiv j \pmod{p^m}\}$ for $j = 0, 1, \ldots, p^m(p-1)$. Then $U_j$ is a coset modulo the subgroup $U$. By definition,

$$\eta_j = \sum_{\substack{0 \leq s \leq (p-1)p^m \\ \bar{s} \in U_j}} \rho^{h^s}.$$

We keep the notations preceding Lemma 4.3:

$$\mathcal{S}_i = \{U_i, U_{i+p^{m-k}}, U_{i+2p^{m-k}}, \ldots, U_{i+(p^k-1)p^{m-k}}\},$$
$$\mathcal{S}_i' = \{\overline{(1+ip)} \cdot U, \overline{[(1+ip^{m-k})p]} \cdot U,$$
$$\ldots, \overline{[1+(i+(p^k-1)p^{m-k})p]} \cdot U\},$$

and let $\pi$ be the permutation of $\{0, 1, \ldots, p^{m-k} - 1\}$ such that $\mathcal{S}_i' = \mathcal{S}_{\pi(i)}$ (see Lemma 4.3). Then

$$\sum_{x=0}^{p^k-1} \eta_{i+xp^{m-k}}^2 = \sum_{U_j \in \mathcal{S}_i} \eta_j^2 = \sum_{U_j \in \mathcal{S}_{\pi^{-1}(i)}} \eta_j^2.$$

Thus we need only to evaluate the sums

$$\mathcal{S}_i = \sum_{x=0}^{p^k-1} \left( \sum_{a=1}^{p-1} \rho^{(1+ip+xp^{m+1-k})a^{p^m}} \right)^2.$$

We have

$$\mathcal{S}_i = \sum_{x=0}^{p^k-1} \sum_{a=1}^{p-1} \sum_{b=1}^{p-1} \rho^{(1+ip+xp^{m+1-k})}(a^{p^m} + b^{p^m})$$

$$= \sum_{a=1}^{p-1} \sum_{b=1}^{p-1} \rho^{a^{p^m}+b^{p^m}} \zeta^{i(a^{p^{m-1}}+b^{p^{m-1}})} \sum_{x=0}^{p^k-1} \xi^{x(a^{p^{k-1}}+b^{p^{k-1}})},$$

where $\zeta = \rho^p$, $\xi = \rho^{p^{m+1-k}}$ (so $\xi^{p^k} = 1$), noting also that $a^{p^m} \equiv a^{p^{m-1}} \pmod{p^m}$, $a^{p^m} \equiv a^{p^{k-1}} \pmod{p^k}$, and similarly for $b$.

By Lemma 4.1,

$$S_i = p^k \sum_{a=1}^{p-1} \rho^{a^{p^m} - a^{p^m}} \zeta^{i(a^{p^{m-1}} - a^{p^{m-1}})} = p^k(p-1),$$

and this concludes the proof.  □

In particular, taking $k = m$ we have

(4.12)
$$\sum_{j=0}^{p^m-1} \eta_j^2 = p^m(p-1).$$

We specialize now the above results to the case where $m = 1$. Let $a$, $1 \le a \le p-1$ be defined as follows. If $s = \text{ind}_h(1+p)$ then $h^s \equiv 1 + p \pmod{p^2}$ hence $h^{sp} \equiv (1+p)^p \equiv 1 \pmod{p^2}$, hence $p(p-1)$ divides $sp$, so $p-1$ divides $s$. We define $a$ by $a(p-1) = s$.

LEMMA 4.8. *If $m = 1$, $a(p-1) = \text{ind}_h(1+p)$ then $\text{ind}_h(1+cp) \equiv -ca$ (mod $p$) for every $c = 0, 1, \dots, p-1$.*

PROOF. From $h^{a(p-1)} \equiv 1+p \pmod{p^2}$ we have $h^{ca(p-1)} \equiv (1+p)^c \equiv 1 + cp \pmod{p^2}$. Hence $\text{ind}_h(1 + cp) \equiv ca(p-1) \pmod{p(p-1)}$. In particular, $\text{ind}_h(1 + cp) \equiv cap - ca \equiv -ca \pmod{p}$.  □

LEMMA 4.9. *If $m = 1$, the Jacobi sums (for $i = 0, 1, \dots, p-1$) are given by*
$$T_i = p\rho^{(ai)^p},$$
*where $a(p-1) = \text{ind}_h(1+p)$.*

PROOF.

$$T_i = \sum_{x=0}^{p^2-1} \zeta^{i\,\text{ind}_h(x)} \rho^x.$$

By (4.3) we may write in a unique way

$$x \equiv h^{pb}(1 + cp) \pmod{p^2},$$

where $0 \le b, c \le p-1$. Then by Lemma 4.8, $\text{ind}_h(x) \equiv pb + \text{ind}_h(1+cp) \equiv -ca \pmod{p}$. Therefore,

$$T_i = \sum_{c=0}^{p-1} \zeta^{-ica} \sum_{b=0}^{p-1} \rho^{h^{pb}(1+cp)} = \sum_{b=0}^{p-1} \rho^{h^{pb}} \sum_{c=0}^{p-1} \zeta^{c(h^{pb} - ia)}$$

(noting that $\rho^p = \zeta$). Since $ia \not\equiv 0 \pmod{p}$ then $(ia)^p \pmod{p^2} \in U$. So there exists $b_0$, $1 \le b_0 \le p-1$ such that $(ia)^p \equiv h^{pb_0} \pmod{p^2}$. By Lemma 4.1(1),

$$\sum_{c=0}^{p-1} \zeta^{c(h^{pb} - ia)} = \begin{cases} p & \text{when } b = b_0, \\ 0 & \text{when } b \ne b_0. \end{cases}$$

We conclude that

$$\tau_i = \rho^{h^{pb_0}} = \rho^{(ia)^p}. \qquad \square$$

In analogy with (2B) we indicate an inductive expression for $N(p, m; k)$ and $F(p, m, k)$ in terms of the periods (see Klösgen (1970)):

**(4L)**

(1) $F(p, m, k) = \dfrac{1}{p} F(p, m-1, k) + \dfrac{p-1}{p^{m+1}} \displaystyle\sum_{i=0}^{p^m - 1} \eta_i^k;$    and

(2) $N(p, m; k) = \dfrac{1}{p} N(p, m-1, k) + \dfrac{1}{p^{m+1}} \displaystyle\sum_{i=0}^{p^m - 1} \eta_i^k.$

PROOF. (1)   By Lemma 4.1(1), we have

$$p^{m+1} F(p, m, k) = \sum_{x_1=1}^{p-1} \sum_{x_2=1}^{p-1} \cdots \sum_{x_k=1}^{p-1} \sum_{y=0}^{p^{m+1}-1} \rho^{y(x_1^{p^m} + \cdots + x_k^{p^m})}$$

$$= \sum_{y=0}^{p^{m+1}-1} \sum_{x_1=1}^{p-1} \rho^{yx_1^{p^m}} \cdots \sum_{x_k=1}^{p-1} \rho^{yx_k^{p^m}}$$

$$= \sum_{y=0}^{p^{m+1}-1} \left( \sum_{x=1}^{p-1} \rho^{yx^{p^m}} \right)^k$$

$$= \sum_{\substack{y=0 \\ p|y}}^{p^{m+1}-1} \left( \sum_{x=1}^{p-1} \rho^{yx^{p^m}} \right)^k + \sum_{\substack{y=0 \\ p \nmid y}}^{p^{m+1}-1} \left( \sum_{x=1}^{p-1} \rho^{yx^{p^m}} \right)^k$$

$$= \sum_{t=0}^{p^m-1} \left( \sum_{x=1}^{p-1} \zeta^{tx^{p^{m-1}}} \right)^k + \sum_{\substack{y=0 \\ p \nmid y}}^{p^{m+1}-1} \left( \sum_{x=1}^{p-1} \rho^{yx^{p^m}} \right)^k,$$

since $\rho^p = \zeta$ and $x^{p^m} \equiv x^{p^{m-1}} \pmod{p^m}$. Since

$$p^m F(p, m-1, k) = \sum_{t=0}^{p^m-1} \left( \sum_{x=1}^{p-1} \zeta^{tx^{p^{m-1}}} \right)^k,$$

by the same computation as above, then

$$p^{m+1} F(p, m, k) = p^m F(p, m-1, k) + \sum_{\substack{y=0 \\ p \nmid y}}^{p^{m+1}-1} \left( \sum_{x=1}^{p-1} \rho^{yx^{p^m}} \right)^k.$$

Each $y$, $0 \le y \le p^{m+1}-1$, $p \nmid y$, may be uniquely written in the form $y \equiv h^i \pmod{p^{m+1}}$, with $0 \le i \le p^m(p-1)$. Also, each $x^{p^m}$ (with $1 \le x \le p-1$) may be written uniquely as $x^{p^m} \equiv h^{cp^m} \pmod{p^{m+1}}$ with $0 \le c \le p-2$. Hence

$$\sum_{\substack{y=0 \\ p \nmid y}}^{p^{m+1}-1} \left( \sum_{x=1}^{p-1} \rho^{yx^{p^m}} \right)^k = \sum_{i=0}^{p^m(p-1)-1} \left( \sum_{c=0}^{p-2} \rho^{h^{i+cp^m}} \right)^k$$

$$= \sum_{i=0}^{p^m(p-1)-1} \eta_i^k$$

$$= (p-1) \sum_{i=0}^{p^m-1} \eta_i^k.$$

Thus

$$F(p, m, k) = \frac{1}{p} F(p, m-1, k) + \frac{p-1}{p^{m+1}} \sum_{i=0}^{p^m-1} \eta_i^k.$$

(2)   Dividing by $p-1$ and taking (4G) into account, we deduce the recurrence relation for $N(p, m; k)$.   $\square$

As a corollary, it follows from (4F) that

$$(4.13) \qquad F(p) = \frac{(p-1)(p-2)}{p} + \frac{p-1}{p^2} \sum_{i=0}^{p-1} \eta_i^3,$$

$$(4.14) \qquad N(p) = \frac{p-2}{p} + \frac{1}{p^2} \sum_{i=0}^{p-1} \eta_i^3.$$

Let $S(p^m, k) = \sum_{i=0}^{p-1} [\eta_i(p, m)]^k$. From the recurrence formulas in (4L) we obtain the expressions:

**(4M)**

(1) $F(p, m, k) = \dfrac{p-1}{p^{m+1}}[(p-1)^k + (-1)^k + S(p^m, k) + S(p^{m-1}, k) + \cdots + S(p, k)].$

(2) $N(p, m, k) = \dfrac{1}{p^{m+1}}[(p-1)^{k-1} + (-1)^k + S(p^m, k) + S(p^{m-1}, k) + \cdots + S(p, k)].$

PROOF. (1)  We have

$$F(p, m, k) = \frac{1}{p}F(p, m-1, k) + \frac{p-1}{p^{m+1}}S(p^m, k),$$

$$\frac{1}{p}F(p, m-1, k) = \frac{1}{p^2}F(p, m-2, k) + \frac{p-1}{p^{m+1}}S(p^{m-1}, k),$$

$$\vdots \quad \vdots \quad \vdots$$

$$\frac{1}{p^{m-1}}F(p, 1, k) = \frac{1}{p^m}F(p, 0, k) + \frac{p-1}{p^{m+1}}S(p, k).$$

Adding these equalities and taking into account (4G), we deduce that

$$F(p, m, k) = \frac{p-1}{p^{m+1}}[(p-1)^k + (-1)^k + S(p^m, k) + S(p^{m-1}, k) + \cdots + S(p, k)].$$

(2)  This formula is obtained from the preceding one by dividing by $p - 1$.  □

Taking $k = 2$ in the above formula and noting that $N(p, m, 2) = 1$ (trivially) then (4M) yields again the relation of Lemma 4.7 for $k = m$. Indeed, if $m = 1$,

$$1 = N(p, 1, 2) = \frac{1}{p^2}[(p-1) + 1 + S(p, 2)],$$

hence

$$S(p, 2) = \sum_{j=0}^{p^2-1}[\eta_j(p, 1)]^2 = p(p-1).$$

Assuming by induction that for $r < m$ we have

$$S(p^r, 2) = \sum_{j=0}^{p^r-1}[\eta_i(p, r)]^2 = p^r(p-1),$$

then

$$1 = N(p, m, 2)$$
$$= \frac{1}{p^{m+1}}[(p-1) + 1 + S(p^m, k) + p^{m-1}(p-1) + \cdots + p(p-1)].$$

Hence $S(p^m, k) = p^m(p-1)$. So $S(p^m, k)$ is an integer which is a multiple of $p^m$.

For the next result, which is about sums of powers of the Gaussian periods, we shall require a lemma about the $q$-adic values of products of factorials.

LEMMA 4.10. *Let $q$ be a prime, $s \geq 1$, $\mu > 1$ and let $r_1, r_2, \ldots, r_\mu$ be integers greater than $0$ such that $q^s = r_1 + r_2 + \cdots + r_\mu$. Then*

$$v_q(r_1! \, r_2! \cdots r_\mu!) \leq q v_q(q^{s-1}!),$$

*and if $\mu = q$, $r_1 = r_2 = \cdots = r_q = q^{s-1}$ then $v_q(q^{s-1}! \cdots q^{s-1}!) = q v_q(q^{s-1}!)$.*

PROOF. If $s = 1$ the statement is trivial (since $\mu > 1$). We proceed by induction on $s$.

We have $[r_1/q] + \cdots + [r_\mu/q] \leq q^{s-1}$, so for some integer $r_0' \geq 0$ we have $r_0' + [r_1/q] + \cdots + [r_\mu/q] = q^{s-1}$. By induction,

$$v_q(r_0'!) + v_q\left(\left[\frac{r_1}{q}\right]!\right) + \cdots + v_q\left(\left[\frac{r_\mu}{q}\right]!\right) \leq q v_q(q^{s-2}).$$

By Chapter II, (1A), if $t \geq 1$ is an integer then

$$v_q(t!) = \left[\frac{t}{q}\right] + \left[\frac{t}{q^2}\right] + \left[\frac{t}{q^3}\right] + \cdots$$

and if $x \geq 0$ is any real number, $a \geq 1$ any integer, then $[x/a] = [[x]/a]$, so

$$v_q(r_1!) + v_q(r_2!) + \cdots + v_q(r_\mu!)$$
$$= \sum_{i=1}^{r} \left(\left[\frac{r_i}{q}\right] + \left[\frac{r_i}{q^2}\right] + \left[\frac{r_i}{q^3}\right] + \cdots\right)$$

$$= \sum_{i=1}^{\mu} \left[ \frac{r_i}{q} \right] + \sum_{i=1}^{\mu} \left( \left[ \frac{[r_i/q]}{q} \right] + \left[ \frac{[r_i/q]}{q^2} \right] + \cdots \right)$$

$$\leq q^{s-1} + v_q(r_0'!) + \sum_{i=1}^{\mu} v_q \left( \left[ \frac{r_i}{q} \right]! \right)$$

$$\leq q^{s-1} + q v_q(q^{s-2}!)$$

$$= q^{s-1} + q(q^{s-3} + \cdots + q + 1)$$

$$= q(q^{s-2} + q^{s-3} + \cdots + q + 1) = q v_q(q^{s-1}!).$$

The last assertion is of course trivial. $\square$

Now we may derive the following properties of the sums $S(p^m, k)$:

**(4N)**

(1)  $S(p^m, k) = p^m[pN(p, m, k) - N(p, m-1, k)] = p^m[F(p, m, k) - F^*(p, m, k)]$.

(2)  If $q$ is a prime, $k = q^s$, then $S(p^m, q^s)$ is a multiple of $q$.

PROOF. (1)  By (3L) and (4H),

$$\begin{aligned} S(p^m, k) &= p^m[pN(p, m, k) - N(p, m-1, k)] \\ &= p^m[(p-1)N(p, m, k) - F^*(p, m, k)] \\ &= p^m[F(p, m, k) - F^*(p, m, k)]. \end{aligned}$$

(2)  By virtue of (1) we may assume $q \neq p$ and it suffices to show that $q$ divides $F(p, m, q^s)$ and $F^*(p, m, q^s) = F(p, m, q^s; c)$, where $1 \leq c \leq p-1$, $c \equiv q \pmod{p}$ (by (4H)). Let $x_1, x_2, \ldots, x_k \in \mathbb{Z}$ be such that $x_1^{p^m} + x_2^{p^m} + \cdots + x_k^{p^m} \equiv 0 \pmod{p^{m+1}}$, respectively, $x_1^{p^m} + x_2^{p^m} + \cdots + x_k^{p^m} \equiv c \pmod{p^{m+1}}$, where $1 \leq x_i \leq p^{m+1} - 1$ for $i = 1, \ldots, k$, $k = q^s$. Each $k$-tuple obtained by a permutation from $(x_1, x_2, \ldots, x_k)$ is still a solution of the congruence. So the set of solutions is organized into disjoint classes of solutions equivalent under permutation. Therefore it suffices to show that the number of solutions equivalent to any given one is a multiple of $q$.

Let the $k$-tuple $(x_1, x_2, \ldots, x_k)$ have exactly $\mu$ distinct components, repeated, respectively, $r_1, r_2, \ldots, r_\mu$ times (with $r_i \geq 1$). We note that $\mu \neq 1$; otherwise from $x_1 = x_2 = \cdots = x_k$ we deduce $qx_1^{p^m} \equiv 0$, respectively, $q \pmod{p}$, which is impossible. It follows that $r_1 < k, \ldots, r_\mu < k$.

The number of solutions obtained by permutation from $(x_1, x_2, \ldots, x_k)$ is equal to $k!/(r_1! r_2! \cdots r_\mu!)$. So we need to prove that $q$ divides this number. We have

$$v_q(k!) = q^{s-1} + q^{s-2} + \cdots + q + 1.$$

Since $r_1 + r_2 + \cdots + r_\mu = q^s$ and $\mu > 1$ then $v_q(r_1! r_2! \cdots r_\mu!) \leq q(q^{s-2} + q^{s-3} + \cdots + q + 1) = q^{s-1} + q^{s-2} + \cdots + q^2 + q$.

Therefore the $q$-adic valuation of $k!/(r_1! r_2! \cdots r_\mu!)$ is at least equal to 1, which was to be proved. $\square$

We now indicate an upper bound for $N(p)$ in terms of the periods. In view of (4.13) we are led to find an upper bound for the sum $\sum_{i=0}^{p-1} \eta_i^3$. For this purpose we establish the following lemma:

LEMMA 4.11. *Let $n \geq 3$, let $f$ be the function of $n$ real variables*

$$f(y_1, \ldots, y_n) = \sum_{i=1}^{n} y_i^3$$

*defined on the set $D$ of all points $(y_1, \ldots, y_n)$ such that $\sum_{i=1}^{n} y_i = 0$ and $\sum_{i=1}^{n} y_i^2 = n(n-1)$.*

(1) *If $(y_1, \ldots, y_n)$ is a point where the function assumes a maximum or a minimum, then there exists an integer $T$, $1 \leq T \leq n-1$, such that (up to a permutation of $\{1, \ldots, n\}$)*

$$\begin{cases} y_1 = \cdots = y_T = (n-T)\sqrt{\dfrac{n-1}{T(n-T)}}, \\[2ex] y_{T+1} = \cdots = y_n = -T\sqrt{\dfrac{n-1}{T(n-T)}}. \end{cases}$$

*Let $y^T$ be the point with the above coordinates.*

(2) *$f(y^T) = n(n-1)(n-2T)\sqrt{(n-1)/(T(n-T))}$, $f(y^{n-T}) = -f(y^T)$, $f(y^T) > 0$ for $1 \leq T < (n-1)/2$.*

(3) *If $T = 1$ then $f(y^T) = n(n-1)(n-2)$ is the absolute maximum of $f$ on the given domain $D$.*

(4) *If $1 \leq T \leq n-1$ then $f(y^T)$ is the absolute maximum of $f$ on the points $y = (y_1, \ldots, y_n) \in D$ such that $y_i \leq (n-T)\sqrt{(n-1)/(T(n-T))}$.*

PROOF. (1)    To find the points of maximum or minimum of the function $F$ on the domain $D$, we employ the method of Lagrange multipliers. Let $\lambda, \mu$ be parameters to be determined and

$$F(y_1, \ldots, y_n) = f(y_1, \ldots, y_n) + \lambda \left( \sum_{i=1}^{n} y_i \right) + \mu \left( \sum_{i=1}^{n} y_i^2 - n(n-1) \right).$$

If $(y_1, \ldots, y_n)$ is an extreme point then $(\partial F / \partial y_1)(y_1, \ldots, y_n) = 0$ (for $i = 1, \ldots, n$), that is,

(4.15)        $3y_i^2 + \lambda + 2\mu y_i = 0$        (for $i = 1, \ldots, n$).

Adding up these relations, we obtain $3n(n-1) + n\lambda = 0$, hence $\lambda = -3(n-1)$. Substituting this value in the above condition (4.15), we have $3y_i^2 + 2\mu y_i - 3(n-1) = 0$, hence

(4.16)        $$y_i = \frac{-\mu \pm \sqrt{\mu^2 + 9(n-1)}}{3}.$$

Since there are only two possible values for the coordinates then, up to a permutation of $\{1, \ldots, n\}$, there exists an integer $T$, $0 \leq T \leq n$, such that

$$y_1 = \cdots = y_T = \frac{-\mu + \sqrt{\mu^2 + 9(n-1)}}{3},$$

$$y_{T+1} = \cdots = y_n = \frac{-\mu - \sqrt{\mu^2 + 9(n-1)}}{3}.$$

We note that not all coordinates can be equal since $\sum_{i=1}^{n} y_i = 0$. Hence $1 \leq T \leq n-1$. Moreover,

$$0 = -\frac{\mu}{3}(T + n - T) + \frac{1}{3}\sqrt{\mu^2 + 9(n-1)}(T - n + T),$$

hence

$$n\mu = (2T - n)\sqrt{\mu^2 + 9(n-1)};$$

therefore

$$n^2\mu^2 = (2T - n)^2[\mu^2 + 9(n-1)]$$

and finally

$$\mu = \frac{3(2T - n)}{2}\sqrt{\frac{n-1}{T(n-T)}},$$

$$\mu^2 + 9(n-1) = \frac{9(n-1)n^2}{4T(n-T)}.$$

Substituting into (4.16) we obtain

(4.17)
$$y_i = \begin{cases} (n-T)\sqrt{\dfrac{n-1}{T(n-T)}}, & \text{or} \\[2ex] -T\sqrt{\dfrac{n-1}{T(n-T)}}. \end{cases}$$

(2)   For the point $y^T$ with above coordinates, we have

$$\sum_{i=1}^{n} y_i^3 = T(n-T)^3 \frac{n-1}{T(n-T)} \sqrt{\frac{n-1}{T(n-T)}}$$

$$-(n-T)T^3 \frac{n-1}{T(n-T)} \sqrt{\frac{n-1}{T(n-T)}}$$

$$= n(n-1)(n-2T)\sqrt{\frac{n-1}{T(n-T)}}.$$

Let

$$\tilde{f}(t) = n(n-1)(n-2t)\sqrt{\frac{n-1}{t(n-t)}}$$

for $0 < t < n$. Then $\tilde{f}(n-t) = -\tilde{f}(t)$; if $0 < t < n/2$ then $\tilde{f}(t) > 0$ and $f(y^T) = \tilde{f}(T)$ for $T = 1, 2, \ldots, n-1$. If $0 < t \le (n-1)/2$ then $t(n-t) \le (t+1)(n-t-1)$, as seen at once. Hence

$$\frac{n-2t}{\sqrt{t(n-t)}} > \frac{n-2t-2}{\sqrt{(t+1)(n-t-1)}}$$

and therefore $f(y^T) > f(y^{T+1}) > 0$ for $T = 1, 2, \ldots, [n/2] - 1$.

(3)   For $T = 1$ we have $f(y^1) = n(n-1)(n-2)$. Since the function is continuous and defined on a closed and bounded domain, it has a maximum and a minimum. By (1) and (2) it follows that $f$ has the absolute maximum at $y^1$.

(4)   Let $1 \le T \le n-1$ and let $D_T$ consist of those points of $D$ with coordinates $y_i$ satisfying $y_i \le (n-T)\sqrt{(n-1)/(T(n-T))}$. If $1 \le T' \le n-1$ then $y^{T'} \in D_T$ exactly when $T \le T'$. Hence from $f(y^{T'-1}) > f(y^{T'}) > 0$ for $2 \le T' \le [n/2]$ and $f(y^{T'}) < 0$ for $T' > [n/2]$, we conclude that $f(y^T)$ is the absolute maximum of $f$ on the domain $D_T$.   □

With the above notations we have:

**(4O)**    *Let $M = \max\{\eta_i \mid i = 0, \dots, p - 1\}$, let $T$ be the largest integer such that*

$$M \leq (p - T)\sqrt{\frac{p - 1}{T(p - T)}}.$$

*Then*

$$N(p) < 1 + (p - 2T)\sqrt{\frac{p - 1}{T(p - T)}}.$$

PROOF. Since $\sum_{i=0}^{p-1} \eta_i = 0$ then $0 < M$. By (4.6), $M < p - 1$. The function

$$g(t) = (p - t)\sqrt{\frac{p - 1}{t(p - t)}}$$

is decreasing, $f(1) = p - 1$, $f(p - 1) = 1$. So there exists the largest integer $T$, $1 \leq T \leq p - 1$, such that

$$M \leq (p - T)\sqrt{\frac{p - 1}{T(p - T)}}.$$

Since

$$\sum_{i=0}^{p-1} \eta_i = 0, \qquad \sum_{i=0}^{p-1} \eta_i^2 = p(p - 1),$$

as seen before, then $(\eta_0, \eta_1, \dots, \eta_{p-1}) \in D_T$. By Lemma 4.11, $f(y_1^T)$ is the absolute maximum of $f$ on $D_T$ and we have

$$\sum_{i=0}^{p-1} \eta_i^3 = f(\eta_0, \dots, \eta_{p-1}) \leq f(y^T) = p(p - 1)(p - 2T)\sqrt{\frac{p - 1}{T(p - T)}}.$$

By (4.13),

$$\begin{aligned}
N(p) &= \frac{p - 2}{p} + \frac{1}{p^2}\sum_{i=0}^{p-1} \eta_i^3 \\
&\leq \frac{p - 2}{p} + \frac{(p - 1)(p - 2T)}{p}\sqrt{\frac{p - 1}{T(p - T)}} \\
&< 1 + (p - 2T)\sqrt{\frac{p - 1}{T(p - T)}}. \quad \square
\end{aligned}$$

We note that if

$$M = (p - T)\sqrt{\frac{p-1}{T(p-T)}},$$

then $N(p) < 1 + M$. Indeed, in the above proof we obtained

$$N(p) \le \frac{p-2}{p} + \frac{(p-1)(p-2T)}{p}\sqrt{\frac{p-1}{T(p-T)}}$$

$$= \frac{p-2}{p} + \frac{p-1}{p}\frac{p-2T}{p-T}M$$

$$< 1 + M.$$

Now we study the asymptotic behavior of $F(p, m, k)$, $N(p, m, k)$ when $k$ tends to infinity.

**(4P)**

$$\lim_{k \to \infty} \frac{N(p, m, k)}{(p-1)^{k-1}/p^{m+1}} = 1.$$

PROOF. From (4M) we have

$$\frac{N(p, m, k)}{(p-1)^{k-1}/p^{m+1}} = 1 + \frac{(-1)^k}{(p-1)^{k-1}} + (p-1)\sum_{j=1}^{m} \frac{S(p^j, k)}{(p-1)^k}$$

$$= 1 + \frac{(-1)^k}{(p-1)^{k-1}} + (p-1)\sum_{j=1}^{m}\sum_{i=1}^{p^j-1}\left\{\frac{\eta_i(p,j)}{p-1}\right\}^k.$$

Since $|\eta_i(p, j)| < p - 1$, by (4.6) it follows that

$$\lim_{k \to \infty}\left\{\frac{\eta_i(p,j)}{(p-1)}\right\}^k = 0,$$

hence $\lim_{k \to \infty} N(p, m, k)/((p-1)^{k-1}/p^{m+1}) = 1$. $\square$

We consider the existence of $p$-adic solutions of a certain congruence. Let $\hat{U}_p$ denote the multiplicative group of $(p-1)$th roots of 1 in the ring $\hat{\mathbb{Z}}_p$ of $p$-adic integers. We observe:

**(4Q)**

(1) *There exists an integer $m_0 = m_0(k, p) \geq 1$ such that for every $m \geq m_0$ we have $N(p, m, k) = N(p, m_0, k)$. Let $N(p, k)$ denote this number; $N(p, k) \geq 0$.*

(2) $N(p, k)$ *is the number of solutions of the equation $1 + X_2 + \cdots + X_k = 0$ by elements in $\hat{U}_p$.*

PROOF. (1)   We have seen in (4H) that $0 \leq N(p, m, k) \leq N(p, m - 1, k)$. Hence there exists $m_0$ with the property indicated.

(2)   Let $m \geq m_0$ and let

$$1 + \sum_{i=2}^{k} x_i^{p^m} \equiv 0 \pmod{p^{m+1}},$$

where $p \nmid x_i$. Since $x_i^{p^{m+1}} \equiv x_i^{p^m} \pmod{p^{m+1}}$ then

$$1 + \sum_{i=2}^{k} x_i^{p^{m+1}} \equiv 0 \pmod{p^{m+1}},$$

so

$$1 + \sum_{i=2}^{k} x_i p^{m+1} \equiv a p^{m+1} \pmod{p^{m+2}}$$

for some $a$, $0 \leq a \leq p - 1$. Since $m + 1 > m_0$, by (4H), $F^*(p, m + 1, k) = 0$; therefore necessarily $a = 0$, so

$$1 + \sum_{i=2}^{k} x_i^{p^{m+1}} \equiv 0 \pmod{p^{m+2}}.$$

In this way we have established the mapping

$$(x_2 \bmod p^{m+1}, \ldots, x_k \bmod p^{m+1})$$
$$\rightarrow (x_2 \bmod p^{m+2}, \ldots, x_k \bmod p^{m+2})$$

from the set $S_m$ of solutions of

$$1 + X_2^{p^m} + \cdots + X_k^{p^m} \equiv 0 \pmod{p^{m+1}}$$

to the corresponding set $S_{m+1}$. This is clearly an injective mapping.

Starting therefore with any $(x_2 \bmod p^{m_0+1}, \ldots, x_k \bmod p^{m_0+1}) \in S_{m_0}$ we obtain the sequences $(\alpha_{m,i})_{m \geq m_0}$, where $\alpha_{m,i} = x_i^{p^m}$ (for $i = 2, \ldots, k$). Since $x_i^{p^{m+1}} \equiv x_i^{p^m} \pmod{p^{m+1}}$ then $\alpha_{m+1,i} \equiv \alpha_{m,i} \pmod{p^m}$, so the sequence $(\alpha_{m,i})_m$ converges $p$-adically.

Let $\alpha_i = \lim_{m \to \infty} \alpha_{m,i}$. Then $1 + \alpha_2 + \cdots + \alpha_k = 0$ in $\hat{\mathbb{Z}}_p$. Since $\alpha_{m,i}^{p-1} \equiv 1 \pmod{p^{m+1}}$ for $m \geq m_0$ then $\alpha_i^{p-1} = 1$, so $\alpha_i \in U_p$ and $(\alpha_2, \ldots, \alpha_k) \in \hat{S}$, the set of solutions in $\hat{U}_p$ of the equation $1 + X_2 + \cdots + X_k = 0$. So we have an injective mapping $\sigma$ from $S_{m_0}$ into $\hat{S}$.

On the other hand, if $\alpha_i \in \hat{U}_p$, $\alpha_i = (y_{m,i})_{m \geq 0}$ (for $i = 2, \ldots, k$) and $1 + \alpha_2 + \cdots + \alpha_k = 0$, since $\alpha_i^{p-1} = 1$ then $y_{m,i}^{p-1} \equiv 1 \pmod{p^{m+1}}$ (for $i = 2, \ldots, k$ and all $m$ sufficiently large; we may take $m \geq m_0$). Thus $y_{m,i} \equiv x_{m,i}^{p^m} \pmod{p^{m+1}}$ and

$$1 + x_{m,2}^{p^m} + \cdots + x_{m,k}^{p^m} \equiv 0 \pmod{p^{m+1}}.$$

In particular,

$$1 + x_{m,2}^{p^{m_0}} + \cdots + x_{m,k}^{p^{m_0}} \equiv 0 \pmod{p^{m_0+1}},$$

so $(x_{m,2}, \ldots, x_{m,k}) \in S_{m_0}$. It is now immediate that the mapping $\sigma$ associates with this solution in $S_{m_0}$ the given solution in $\hat{S}$, thus $\sigma$ is surjective. We conclude that the number of elements in $\hat{S}$ is equal to $N(p, k) = \#S_{m_0}$.   $\square$

As a corollary, we have:

**(4R)**    *If $p \equiv 1 \pmod{k}$ then there exist $(p - 1)$th roots of $1$, $\alpha_2, \ldots, \alpha_k \in \hat{U}_p$, such that $1 + \alpha_2 + \cdots + \alpha_k = 0$.*

PROOF. By (4A), for every $m \geq 1$ the congruence $1 + X_2^{p^m} + \cdots + X_k^{p^m} \equiv 0 \pmod{p^{m+1}}$ has a nontrivial solution. In particular, $N(p, k) \geq 1$, hence by (4Q), there exist $\alpha_2, \ldots, \alpha_k \in \hat{U}_p$ such that $1 + \alpha_2 + \cdots + \alpha_k = 0$.   $\square$

We reproduce now the following tables, computed by Klösgen, which give $N(p, m, k)$ for low values of the argument. Table for $N(p, 1, k)$:

|    | 3 | 4  | 5   | 6    | 7      |
|----|---|----|-----|------|--------|
| 5  | 0 | 9  | 0   | 100  | 35     |
| 7  | 2 | 15 | 60  | 340  | 1680   |
| 11 | 0 | 31 | 24  | 1600 | 5250   |
| 13 | 2 | 33 | 200 | 2260 | 21630  |
| 17 | 0 | 57 | 140 | 6220 | 50120  |
| 19 | 2 | 51 | 390 | 6880 | 101430 |

Table for $N(p, 2, k)$ (italicized are the values which are not the same as in the table above):

|    | 3 | 4  | 5   | 6    | 7      |
|----|---|----|-----|------|--------|
| 5  | 0 | 9  | 0   | 100  | *0*    |
| 7  | 2 | 15 | 60  | 340  | 1680   |
| 11 | 0 | *27* | 24 | *1090* | *2520* |
| 13 | 2 | 33 | *180* | *1930* | *15540* |
| 17 | 0 | *45* | 0  | *3160* | *945*  |
| 19 | 2 | 51 | *300* | *4600* | *44520* |

To conclude this section, we shall discuss a heuristic method to indicate the probability for the congruence

$$1 + Y^p + Z^p \equiv 0 \pmod{p^2}$$

to have a given number of equivalent classes of nontrivial solutions. We exclude also the cyclic solutions from these considerations.

Let $1 \le a \le p - 2$ and let

$$M_a = \left\{ a \bmod p, \frac{1}{a} \bmod p, -(1 + a) \bmod p, -\frac{1}{(1 + a)} \bmod p, \right.$$
$$\left. -\frac{(a + 1)}{a} \bmod p, -\frac{a}{(a + 1)} \bmod p \right\}.$$

$M_a$ consists of six distinct elements (all different from $\bar{0}, -\bar{1}$), except in the following cases:

(a) $a = 1, p - 2$ or $(p - 1)/2$; then $M_a = \{1 \bmod p, (p - 2) \bmod p, (p - 1)/2 \bmod p\}$; and

(b) $a \neq 1, a^3 \equiv 1 \pmod{p}$; then $M = \{a \bmod p, a^2 \bmod p\}$.

This latter case happens if and only if $p \equiv 1 \pmod 6$.

Thus, we have a partition of $\{\bar{1}, \bar{2}, \ldots, \overline{p-2}\}$ into disjoint classes. If $p = 6n \pm 1$, the number of such classes $M_a$ (with more than two elements) is equal to $n$. Indeed, if $p = 6n + 1$, this number is $1 + (p - 2 - 2 - 3)/6 = n$. If $p = 6n - 1$ then the number is again $1 + (p - 2 - 3)/6 = n$. In each class $M_a$ (with more than two elements) let $\tilde{a}$ be the smallest integer, $1 \leq \tilde{a} \leq p - 2$, such that $\tilde{a} \in M_a$. Clearly $\tilde{a} \leq (p - 1)/2$.

If $1 + y^p + z^p \equiv 0 \pmod{p^2}$ then necessarily $z \equiv -(1 + y) \pmod p$. With this solution we associate $\tilde{y}$, and we note that $(1 + \tilde{y})^p - 1 - \tilde{y}^p \equiv 0 \pmod{p^2}$. If $1 + y'^p + z'^p \equiv 0 \pmod{p^2}$, this is an equivalent solution to the above one, if and only if $y' \bmod p \in M_y$, that is, $\tilde{y}' = \tilde{y}$.

By Fermat's little theorem, if $t \geq 1$ then $(1 + t)^p - 1 - t^p \equiv 0 \pmod p$ hence $(1 + t)^p - 1 - t^p = t(p)p \pmod{p^2}$ where $0 \leq t(p) \leq p - 1$. So the solutions of the congruence correspond to the integers $t$, $1 \leq t \leq (p - 1)/2$ such that $t(p) = 0$.

We consider the sequence $(t(p))_{t \leq (p-1)/2}$.

**(4S)**    *Assuming that the sequences $(t(p))_{t \leq (p-1)/2}$ are random, for all primes $p = 6n \pm 1$ we have:*

(1) *the probability that $1 + Y^p + Z^p \equiv 0 \pmod{p^2}$ have only the trivial or cyclic solution is equal to $((p - 1)/p)^n$.*

(2) *The probability that the above congruence have $r$ (nontrivial, noncyclic) equivalence classes of solutions is equal to*

$$\frac{1}{p^r} \binom{n}{r} \left(\frac{p - 1}{p}\right)^{n - r}.$$

(3) *The density of primes for which there are exactly $r$ (nontrivial, noncyclic) equivalence classes of solutions is equal to*

$$\frac{1}{r! \, 6^r} \frac{1}{\sqrt[6]{e}}.$$

PROOF. (1)   As already indicated, there are $n$ equivalence classes in $\{\bar{1}, \bar{2}, \ldots, \overline{p-2}\}$ consisting of at least three elements. Let us denote them by $M_{\tilde{x}_1}, M_{\tilde{x}_2}, \ldots, M_{\tilde{x}_n}$.

$M_{\tilde{x}_i}$ consists of solutions of $(1 + t)^p \equiv 1 + t^p \pmod{p^2}$ exactly when $\tilde{x}_i(p) = 0$. Since the sequence $(\tilde{x}_i(p))_{\tilde{x}_i \leq (p-1)/2}$ is random, the probability is $1/p$. Hence, the probability that none of the $n$ classes $M_{\tilde{x}_i}(p)$ consist of solutions is $(1 - 1/p)^n$.

(2)   In the same way the probability that $r$ among the $n$ classes $M_{\tilde{x}_i}$ consist of solutions is

$$\binom{n}{r} \frac{1}{p^r} \left(1 - \frac{1}{p}\right)^{n-r}.$$

(3)   The density in question is equal to

$$D = \lim_{p \to \infty} \frac{1}{p^r} \binom{n}{r} \left(\frac{p-1}{p}\right)^{n-r}$$

$$= \lim_{n \to \infty} \frac{n^r \left(1 - \frac{1}{n}\right)\left(1 - \frac{2}{n}\right) \cdots \left(1 - \frac{r-1}{n}\right)}{6^r r! n^r \left(1 \pm \frac{1}{6n}\right)^r} \left(\frac{p-1}{p}\right)^{n-r}.$$

If $p = 6n + 1$ then

$$D = \frac{1}{6^r r!} \lim_{n \to \infty} \left(\frac{6n}{6n+1}\right)^{6n/6} = \frac{1}{6^r r!} \frac{1}{\sqrt[6]{e}}.$$

Similarly, if $p = 6n - 1$ then

$$D = \frac{1}{6^r r!} \lim_{n \to \infty} \left(\frac{6n-2}{6n-1}\right)^{(6n-2)/6} \times \left(1 - \frac{1}{6n-1}\right)^{1/3}$$

$$= \frac{1}{6^r r!} \frac{1}{\sqrt[6]{e}}. \qquad \square$$

Klösgen computed the solutions of the congruence $1 + Y^p + Z^p \equiv 0$ (mod $p^2$) for all primes $p < 20\,000$.

If $0 \le r$ let $v_r^+$ (respectively, $v_r^-$) be the number of primes $p < 20\,000$, $p \equiv 1$ (mod 6) (respectively, $p \equiv -1$ (mod 6)) for which the above congruence has exactly $r$ (nontrivial, noncyclic) equivalence classes of solutions.

There are 1124 primes $p$ such that $p \equiv 1$ (mod 6) and $p < 20\,000$. Klösgen found that

$$
\begin{array}{lll}
v_0^+ = 970, & v_0^+/1124 = 86.30\%, & \text{probability } 84.35\%. \\
v_1^+ = 144, & v_1^+/1124 = 12.81\%, & \text{probability } 14.11\%. \\
v_2^+ = 9, & v_2^+/1124 = 0.80\%, & \text{probability } 1.18\%. \\
v_3^+ = 1, & v_3^+/1124 = 0.09\%, & \text{probability } 0.07\%.
\end{array}
$$

In a similar way for $p \equiv -1 \pmod 6$, $p < 20\,000$:

$$v_0^- = 957, \qquad v_0^-/1136 = 84.24\%, \quad \text{probability } 84.35\%.$$
$$v_1^- = 166, \qquad v_1^-/1136 = 14.61\%, \quad \text{probability } 14.11\%.$$
$$v_2^- = 13, \qquad v_2^-/1136 = 1.15\%, \quad \text{probability } 1.18\%.$$

## Bibliography

1965 Peschl, E., *Remarques sur la résolubilité de la congruence* $x^p + y^p + z^p \equiv 0 \pmod{p^2}$, $xyz \not\equiv 0 \pmod p$ *pour un nombre premier impair p*, Mém. Acad. Sci. Inscriptions Belles Lettres Toulouse, $14^e$ série, **6** (1965), 121–127.

1970 Klösgen, W., *Untersuchungen über Fermatsche Kongruenzen*, Gesellschaft Math. Datenverarbeitung, No. 36, 1970, 124 pp., Bonn.

1999 Ribenboim, P., *Classical Theory of Algebraic Numbers*, Springer-Verlag, New York, 1999.

# XI
# Epilogue

This book about Fermat's last theorem was written for the enjoyment of amateurs. Most of the proofs are given in full detail and use only elementary and easily understandable methods. For this reason, it was imperative to exclude developments depending on the study of ideals of number fields or on more sophisticated theories. However, in this final part we indicate the more important achievements which could not be dealt with using elementary methods. We also give a succinct description of the approach to the proof of Fermat's last theorem. To help the reader who wants to know more about these matters, a bibliography of important articles is also included.

## XI.1. Attempts

In this section, we give a brief overview of various approaches to the proof of Fermat's last theorem. They were not quite successful but should not be dismissed. At their time, these results raised hopes for the proof of Fermat's last theorem and led to new research problems of independent interest.

## A. The Theorem of Kummer.

In 1847, Kummer proved the following important theorem:

*If $p > 2$ is a regular prime, then Fermat's last theorem is true for the exponent $p$.*

The concept of a *regular prime* needs explanation. It may be defined in terms of the class number of cyclotomic fields or by means of *Bernoulli numbers*.

The *Bernoulli numbers* $B_0, B_1, B_2, \ldots$ are defined recursively:

$$B_0 = 1$$

and for $n \geq 1$,

$$\binom{n+1}{1} B_n + \binom{n+1}{2} B_{n-1} + \cdots + \binom{n+1}{n} B_1 + 1 = 0.$$

Thus $B_1 = \frac{-1}{2}$, $B_2 = \frac{1}{6}$, $B_3 = 0, \ldots$. It is easily seen that $B_{2k+1} = 0$ for all $k \geq 1$. The prime number $p$ is *regular* if $p$ does not divide the numerators of the Bernoulli numbers $B_2, B_4, \ldots, B_{r-5}, B_{r-3}$. Let $p$ be an odd prime, let

$$\zeta_p = \cos(2\pi/p) + i \sin(2\pi/p)$$

be a primitive $p$th root of 1. Let $\mathbb{Q}(\zeta_p)$ be the $p$th cyclotomic field; it consists of all complex numbers of the form

$$r_0 + r_1 \zeta_p + \cdots + r_{p-2} \zeta_p^{p-2},$$

with $r_0, r_1, \ldots, r_{p-2} \in \mathbb{Q}$. The *class number* $h_p$ of $\mathbb{Q}(\zeta_p)$ is a certain positive integer attached to $\mathbb{Q}(\zeta_p)$; it is the number of classes of ideals of $\mathbb{Q}(\zeta_p)$, but we shall not explain these concepts any further (see any book on the theory of algebraic numbers, like the one by Borevich and Shafarevich (1966), or even this author's own book (1999)). Kummer showed that the prime $p$ is regular if and only if $p$ does not divide $h_p$.

The smallest irregular prime is 37. It is known that there are infinitely many irregular primes. On the other hand, it is conjectured, but it has never been proved, that there are infinitely many regular primes.

The method of Kummer could be extended to deal also with many irregular primes. However, with these methods it was never possible to establish that Fermat's last theorem is true for infinitely many prime exponents.

To determine if a prime $p$ is regular is not a simple matter as soon as $p$ is large, because the numerators of the Bernoulli numbers become very large. Noting that what is required is to ascertain that the exponent $p$ does not divide the numerators of $B_2, \ldots, B_{p-3}$ (rather than calculating their numerators), Lehmer, Lehmer , and Vandiver gave a criterion which was possible to implement for actual calculations. In this way it was shown (at a time when the proof of FLT was not yet discovered) that FLT is true for all prime exponents up to $4 \times 10^6$ (see Buhler et al. (1993)).

## Bibliography

1847 Kummer, E.E., *Extrait d'une lettre de M. Kummer à M. Liouville*, J. Math. Pures Appl., **12** (1847), 136.

1851 Kummer, E.E., *Mémoire sur les nombres complexes composés de racines de l'unité et de nombres entiers*, J. Math. Pures Appl., **16** (1851), 377–498.

(The above papers are reprinted in *Collected Papers of E. E. Kummer*, Vol. 1 (editor, A. Weil), Springer-Verlag, Berlin, 1975.)

1966 Borevich, Z.I. and Shafarevich, I.R., *Number Theory*, Academic Press, New York, 1966.

1993 Buhler, J., Crandall, R.E., Ernvall, R., and Metsänkyla, T., *Irregular primes and cyclotomic invariants to four million*, Math. Comp., **61** (1993), 151–153.

1999 Ribenboim, P., *Classical Theory of Algebraic Numbers*, Springer-Verlag, New York, 1999.

## B. The Theorem of Wieferich.

In 1909, Wieferich proved:

*If the first case of FLT is false for the exponent $p$ then*

$$2^{p-1} \equiv 1 \bmod p^2.$$

This is a criterion involving only the exponent $p$ and none of the hypothetical nonzero solutions $x, y, z$ of $X^p + Y^p = Z^p$. It was immediately noted that no very small prime $p$ satisfies the above congruence. Before the age of computers Meissner proved in 1913 that $p = 1093$ is the smallest prime with the above-mentioned property. It is difficult to imagine the amount of calculations which was required. A further example, $p = 3511$, was found by Beeger in 1921. Further computations by Lehmer, Keller, Clark and lately by Crandall, Dilcher, and Pomerance have shown that no other prime $p < 4 \times 10^{12}$ satisfies the congruence.

Other criteria of a similar kind were discovered by Mirimanoff, Vandiver, Frobenius, Pollaczek, Rosser, and Granville and Monagan, namely,

*If the first case of FLT is false for the exponent $p$ then*

$$l^{p-1} \equiv 1 \bmod p^2$$

*for all primes $l \leq 89$.*

A clever combinatorial combination of these criteria, by Gunderson and Coppersmith, followed by extensive calculations (Granville and Monagan, Tanner and Wagstaff), allowed us to show that the first case of FLT is true for every exponent $p < 6.93 \times 10^{17}$. All this was done before the discovery of the proof of FLT for all exponents.

## Bibliography

1909 Wieferich, A., *Zum letzten Fermat'schen Theorem*, J. Reine Angew. Math., **136** (1909), 293–302.

1911 Mirimanoff, D., *Sur le dernier théorème de Fermat*, J. Reine Angew. Math., **139** (1911), 309–324.

1917 Pollaczek, F., *Über den grossen Fermat'schen Satz*, Sitzungsber. Akad. Wiss. Wien, Abt. IIa, **126** (1917), 45–59.

1988 Granville, A. and Monagan, M.B., *The First Case of Fermat's last theorem is true for all prime exponents up to 714,-591,116,091,389*, Trans. Amer. Math. Soc., **306** (1988), 329–359.

1989 Tanner, J.W. and Wagstaff, S.S., Jr., *New bound for the first case of Fermat's last theorem*, Math. Comp., **53** (1989), 743–750.

1990 Coppersmith, D., *Fermat's last theorem (case 1) and the Wieferich criterion*, Math. Comp., **54** (1990), 895–902.

## C. The First Case of Fermat's Last Theorem for Infinitely Many Prime Exponents.

Using methods from sieve theory, Adleman, Heath-Brown, and Fouvry proved in 1985:

*There exists an infinite set S of prime numbers, such that the first case of Fermat's last theorem is true for every exponent $p \in S$.*

A stronger result, valid not only for the first case, could not be established with the same methods. This theorem represented an important advance at that time. The method of proof was inspired from the old ideas of Sophie Germain and was connected with the estimation of the size of the smallest prime in arithmetic progressions; the use of refined sieve theory was essential.

The infinite set $S$, guaranteed by the theorem, is not effectively defined, so it is not possible, with the method of the proof, to deduce for any given $p$ that the first case of FLT holds for $p$.

Once again this substantial theorem is obsolete, due to the proof of FLT for all exponents $n > 2$.

### Bibliography

1985 Fouvry, E., *Théorème de Brun–Titchmarsh. Application au théorème de Fermat*, Invent. Math., **79** (1985), 383–407.

1985 Adleman, L.M. and Heath-Brown, D.R., *The first case of Fermat's last theorem*, Invent. Math., **79** (1985), 409–416.

## D. The Theorem of Faltings.

Mordell observed and conjectured that irreducible curves defined by homogeneous polynomials of high degree in three variables with rational coefficients should have only finitely many rational points, when they have few singularities, all of lower order. The exact conjecture is expressed in terms of the *genus* of the curve, a concept which will not be explained here. In a remarkable paper Faltings proved, among many other theorems, Mordell's conjecture. In the particular case of Fermat's equation his result becomes:

*For every $n > 3$, there exist at most finitely many triples $(x, y, z)$
where $x, y, z$ are integers, not all equal to $0$ and such that $\gcd(x, y, z)$
$= 1$ and $x^n + y^n = z^n$.*

Despite its importance, this result could not lead to the proof of
FLT. However it was used, independently by Granville and Heath-
Brown, to deduce that the set of exponents $n \geq 3$ for which FLT is
true has density one. This method of Granville or Heath-Brown is
also applicable to a very wide class of exponential diophantine equations (see Ribenboim, 1993), the conclusion being the zero density
for the exponents for which the equations have nontrivial solution.

## Bibliography

1983 Faltings, G., *Endlichkeitssätze für Abelsche Varietäten über
    Zahlkörpern*, Invent. Math., **73** (1983), 349–366.
1984 Filaseta, M., *An application of Faltings' results to Fermat's
    last theorem*, C. R. Math. Rep. Acad. Sci. Canada, **6** (1984),
    31–32.
1985 Granville, A., *The set of exponents for which Fermat's last
    theorem is true, has density one*, C. R. Math. Rep. Acad.
    Sci. Canada, **7** (1985), 55–60.
1985 Heath-Brown, D.R., *Fermat's last theorem is true for almost
    all exponents*, Bull. London Math. Soc., **17** (1985), 15–16.
1989 Tzermias, P., *A short note on Fermat's last theorem*, C. R.
    Math. Rep. Acad. Sci. Canada, **11** (1989), 259–260.
1990 Brown, T.C. and Friedman, A.R., *The uniform density of
    sets of integers and Fermat's last theorem*, C. R. Math. Rep.
    Acad. Sci. Canada, **12** (1990), 1–6.
1993 Ribenboim, P., *Density results on families of diophantine
    equations with finitely many solutions*, Enseign. Math., **39**
    (1993), 3–23.

## E. The $(abc)$ Conjecture.

The $(abc)$ conjecture, attributed to Masser and Oesterlé, was inspired by a result about polynomials, due to Mason. The conjecture
is stated as follows:

*For any $\epsilon > 0$ there exists a number $C(\epsilon) > 0$ such that if $a, b, c$ are integers, $1 \leq a < b < c$, with $c = a + b$ and $\gcd(a, b, c) = 1$ then $c < C(\epsilon)r^{1+\epsilon}$, where $r$ is the product of the distinct prime factors of $abc$.*

Intuitively, as an example, if $a = 2^m$, $b = 3^n$ (with $m, n$ large) then $c = a + b$ is large, so the conjecture states that $c$ must have a large prime factor or a large number of prime factors, so that $r$ is large.

It is easy to show that the $(abc)$ conjecture implies:

*FLT is true for all sufficiently large exponents.*

Indeed, let $n > 3$ and assume that $x, y, z$ are positive integers, such that $\gcd(x, y, z) = 1$ and $x^n + y^n = z^n$. Let $\epsilon = \frac{1}{2}$, so by the $(abc)$ conjecture $z^n < C(\frac{1}{2})r^{3/2}$ where

$$r = \prod_{p \mid x^n y^n z^n} p = \prod_{p \mid xyz} p \leq xyz \leq z^3,$$

so $z^n < C(\frac{1}{2})z^{9/2}$. This shows that there exists $n_0$ such that $n \leq n_0$, in other words, FLT is true for every exponent $n > n_0$, or in short, FLT is asymptotically true.

The $(abc)$ conjecture is known to imply many other statements in number theory which have never been proved, as well as Mordell's conjecture which was proved by Faltings. The proof of the $(abc)$ conjecture should be very difficult and this is presently the object of intense research.

## Bibliography

1984 Mason, R.C., *Diophantine Equations over Function Fields*, London Math. Soc. Lecture Notes Ser., No. 96, Cambridge University Press, Cambridge, 1984.

1985 Masser, D.W., *Some open problems*, Symp. Analytic Number Theory, Imperial College, London, 1985 (unpublished).

1988 Oesterlé, J., *Nouvelles approches au théorème de Fermat*, Sém. Bourbaki, 40ème année, No. 694, Février 1988, 1987–1988.

1990 Masser, D.W., *More on a conjecture of Szpiro*, Astérisque, **183** (1990), 19–24.

1991 Elkies, N.D., *ABC implies Mordell*, Duke Math. J., Intern. Math. Res. Notes, **7** (1991), 99–109.

## XI.2. Victory, or the Second Death of Fermat

Mathematicians have the obligation of solving problems. When a long-sought proof, like the one for Fermat's last theorem, is finally discovered, it is the moment of crying VICTORY.

On June 23, 1993, in the third of his lectures at the Newton Institute in Cambridge, England, Wiles announced the proof of Fermat's last theorem. His manuscript, scrutinized by various experts, revealed flaws which needed corrections. Undeterred and with the help of Taylor, Wiles found a way out of the difficulties and in October 1994, he made public two manuscripts, one co-authored by Taylor. They contain the proof of the conjecture of Shimura-Taniyama, for the case of semistable elliptic curves. According to the previous work of Ribet this entails that FLT is true. For most mathematicians this represents the end of the saga. Wiles deserves the admiration of all mathematicians for his achievement. The method used has already been applied to other diophantine equations. Wiles' work was the final step in a new strategy which will be evoked shortly.

There are some mathematicians who are not satisfied with the method of proof using elliptic curves and modular forms, considered — perhaps wrongly? or rightly? — to be extraneous to the problem. It is a legitimate task to try to find another, simpler, proof of FLT. But the solution of Fermat's problem also harbors a negative aspect and a tear of regret is unavoidable, because mathematicians also like unsolved problems to stimulate their research, just like night butterflies are attracted by intense sources of light. The study of Fermat's theorem led to the creation of the theory of algebraic numbers, in the same way as the study of quadratic fields was prompted by Gauss' theory of quadratic forms. The branch of mathematics which is the confluence of number theory and algebraic geometry, called Arithmetic Algebraic Geometry, developed not only by its internal problems, but also in view of solving Fermat's last theorem. The attempts to prove Fermat's theorem, the old and the new, show a myriad of interesting ideas in many directions of number theory, by illustrious names. Will this stimulation disappear now that FLT is proved? Not at all. Variants of the problem, generalizations to higher dimensions, will continue tantalizing mathematicians. So, we celebrate this striking victory and admire our colleagues who, through effort and ingenuity, succeeded in solving the problem.

The proof of Fermat's last theorem must be indirect. We assume that there exists $n \geq 3$ and positive integers $a, b, c$ such that $a^n + b^n = c^n$. The aim is to deduce a statement which is known to be false. No contradictions were found with statements in elementary number theory, nor with statements about number fields, nor for that matter, for any other statements until the expression of FLT in terms of elliptic curves. The proof of FLT was established with the following steps:

(I) To associate an elliptic curve to a hypothetical nontrivial solution of Fermat's equation, with arbitrary exponent $n \geq 5$.

(II) To obtain a contradiction to the assumption of validity of a certain conjecture about elliptic curves and modular forms.

(III) To prove the validity of the conjecture.

These steps require sophisticated concepts and theories, far beyond the level of this book and the knowledge usually expected from amateurs — and also from professional mathematicians working in other disciplines. My task is difficult if not hopeless. What will follow is simple-minded and superficial, but still mysterious and perhaps out of grasp for anyone who is not yet familiar with the concepts involved. The key notions needed are elliptic curves, modular forms, and Galois representations.

## A. The Frey Curves.

For relatively prime positive integers $A$, $B$ and $A$ divisible by 16, Frey considered the elliptic curve of equation

$$(2.1) \qquad\qquad Y^2 = X(X - A)(X + B)$$

(see Chapter VIII, §1, (A6)) and studied its properties.

If Fermat's last theorem is false for the prime exponent $q \geq 5$, let $a, b, c$ be positive pairwise relatively prime integers, with $a$ even, such that $a^q + b^q = c^q$. Let $A = a^q$, $B = b^q$. The associated Frey curve displayed properties in sharp contrast with those of other elliptic curves. Frey became convinced that such a situation was not possible and envisioned a method to derive a contradiction with the, by then well-known, conjecture of Shimura-Taniyama (see below). But there were serious obstacles to overcome, which would require many years of work (see below).

Here are some propeties of the Frey curves. The minimal discriminant of the Frey curve is

$$\Delta = \frac{a^{2q}b^{2q}(a^q + b^q)^2}{2^8} = \frac{(abc)^{2q}}{2^8}.$$

Since $\Delta \neq 0$, the curve is nonsingular, so it is an elliptic curve.

For every prime $p$ not dividing $\Delta$, we consider the congruence

$$(2.2) \qquad Y^2 \equiv X(X - a^q)(X + b^q) \pmod{p}.$$

It defines a curve in the two-dimensional space over the finite field $\mathbb{F}_p$. Since $p$ does not divide $\Delta$, the curve is nonsingular, so it is an elliptic curve. On the other hand, if $p$ divides $\Delta$, the curve is singular. The type of singularities is encoded in the invariant called the *conductor*. The primes $p$ dividing the conductor are exactly those dividing the discriminant, that is, the primes $p$ for which the curve in $\mathbb{F}_p \times \mathbb{F}_p$ has singularity. The exponent of $p$ indicates the type of singularity. In the present case, where the singularities are nodes, the conductor $N$ is square-free, so it is equal to

$$N = \prod_{p \mid \Delta} p.$$

Elliptic curves with square-free conductor are said to be *semistable*. Thus, Frey curves are semistable. As it was known, if Fermat's last theorem is assumed false for the prime exponent $q$, then $q$ has to be very large; moreover, since Fermat's equation is homogeneous, the discriminant is a power — and this seemed unlikely to be possible.

We shall count the number of points of Frey's curve modulo $p$ (for every $p$ not dividing $\Delta$). To this count we add 1, which corresponds to the point at infinity in the associated projective curve. Let $\nu_p$ be the number of points and let $a_p = p + 1 - \nu_p$ ($a_p$ need not be positive). We pause to recall that in Chapter I, §1, we studied the Pythagorean equation $X^2 + Y^2 = 1$ modulo all odd primes; we proved that the numbers $a_p$ defined there are easily determined by a simple congruence for the prime $p$. Similar considerations are important for all elliptic curves (not only for Frey curves). The discriminant, the conductor, and the integers $a_p$ (for $p$ not dividing the discriminant) are defined and studied in the same spirit. Elliptic curves which can be given by an equation with coefficients in $\mathbb{Q}$ are said to be defined over $\mathbb{Q}$. The rule for determination of the integers $a_p$ involves modular forms.

## B. Modular Forms and the Conjecture of Shimura-Taniyama.

Let $N \geq 1$ be an integer. Let $\Gamma_0(N)$ be the set of all $2 \times 2$ matrices

$$\begin{pmatrix} a & b \\ c & d \end{pmatrix},$$

where $a, b, c, d$ are integers, $N$ divides $c$ and $ad - bc = 1$. $\Gamma_0(N)$ is a multiplicative group called the *congruence group of level $N$*. Let $H$ denote the upper half-plane, that is, $H = \{z = x + iy \in \mathbb{C} \mid y > 0\}$. $\Gamma_0(N)$ acts on $H$ as follows:

$$(2.3) \qquad \begin{pmatrix} a & b \\ c & d \end{pmatrix} z = \frac{az + b}{cz + d}$$

for all matrices of $\Gamma_0(N)$ and $z \in H$. Associated to the group $\Gamma_0(N)$ there are finitely many special points (which we do not define here), called *cusps*; these are the point at infinity of the half-line $\{iy \mid y \geq 0\}$ and other points in $H \cup \mathbb{Q}$ (when $N > 1$).

A *modular form of level $N$* (and weight 2 — the only ones we wish to consider) is a map $f$ from $H^* = H \cup \{\text{cusps of } \Gamma_0(N)\}$ to $\mathbb{C}$ such that:

(i) for all $\begin{pmatrix} a & b \\ c & d \end{pmatrix} \in \Gamma_0(N)$ and $z \in H^*$:

$$(2.4) \qquad f\left(\frac{az + b}{cz + d}\right) = (cz + d)^2 f(z);$$

(ii) $f$ is holomorphic at every point of $H^*$ (this requires an appropriate definition at the cusps).

A modular form which vanishes at all cusps is called a *cusp form*.

The theory of modular forms is very rich. Here are some relevant facts (for which we give no hint of proof):

(1) The set $\mathcal{M}_2(N)$ of modular forms of level $N$ and weight 2 is a finite-dimensional vector space over $\mathbb{C}$ and the subset of cusp forms is

a subspace. For the level $N = 2$ the subspace of cusp forms consists only of the form 0.

(2) There is a natural inner product on $\mathcal{M}_2(N)$, so it is possible to consider orthogonality in $\mathcal{M}_2(N)$.

(3) Let $N \geq 1$. If $M$ divides $N$ then $\mathcal{M}_2(M) \subseteq \mathcal{M}_2(N)$. There is also the embedding from $\mathcal{M}_2(M)$ into $\mathcal{M}_2(N)$ given as follows: if $f \in \mathcal{M}_2(N)$ let $\tilde{f}(z) = f((N/M)z)$ for every $z \in H^*$; then $\tilde{f} \in \mathcal{M}_2(N)$.

(4) A form $f \in \mathcal{M}_2(N)$ is called an *old form* if $f$ is in the subspace of $\mathcal{M}_2(N)$ generated by the images of the mappings considered in (3), for all $M$ dividing $N$. A form $f \in \mathcal{M}_2(N)$ is called a *new form* if it is in the subspace which is orthogonal to the subspace of old forms.

(5) Since

$$\begin{pmatrix} 1 & 1 \\ 0 & 1 \end{pmatrix} \in \Gamma_0(N)$$

then $f(z + 1) = f(z)$ for each modular form and every $z$. Thus $f$ has a Fourier expansion, which is of the form

$$(2.5) \qquad f(z) = \sum_{n=0}^{\infty} c_n e^{2\pi i n z}.$$

For cusp forms, $c_0 = 0$.

(6) Hecke defined for each $n \geq 1$ coprime to the level $N$, a linear operator $T_n$ of $\mathcal{M}_2(N)$. The Hecke operators commute: $T_m \circ T_n = T_n \circ T_m$ for all $m, n$ coprime to the level. A modular form which is an eigenvalue for all Hecke operators $T_n$ is called an *eigenform*.

Other operators associated to the integers $n$, not coprime with $N$, have also been introduced and, together with the above Hecke operators $T_n$, they generate a larger Hecke algebra, whose properties are of essential importance (see Wiles and Taylor, and also Lenstra). New forms of level $N$, which are eigenforms for each $T_n$ (with $n$ coprime to $N$) are also eigenforms of the operators of the larger Hecke algebra.

Now we discuss the relationship between elliptic curves and modular forms. For a given elliptic curve, the numbers $a_p$ (for all primes $p$ not dividing the discriminant) contain very important "local" information about the curve (for each $p$). It is crucial to relate these local data by means of some "global" invariant.

This important idea is a sophisticated transfiguration of the fact that every natural number is the product of powers of primes in a unique way. Thus Euler already introduced this relation between an infinite product extended over all primes and infinite Dirichlet series, summed over all integers:

$$\prod_p \left(1 - \frac{1}{p^s}\right)^{-1} = \sum_{n=1}^{\infty} \frac{1}{n^s}.$$

First $s$ was restricted to be a real number $s > 1$, for which both sides converge and are equal. Riemann had the idea, courageous and deep, of allowing $s$ to be any complex number with $\text{Re}(s) > 1$. The above series is the Riemann zeta function. To prove the existence of infinitely many primes in arithmetic progressions, Dirichlet considered "twisted" $L$-series, where the numerators are no longer 1, but values of characters of appropriate finite Abelian groups; each series has also an abscissa of convergence and admits an Euler product, reflecting also the multiplicative property of characters.

As for the Riemann zeta function, $L$-series of characters have only poles but no essential singularities at the boundary of the domain of convergence. Riemann proved that the functions definded above could be extended to the whole plane by analytic continuation, and even more remarkably, the values to the right and left of the boundary line are linked by a functional equation involving the gamma function. A great discovery and the royal road for analytical methods to enter into number theory.

In great analogy with number fields, elliptic curves also display very important analytical properties of the same kind. The local numbers $a_p$, defined above (not forgetting finitely many factors attached to the primes dividing the discriminant), combine together multiplicatively to define numbers $a_n$ (for every $n \geq 1$), thus leading to a Dirichlet series, called the $L$-series of the elliptic curve; they converge for $\text{Re}(s) > \frac{3}{2}$. In computed examples it was observed that these $L$-series admit analytic continuations and functional equations. Hasse conjectured that this should be true for every elliptic curve. Deuring proved it for the elliptic curves admitting more "symmetries," namely those with complex multiplication.

For a certain time it had been observed by numerical calculations that for many specific elliptic curves the numbers $a_p$ coincide with the coefficients $c_p$ of the Fourier series of some modular form. Elliptic curves with the above property have been called modular elliptic curves, or also Weil elliptic curves.

In 1955, during the Tokyo–Nikko conference on number theory, Taniyama proposed problems, two of which concerned — if still somewhat imprecisely — the above question. If Hasse's conjecture were true, would the $L$-series be associated to some automorphic function, or even to a modular form? These problems were discussed with Shimura and Weil. By 1964, Shimura made known in his lectures a very specific conjecture (which however did not appear in print on that occasion). Weil contributed in an important way to the investigation of the modularity of elliptic curves. His paper (of 1967) acknowledges previous communications by Shimura but does not contain a statement of the conjecture which he considered, even later, to be problematic. According to a well-documented study by Lang (1995), we shall adopt the name "Shimura–Taniyama conjecture" for this penetrating statement.

(7) **The Shimura–Taniyama Conjecture:** *Every elliptic curve is modular.*

This is a short way of expressing the following:

*If $E$ is any elliptic curve defined over $\mathbb{Q}$, if $N$ is its conductor, then there is a new cusp eigenform $f$ of level $N$, whose Fourier coefficients $c_n$ are integers and such that for every prime $p$ not dividing $N$, $c_p = a_p$ (where $a_p$ is defined by counting the number of points of $E$ in $\mathbb{F}_p$).*

This conjecture says that the rule of determination of the integers $a_p$ is given by some modular form.

(8) Shimura proved the converse of the Shimura–Taniyama conjecture. Let $f \in \mathcal{M}_2(N)$ be such that its Fourier coefficients are in $\mathbb{Z}$. We explain how it is possible to associate an elliptic curve. Let $z_0 \in H$. For each $\gamma \in \Gamma_0(N)$ consider the integral

$$w_{z_0}(\gamma) = \int_{z_0}^{\gamma(z_0)} f(z)\, dz;$$

it is independent of the path. The set $\{w_{z_0}(\gamma) \mid \gamma \in \Gamma_0(N)\}$ is independent of $z_0$, so it depends only on $f$. Using the fact that the Fourier coefficients of $f$ are integers, the above set is a lattice in $H$, that is, it is the set of all linear combinations, with integral coefficients, of two numbers in $H^*$ (the *periods* of $f$). This lattice gives rise in the usual manner to an analytic torus, hence to an elliptic curve $E$ having an equation with coefficients in $\mathbb{Z}$ (thus $E$ is defined over $\mathbb{Q}$). Let $\mathcal{C}_2(N) = \{f \in \mathcal{M}_2(N) \mid f$ is a cusp eigenform whose Fourier coefficients are integers$\}$. The above construction associates to each $f \in \mathcal{C}_2(N)$ an elliptic curve $E$ defined over $\mathbb{Q}$. Moreover, the conductor of $E$ is the level $N$ of $f$ and for each prime $p$ not dividing the discriminant of $E$, the Fourier coefficient $c_p$ of $f$ is equal to the number $a_p$ (associated to $E$ and $p$ as was already indicated).

Analytical methods involving the $L$-series of elliptic curves, their Euler product, analytic continuation, and functional equation play a fundamental role.

## C. The Work of Ribet and Wiles.

The work of Ribet involved an argument of descent concerning Galois representations and modular forms. We need to explain how Serre attached Galois representations to any elliptic curve $E$ defined over $\mathbb{Q}$, that is, having an equation with integral coefficients. The set of points with complex coordinates (to which is added the point at infinity) constitutes an Abelian additive group, well defined by the stipulation that the point at infinity should be the zero for the addition. The addition is defined by the following rule: if $P, Q, R$ are points on the curve, then $P + Q + R = 0$ when $P, Q, R$ lie on one line (clarifications are needed when $P = Q$ or in some other special cases). If $K$ is a subfield of $\mathbb{C}$, let $E(K)$ be the set of pairs of elements of $K$ which satisfy the equation of $E$; then $E(K)$ is a subgroup of $E(\mathbb{C})$.

For each prime $p$, it is equally possible to define the additive group $E(\mathbb{F}_p)$. In the Abelian group $E(\mathbb{C})$ we consider the set $E(\mathbb{C})[p]$ of all elements of order dividing $p$. These are the point $0$ and the points $P$ such that $P + P + \cdots + P$ ($p$ times) is equal to $0$. Then $E(\mathbb{C})[p]$ is a subgroup of order $p^2$, which is isomorphic to $\mathbb{Z}/p \times \mathbb{Z}/p$. The coordinates of the points in $E(\mathbb{C})[p]$ are in some Galois extension $K$ of finite degree over $\mathbb{Q}$. The elements of the Galois group of $K|\mathbb{Q}$ act linearly on $E(K)$ and permute among themselves the elements of $E(\mathbb{C})[p]$; by isomorphism with $\mathbb{Z}/p \times \mathbb{Z}/p$ this gives rise to linear transformations of $\mathbb{Z}/p \times \mathbb{Z}/p$. Thus, we obtain a representation, associated to $E$, of the Galois group of $K|Q$. It is usual to consider the field $\overline{\mathbb{Q}}$ of all algebraic numbers; it has infinite degree and contains $K$. The Galois group of $K|\mathbb{Q}$ is a quotient of the Galois group $G$ of $\overline{\mathbb{Q}}|\mathbb{Q}$. So we obtain a representation $\rho_{E,p}$ from $G$ by a group of $2 \times 2$ matrices with entries in $\mathbb{Z}/p = \mathbb{F}_p$ (attention is also paid to the natural Krull topology of $G$). Similar considerations lead to representations $\rho_{E,p^n}$ by means of $2 \times 2$ matrices with entries in $\mathbb{Z}/p^n$ (for all $n \geq 1$). All the representations $\rho_{E,p^n}$ (for $n \geq 1$) fit together to produce a representation $\rho_{E,\mathbb{Q}_p}$ of $G$ by $2 \times 2$ matrices with entries in the field $\mathbb{Q}_p$ of $p$-adic numbers.

It is also possible to attach to any eigenform $f$ with Fourier coefficients in $\mathbb{Z}$, a representation $\rho_{f,\mathbb{Q}_p}$ of $G$ by $2 \times 2$ matrices with entries in $\mathbb{Q}_p$. If $E$ is the elliptic curve associated to $f$, as indicated before, then the representations $\rho_{f,\mathbb{Q}_p}$ and $\rho_{E,\mathbb{Q}_p}$ are isomorphic.

We outline the proof of Ribet. Assume that FLT is false for the exponent $q$, let $E$ be the Frey curve associated to a hypothetical solution; $E$ is a semistable elliptic curve. Assuming that the conjecture of Shimura-Taniyama is valid, there exists a new cusp eigenform $f$ of weight 2 and level equal to the conductor $N$ of $E$. Then $\rho_{f,\mathbb{Q}_q} \cong \rho_{E,\mathbb{Q}_q}$.

Ribet proved that if $p$ is an odd prime dividing $N$, $N_1 = N/p$, then there is a new cusp eigenform $f_1$ of weight 2 and level $N_1$, such that $\rho_{f_1,\mathbb{F}_p} = \rho_{E,\mathbb{F}_p}$. The argument may be repeated, leading to a nonzero cusp form of weight 2 and level 2 — which is impossible.

Wiles proved that the Shimura–Taniyama conjecture is valid for semistable elliptic curves, in particular for the Frey curve. The theory of deformation of representations, created by Mazur, plays a great role; so does the result (proved with Taylor, and also later by Lenstra) on the structure of the commutative algebra generated by the Hecke operators. Cohomological results were developed and used in essential ways. The proof is at a maximal level of sophistication, so it is impossible to report in an intelligent way in this book. Among the expository papers listed in the Bibliography, we may recommend the one by Gouvêa which delineates the proof, avoiding technical details, making it accessible for the courageous amateur.

# XI.3. A Guide for Further Study

For my readers who are still courageous, I include a bibliography which lists not only research papers, but also expository material and should be explored by readers wishing to enter deeper into the proof of Wiles.

For the convenience of the reader, the references are organized as follows:

## A. Elliptic Curves, Modular Forms: Basic Texts.

1962 Gunning, R.C., *Lectures on Modular Forms*, Princeton University Press, Princeton, NJ, 1962.

1971 Shimura, G., *Introduction to the Theory of Automorphic Functions*, Princeton University Press, Princeton, NJ, 1971.

1972 Ogg, A., *Survey of modular functions of one variable*, in: *Modular Functions of One Variable* (editor, W. Kuyk), Springer-Verlag, New York, 1972.

1974 Tate, J., *The arithmetic of elliptic curves*, Invent. Math., **23** (1974), 179–206.

1976 Lang, S., *Introduction to Modular Forms*, Springer-Verlag, New York, 1976.

1984 Koblitz, N., *Introduction to Elliptic Curves and Modular Forms*, Springer-Verlag, New York, 1984.

1986 Silverman, J.H., *The Arithmetic of Elliptic Curves*, Springer-Verlag, New York, 1986.

1986 Cornell, G. and Silverman, J.H. (editors), *Arithmetic Geometry*, Springer-Verlag, Berlin, 1986.

1989 Miyake, T., *Modular Forms*, Springer-Verlag, New York, 1989.

1989 Hida, H., *Theory of p-adic Hecke algebras and Galois representations*, Sûgaku Expositions, **2** (1989), 75–102.

1989 Gouvêa, F.Q., *Formas Modulares, uma Introdução*, Instituto de Matemática Pura e Aplicada, Rio de Janeiro, 1989.

1991 Cassels, J.W.S., *Lectures on Elliptic Curves*, Cambridge University Press, Cambridge, 1991.

1992 Tate, J. and Silverman, J.H., *Rational Points on Elliptic Curves*, Springer-Verlag, New York, 1992.

## B. Expository.

1988  Cipra, B.A., *Fermat's last corollary?*, Focus, March–April 1988, pp. 2 and 6.

1989  Shimura, G., *Yataku Taniyama and his time. Very personal recollections*, Bull. London Math. Soc., **21** (1989), 186–196.

1990  Ribet, K.A., *From the Taniyama–Shimura conjecture to Fermat's last theorem*, Ann. Fac. Sci. Toulouse Math., (5), **11** (1990), no. 1, 116–139.

1993  Murty, M. Ram, *Fermat's last theorem, an outline*, Gaz. Soc. Math. Québec, **16** (1993), No. 1, 4–13.

1993  Murty, M. Ram, *Topics in Number Theory*, Mehta Res. Inst. Lect. Notes, No. 1, Allahabad, 1993.

1993  Frey, G., *Über A. Wiles' Beweis der Fermatschen Vermutung*, Math. Semesterber., **40** (1993), no. 2, 177–191.

1993  Ribet, K.A., *Modular elliptic curves and Fermat's last theorem*, Videocassette, 100 min., Amer. Math. Soc., Providence, RI.

1994  Gouvêa, F.Q., *A marvelous proof*, Amer. Math. Monthly, **101** (1994), 203–222. (Updated Portuguese translation: Matem. Univ., no. 19, Dec. 1995, pp. 16–43.)

1994  Cox, D.A., *Introduction to Fermat's last theorem*, Amer. Math. Monthly, **101** (1994), 3–14.

1994  Rubin, K. and Silverberg, A., *Wiles' Cambridge lecture*, Bull. Amer. Math. Soc., **11** (1994), 15–38.

1994  Ribet, K.A. and Hayes, B., *Fermat's last theorem and modern arithmetic*, American Scientist, March–April 1994, pp. 146–156.

1994  Ribet, K.A., *Wiles proves Taniyama's conjecture; Fermat's last theorem follows*, Notices Amer. Math. Soc., **40** (1993), no. 6, 575–576.

1995  Ribenboim, P., *Fermat's last theorem before June 23, 1993*, in: *Proc. Fourth Conference Canad. Number Theory Assoc.*, Halifax, July 1994 (editor, K. Dilcher), Amer. Math. Soc., Providence, RI, 1995, pp. 279–293.

1995  Schoof, R., *Wiles' proof of Taniyama–Weil conjecture for semi-stable elliptic curves over* $\mathbb{Q}$, Gaz. Math., Soc. Math. France, No. 66 (1995), 7–24.

1995  Edixhoven, B., *Le rôle de la conjecture de Serre dans la démonstration du théorème de Fermat*, Gaz. Math., Soc.

Math. France, No. 66 (1995), 25–41. (Erratum and addendum: Gaz. Math., Soc. Math. France, No. 67 (1996), 19.)

1995 Lang, S., *Some history of the Shimura–Taniyama conjecture*, Notices Amer. Math. Soc., **42** (1995), no. 11, 1301–1307.

1995 Faltings, G., *The proof of Fermat's last theorem by R. Taylor and A. Wiles*, Notices Amer. Math. Soc., **42** (1995), no. 7, 743–746.

1995 Serre, J.-P., *Travaux de Wiles (et Taylor, ...), Partie I*, Séminaire Bourbaki, Vol. 1994/95. Astérisque, No. 237 (1996), Exp. No. 803, 5, 319–332.

1995 Oesterlé, J., *Travaux de Wiles (et Taylor, ...), Partie II*, Séminaire Bourbaki, Vol. 1994/95. Astérisque, No. 237 (1996), Exp. No. 804, 5, 333–355.

1995 Darmon, H., Diamond, F., and Taylor, R., *Fermat's last theorem*. In: *Current Developments in Mathematics*, 1995 (editors, R. Bott, A. Jaffe, and S.T. Yau), pp. 1–154, Internat. Press, Cambridge, MA, 1995. Also in: *Elliptic curves, modular forms & Fermat's last theorem* (Hong Kong, 1993), pp. 2–140, Internat. Press, Cambridge, MA, 1997.

1995 Gouvêa, F.Q., *Deforming Galois representations: a survey*. In: *Seminar on Fermat's Last Theorem* (Toronto, ON, 1993–1994), pp. 179–207, CMS Conf. Proc., No. 17, Amer. Math. Soc., Providence, RI, 1995.

1995 Mazur, B., *Fermat's last theorem*, Videocassette, 60 min., American Math. Soc., Providence, RI.

1996 Darmon, H. and Levesque, C., *Sommes infinies, équations diophantiennes et le dernier théorème de Fermat*, Gaz. Soc. Math. Québec, **18** (1996), 3–18.

1996 van der Poorten, A., *Notes on Fermat's Last Theorem*, Wiley, New York, 1996.

1997 Cornell, G., Silverman, J.H., and Stevens, G. (editors), *Modular Forms and Fermat's Last Theorem*, Springer-Verlag, New York, 1997.

1997 Singh, S., *Fermat's Enigma*, Viking, London, 1997.

1997 Singh, S. and Ribet, K.A., *Fermat's last theorem*, Scientific American, **277** (1997), no. 5, 36–41.

1997 Kani, E., *Fermat's last theorem*, Queen's Math. Communicator (Queen's University at Kingston, Ontario, Canada),

Summer 1997, pp. 1–8.

1997 Frey, G., *The way to the proof of Fermat's last theorem*, preprint, 20 pp., 1997.

## C. Research.

1958 Shimura, G., *Correspondances modulaires et les fonctions zeta de courbes algébriques*, J. Math. Soc. Japan, **10** (1958), 1–28.

1961 Shimura, G., *On the zeta-functions of the algebraic curves uniformized by certain automorphic functions*, J. Math. Soc. Japan, **13** (1961), 275–331.

1967 Shimura, G., *Construction of class fields and zeta functions of algebraic curves*, Ann. of Math., (2), **85** (1967), 58–159.

1967 Weil, A., *Über die Bestimmung Dirichletscher Reihen durch Funktionalgleichungen*, Math. Ann., **168** (1967), 149–156.

1975 Hellegouarch, Y., *Points d'ordre 2p sur les courbes elliptiques*, Acta Arith., **26** (1975), 253–263.

1977 Mazur, B., *Modular curves and the Eisenstein ideal*, Inst. Hautes Études Sci. Publ. Math., **47** (1977), 33–186 (1978).

1982 Frey, G., *Rationale Punkte auf Fermatkurven und getwisteten Modulkurven*, J. Reine Angew. Math., **331** (1982), 185–191.

1986 Frey, G., *Elliptic curves and solutions of $A - B = C$*, in: *Sém. Th. Nombres*, Paris, 1985–1986 (editor, C. Goldstein), Progress in Mathematics, Birkhäuser, Boston, 1986, pp. 39–51.

1986 Frey, G., *Links between elliptic curves and certain diophantine equations*, Ann. Univ. Sarav. Ser. Math., **1** (1986), No. 1, 1–40.

1987 Frey, G., *Links between elliptic curves and solutions of $A - B = C$*, J. Indian Math. Soc., **51** (1987), 117–145.

1987 Frey, G., *Links between solutions of $A - B = C$ and elliptic curves*, in: *Number Theory* (Ulm, 1987) (editors, H.-P. Schlickewei and E. Wirsing), Springer Lect. Notes in Math., No. 1380, Springer-Verlag, New York, 1989.

1987 Serre, J.-P., *Sur les représentations modulaires de degré 2 de* $\mathrm{Gal}(\overline{\mathbb{Q}}|\mathbb{Q})$, Duke Math. J., **54** (1987), 179–230.

1987–1990 Ribet, K.A., *On modular representations of* $\mathrm{Gal}(\overline{\mathbb{Q}}|\mathbb{Q})$, preprint, 1987. Invent. Math., **100** (1990), 115–139.

1989 Mazur, B., *Deforming Galois representations*, in: *Galois Groups over* $\mathbb{Q}$ (editors, Y. Ihara, K.A. Ribet, and J.-P.

Serre), Math. Sci. Res. Inst. Publ., Vol. 16, Springer-Verlag, New York, 1989.

1990 Ribet, K.A., *From the Taniyama–Shimura conjecture to Fermat's last theorem*, Ann. Sci. Univ. Toulouse, (5), **11** (1990), 115–139.

1991 Ribet, K.A., *Lowering the levels of modular representations without multiplicity one*, Internat. Math. Res. Notices **1991**, no. 2, 15–19.

1991 Kolyvagin, V., *Euler systems*, in: *The Grothendieck Festschrift*, Vol. 2, pp. 435–483, Birkhäuser, Boston, 1991.

1992 Flach, M., *A finiteness theorem for the symmetric square of an elliptic curve*, Invent. Math., **109** (1992), 307–327.

1993 Lenstra, H.W. Jr., *Complete intersections and Gorenstein rings*, preprint (September 27, 1993).

1993 Ramakrishna, R., *On a variation of Mazur's deformation functor*, Compositio Math., **87** (1993), 269–286.

1995 Ribet, K.A., *Galois representations and modular forms*, Bull. Amer. Math. Soc., **32** (1995), 375–401.

1995 Wiles, A., *Modular elliptic curves and Fermat's last theorem*, Ann. of Math., (2), **141** (1995), 443–551.

1995 Taylor, R. and Wiles, A., *Ring theoretic properties of certain Hecke algebras*, Ann. of Math., (2), **141** (1995), 553–572.

## XI.4. The Electronic Mail in Action

As a concluding note, here are timely communications by Karl Rubin which circulated widely.

E-mail message no. 1:

*Date: June 23, 1993, 05:52:30*
*Subject: big news*
*Andrew Wiles just announced, at the end of his 3rd lecture here, that he has proved Fermat's Last Theorem. He did this by proving that every semistable elliptic curve over Q (i.e. square-free conductor) is modular. The curves that Frey writes down, arising from counterexamples to Fermat, are semistable and by work of Ribet they cannot be modular, so this does it.*
*It's an amazing piece of work.*
*Karl*

E-mail message no. 2:

*Date: Oct. 25, 1994, 10:24:46*
*Subject: update on Fermat's last theorem*
*As of this morning, two manuscripts have been released:*

> *Modular elliptic curves and Fermat's last theorem,*
> *by Andrew Wiles*
> *Ring theoretic properties of certain Hecke algebras, by*
> *Richard Taylor and Andrew Wiles.*

*The first one (long) announces a proof of, among other things, Fermat's last theorem, relying on the second one (short) for one crucial step.*

*As most of you know, the argument described by Wiles in his Cambridge lectures turned out to have a serious gap, namely the construction of an Euler system. After trying unsuccessfully to repair that construction, Wiles went back to a different approach, which he had tried earlier but abandoned in favor of the Euler systems idea. He was able to complete his proof, under the hypothesis that certain Hecke algebras are local complete intersections. This and the rest of the ideas described in Wiles' Cambridge lectures are written up in the first manuscript. Jointly, Taylor and Wiles establish the necessary property of the Hecke algebras in the second paper.*

*The overall outline of the argument is similar to the one Wiles described in Cambridge. The new approach turns out to be significantly simpler and shorter than the original one, because of the removal of the Euler system. (In fact, after seeing these manuscripts, Faltings has apparently come up with a further significant simplification of that part of the argument.)*

*Versions of these manuscripts have been in the hands of a small number of people for (in some cases) a few weeks. While it is wise to be cautious for a little while longer, there is certainly reason for optimism.*

*Karl Rubin*

Excitement, caution, and amazement in the face of a superlative feat in Mathematics.

# Appendix A
# References to Wrong Proofs

It is well known that there have been literally thousands of wrong proofs of Fermat's last theorem. This can be explained by the fact that the statement of the problem is easily understandable to an amateur. Moreover, there have been important prizes offered by academies and foundations which have stimulated efforts by dilettantes as well as professional mathematicians.

Since the Wolfskehl Prize was established in 1908, in the first years alone, 621 wrong solutions were submitted, and today there are about 3 meters of file correspondence and proposed solutions of Fermat's problem stored in Göttingen.

We indicate below a list — obviously incomplete — of some notoriously wrong published attempts to solve the problem. Even good professional mathematicians have not escaped from being included in the roll.

Remarkable is the case of F. Lindemann, who discovered the transcendency of the number $\pi$. Yet, with respect to Fermat's last theorem, all his attempts failed.

F. Paulet seems to have been one of the most persistent, with twelve submissions to the Academy of Sciences of Paris, spanning the years 1841 to 1862, but not bringing any progress to the investigation.

With only a few exceptions, we do not mention wrong solutions published by the authors as independent books or brochures; some of these have been listed by Fleck and Maennchen (1908–1912), Mirimanoff (1909), and Perron (1916). Instead, we concentrate only on the intended, but failed solutions, published in mathematical journals or proceedings of conferences.

First, we give a selection of books or papers containing references to wrong proofs. This is followed by a list of wrong papers, including an indication of where the mistake is discussed.

## I. Papers or Books Containing Lists of Wrong Proofs

1908 Hoffmann, F., *Der Satz vom Fermat. Sein seit dem Jahr 1658 gesuchter Beweis*, J. Singer, Strasbourg, 1908.

1909 Lampe, E., Jahrbuch Fortschritte Math., **40** (1909), 258–261.

1909/10/11/12/16 Fleck, A. and Maennchen, A., *Vermeintliche Beweise des Fermatschen Satzes*, Arch. Math. Phys., (3),
**14** (1909), 284–286, 370–372;
**15** (1909), 108–111;
**16** (1910), 105–109 and 372–375;
**17** (1911), 108–109 and 370–374;
**18** (1912), 105–109 and 204–206;
**25** (1916), 267–268.

1910 Lind, B., *Über das letzte Fermatsche Theorem*, Abh. Geschichte Math. Wiss., **26** (1910), 23–65.

1920 Dickson, L.E., *History of the Theory of Numbers*, Vol. II, Carnegie Institution, Washington, DC, 1920; reprinted by Chelsea, New York, 1971.

1973 Besenfelder, H.J.,[1] *Das Fermat-Problem*, Diplomarbeit, Universität Karlsruhe, 1973, 61 pp.

## II. Wrong Proofs in Papers

1810 Barlow, P., *Demonstration of a curious numerical proposition*, J. Nat. Phil. Chem. Arts, **27** (1810), 193–205.

---

[1]His family name has changed from "Besenfelder" to "Bentz" since August 1979.

[2]This paper uses an incorrect result of Kapferer (1933).

TABLE 8. Wrong proofs.

| Year | Author | Mistake Pointed Out by | Year |
|---|---|---|---|
| 1810, | Barlow | Smith | 1860 |
| 1811 | | Talbot | 1864 |
| 1845 | Drach | Dickson, p. 738 | 1820 |
| 1847 | Lamé | Liouville | 1847 |
| | | Kummer | 1847 |
| | | Dickson, pp. 739/40 | 1920 |
| 1855 | Calzolari | Lind, p. 48 | 1910 |
| /57/64 | | Dickson, pp. 743, 744, 746 | 1920 |
| 1864 | Gaudin | Dickson, p. 746 | 1920 |
| 1890 | Lefébure | Pepin | 1880 |
| 1889 | Varisco | Landsberg | 1890 |
| | | Dickson, p. 754 | 1920 |
| 1893 | Korneck | Picard and Poincaré | 1894 |
| | | Dickson, p. 756 | 1920 |
| 1901/ | Lindemann | Fleck and Maennchen | 1909 |
| 1907/ | | Furtwängler, Fleck | 1909 |
| 1909 | | Ivanov | 1910 |
| | | Dickson, pp. 759, 762 | 1920 |
| 1908 | Werebrusow | Dickson, Worms de Romilly, | 1908 |
| | | Duran-Loriga, Curjel | |
| | | Dickson, p. 762 | 1920 |
| 1910 | Lind | Fleck | 1910 |
| | | Dickson, p. 760 | 1920 |
| 1913 | Fabry | Mirimanoff | 1913 |
| 1955 | Becker, W. W. | Eggan | 1981 |
| 1956 | Fraga Torrejón | Rodeja, F. | 1956 |
| 1957 | Villaseñor, Z. | *Math. Rev.*, **19** (1958), No. 251f | 1958 |
| 1957 | Noguera | *Math. Rev.*, **19** (1958), No. 16e | 1958 |
| | Barreneche | *Math. Rev.*, **20** (1959), No. 1658 | 1959 |
| 1958/9 | Draeger | Morishima | 1960 |
| 1958/73 | Yahya[2] | Gandhi and Stuff | 1975 |
| 1973/77 | | Inkeri | 1984 |
| 1966 | Sarantopoulos | Garrison | 1967 |
| 1978 | Zinoviev | Kreisel | 1978 |
| 1979 | Clarke and | Oral communication by | 1983 |
| | Shannon | J. H. Ursell | |
| 1980 | Maggu | Eggan | 1980 |
| 1980 | Lallu-Singh | Yamaguchi | 1982 |

1811 Barlow, P., *An Elementary Investigation of Theory of Numbers*, pp. 160–169, J. Johnson, St. Paul's Church-Yard, London, 1811.

1845 Drach, S.M., *Proof of Fermat's undemonstrated theorem that $x^n + y^n = z^n$ is only possible in whole numbers when $n = 1$ or 2*, Phil. Mag., **27** (1845), 286–289.

1847 Lamé, G., *Mémoire sur la résolution en nombres complexes de l'équation $A^n + B^n + C^n = 0$*, J. Math. Pures Appl., **12** (1847), 172–184.

1847 Lamé, G., *Démonstration générale du théorème de Fermat sur l'impossibilité en nombres entiers de l'équation $x^n + y^n = z^n$*, C. R. Acad. Sci. Paris, **24** (1847), 310–314.

1847 Lamé, G., *Note au sujet de la démonstration du théorème de Fermat*, C. R. Acad. Sci. Paris, **24** (1847), 352.

1847 Lamé, G., *Second mémoire sur le dernier théorème de Fermat*, C. R. Acad. Sci. Paris, **24** (1847), 569–572.

1847 Lamé, G., *Troisième mémoire sur le dernier théorème de Fermat*, C. R. Acad. Sci. Paris, **24** (1847), 888.

1847 Kummer, E. E., *Extrait d'une lettre de M. Kummer à M. Liouville*, J. Math. Pures Appl., **12** (1847), 136; reprinted in *Collected Papers*, Vol. I, p. 298. Springer-Verlag, Berlin, 1975.

1847 Liouville, J., *Remarques à l'occasion d'une communication de M. Lamé sur un théorème de Fermat*, C. R. Acad. Sci. Paris, **24** (1847), 315–316.

1855 Calzolari, L., *Tentativo per dimostrare il teorema di Fermat sull'equazione indeterminata $x^n + y^n = z^n$*, Ferrara, 1855.

1857 Calzolari, L., *Dimostrazione dell'ultimo teorema di Fermat*, Annali Sci. Mat. B. Tortolini, **8** (1857), 339–349.

1860 Smith, H.J.S., *Report on the Theory of Numbers, Part II, Art. 61, "Application to the Last Theorem of Fermat"*, Collected Math. Papers, Vol. I, 1894, pp. 131–137 Clarendon Press, Oxford, 1894; reprinted by Chelsea, New York, 1965.

1864 Calzolari, L., *Impossibilità in numeri interi dell'equazione $z^n = x^n + y^n$ quando $n > 2$*, Ann. Mat., **6** (1864), 280–286.

1864 Gaudin, A., *Impossibilité de l'équation $(x + h)^n - x^n = z^n$*, C. R. Acad. Sci. Paris, **59** (1864), 1036–1038.

1864 Talbot, W.H.F., *On the theory of numbers*, Trans. Roy. Soc. Edinburgh, **23** (1864), 45–52.

1880 Lefébure, A., *Sur la résolution de l'équation* $x^n + y^n = z^n$ *en nombres entiers*, C. R. Acad. Sci. Paris, **90** (1880), 1406–1407.

1880 Pepin, T., *Sur diverses tentatives de démonstration du théorème de Fermat*, C. R. Acad. Sci. Paris, **91** (1880), 366–367.

1889 Varisco, D., *Ricerche aritmetiche contenente la dimostrazione generale del teorema di Fermat*, Giorn. Mat., **27** (1889), 371–380.

1890 Landsberg, O., *Lettera al redattore*, Giorn. Mat., **28** (1890), 52.

1893 Korneck, G., *Beweis des Fermatschen Satzes von der Unmöglichkeit der Gleichung* $x^n + y^n = z^n$ *für rationale Zahlen und* $n > 2$, Arch. Math. Phys., (2), **13** (1893), 1–9.

1893 Korneck, G., *Nachtrag zum Beweis des Fermatschen Satzes*, Arch. Math. Phys., (2), **13** (1893), 263–267.

1894 Picard, E. and Poincaré, H., *Rapport verbal sur les articles de M. G. Korneck*, C. R. Acad. Sci. Paris, **118** (1894), 841.

1901 Lindemann, F., *Über den Fermatschen Satz betreffend die Unmöglichkeit der Gleichung* $x^n = y^n + z^n$, Sitzungsber. Akad. Wiss. München, Math., **31** (1901), 185–202; corrigenda, p. 495.

1907 Lindemann, F., *Über das sogenannte letzte Fermatsche Theorem*, Sitzungsber. Akad. Wiss. München, Math., **37** (1907), 287–352.

1908 Dickson, L. E., *Dernier théorème de Fermat*, L'Interm. Math., **15** (1908), 174.

1908 Curjel, H. W., *Dernier théorème de Fermat (Question 612 de Worms de Romilly)*, L'Interm. Math., **15** (1908), 247.

1908 Duran-Loriga, J. J., *Sur le dernier théorème de Fermat (Réponse de M. Werebrusow)*, L'Interm. Math., **15** (1908), 177.

1908 Werebrusow, A. S., *Impossibilité de l'équation* $x^n = y^n + z^n$ *(Question 612 de Worms de Romilly)*, L'Interm. Math., **15** (1908), 79–81.

1908 Worms de Romilly, A. S., *Le dernier théorème de Fermat*, L'Interm. Math., **15** (1908), 175–177.

1909 Lindemann, F., *Über den sogenannten letzten Fermatschen Satz*, Veit, Leipzig, 1909, 83 pp.

1909 Fleck, A. and Maennchen, A.: See in List I.

1909 Furtwängler, P., *Review of Lindemann's "Über den sogenan-*

*nten letzten Fermatschen Satz,"* Jahrbuch Fortschritte Math., **40** (1909), 258.

1910 Ivanov, I.I., *Über den von Prof. F. Lindemann vorgeschlagenen Beweis des Fermatschen Satzes (Brief an die Redaktion)*, Jahrbuch Fortschritte Math., **41** (1910), 238.

1910 Lind, B.: See in List I.

1913 Fabry, E., *Un essai de démonstration du théorème de Fermat*, C. R. Acad. Sci. Paris, **156** (1913), 1814–1816.

1913 Mirimanoff, D., *Remarque sur une communication de M. Eugène Fabry*, C. R. Acad. Sci. Paris, **157** (1913), 491–492.

1920 Dickson, L.E., See in List I.

1933 Kapferer, H., *Über die diophantischen Gleichungen $z^3 - y^2 = 3^3 \cdot 2^\lambda x^{\lambda+2}$ und deren Abhängigkeit von der Fermatschen Vermutung*, Heidelberger Akad., Math. Naturwiss. Klasse, Abh., **2** (1933), 32–37.

1956 Fraga Torrejón, E. de, *Note on Fermat's last theorem*, Las Ciências, **21** (1956), 5–13.

1956 Rodeja, F., E.G., *On Fermat's last theorem*, Las Ciências, **21** (1956), 382–383.

1957 Noguera Barreneche, R., *Solución general de la ecuación algebraico-exponencial $X^\nu + Y^\nu = Z^\nu$*, Studia Rev. Univ. Atlantico, **2** (1957), 119–126.

1957 Noguera Barreneche, R., *Historically the first proof incontrovertible, complete and universal of the grand theorem of Fermat, with the Davidic algebra of the "principle of the amateurs" in mathematical investigation (in Spanish)*, Studia Rev. Univ. Atlantico, **2** (1957), 199–209.

1957 Villaseñor Z., F., *El celebre teorema de Fermat y su demonstración*, Mexico, 1957, 127 pp.

1958 Yahya, Q.A.M.M., *Complete proof of Fermat's last theorem*, Author's publication, Pakistan Air Force, Kohat, Pakistan, 1958, 14 pp.

1958/9 Draeger, M., *Das Fermat-Problem*, Wiss. Z. Techn. Hochsch. Dresden, **8** (1958/9), 941–946.

1960 Morishima, T., *Review of the paper by Draeger "Das Fermat-Problem,"* Math. Rev., **23** (1960), A2375.

1966 Sarantopoulos, S., *Du premier cas du théorème de Fermat*. Bull. Soc. Math. Grèce (N.S.), **10** (1966), 76–115.

1971 Garrison, B., Review of the above paper by Sarantopoulos (with a remark by E.G. Straus), Math. Rev., **42** (1971), No. 4483.

1973 Yahya, Q.A.M.M., *On general proof of Fermat's last theorem*, Portugal. Math., **32** (1973), 157–170.

1975 Gandhi, J.M. and Stuff, M., *Comments on certain results about Fermat's last theorem*, Notices Amer. Math. Soc., **22** (1975), A-502.

1978 Kreisel, G., Letter to Ribenboim (6 June 1978). Atlantis Hotel, Zürich.

1979 Clarke, J.H. and Shannon, A.G., *Some observations on Fermat's last theorem*, New Zealand Math. Mag., **16** (1979), 80–83.

1979 Zinoviev, A. A., *Complete (rigorous) induction and Fermat's great theorem (with a report by G. Kreisel)*, Logique et Anal., **22** (1979), no. 87, 243–263.

1980 Maggu, P.L., *On the proof of Fermat's last theorem*, Pure Appl. Math. Sci., **12** (1980), 1–9.

1981 Eggan, L.C., Review of the above paper by Maggu, Math. Rev., **81g** (1981), No. 10032.

1984 Inkeri, K., *On certain equivalent statements for Fermat's last theorem — with requisite corrections*, Ann. Univ. Turku., Ser. AI, **186** (1984) 12–22; reprinted in *Collected Papers of Kustaa Inkeri* (editor, P. Ribenboim), Queen's Papers in Pure and Applied Mathematics, Vol. 91, Kingston, Ontario, 1992.

## III.  Insufficient Attempts

We add to the above list some publications involving methods which are clearly insufficient to solve the problem.

1951 Natucci, A., *Osservazioni sul problema di Fermat*, Bull. Un. Mat. Ital., (3), **6** (1951), 245–248.

1953 Natucci, A., *Ricerche sistematiche sull'ultimo teorema di Fermat*, Giorn. Mat., (5), **1(81)** (1953), 171–179.

1975 Peiulescu, V., *Teorema lui Fermat*, Ed. Litera, Bucureşti, 1975, 86 pp.

1976 Yahya, Q.A.M.M., *On general proof of Fermat's last theorem—epilogue*, Portugal. Math., **35** (1976), 9–15.

1977 Yahya, Q.A.M.M., *Fermat's last theorem—a topological verification*, Portugal. Math., **36** (1977), 25–31.

1979 De Fermate, J.F. (pseudonym of Guillotte, G.), *A Famous Problem in Number Theory of the First Kind Dating from 1637:* $x^p + y^p = z^p, p > 2$, Cowansville Printing, Cowansville, Quebec; Vol. I, 1979, 22 pp.; Vol. II, 1980, 8 pp.

1980 Singh, L., *The general proof of Fermat's last theorem*, J. Indian Acad. Math., **2** (1980), 43–50.

There have also been quite a number of mistakes in papers related to Fermat's last theorem, aiming to establish partial results, necessary conditions, etc. We have referred to these errors at the appropriate place in the text.

# Appendix B
# General Bibliography

I. The Works of Fermat

After Fermat's death, his collected papers were published under the supervision of his son Samuel de Fermat.

1679      *Varia Opera Mathematica*, D. Petri de Fermat, Senatoris Tolosani. Tolosae, Apud Joannem Pec, Comitiorum Fuzensium Typographum justa Collegium P P. Societatis JESU.

These books were reprinted in four volumes and a supplement:

1891, 1894,      *Oeuvres de Pierre de Fermat,*
1896, 1912,      Publiées par les soins de MM. Paul Tannery et
1922      Charles Henry. Gauthier-Villars, Paris.

Among the letters of Fermat, the following ones are relevant to the subject of this book:

1636, April 26 Lettre à Mersenne,[1] *Correspondance du Père Marin Mersenne*, Vol. 6, p. 50. Commencée par Mme. Paul Tannery, publiée et annotée par Cornelis de Waard. Ed. du C.N.R.S, Paris, 1962.

---

[1] According to Itard (1948) the actual date of this letter is June 1638; it has appeared with this date in Mersenne's Correspondence.

1636, Sept.      Lettre à Mersenne [pour Sainte-Croix]. *Oeuvres de Fermat*, Vol. III, pp. 286–292.

1636, Nov. 4    Lettre à Roberval. *Oeuvres de Fermat*, Vol. II, p. 83.

1638, beg. June Lettre à Mersenne. *Correspondance du Père Marin Mersenne*, Vol. 7, pp. 272–283. Commencée par Mme. Paul Tannery, publiée et annotée par Cornelis de Waard. Ed. du C.N.R.S, Paris, 1962.

1640, May (?)   Lettre à Mersenne. *Oeuvres de Fermat*, Vol. II, pp. 194–195.

1654, Aug. 29   Lettre à Pascal. *Oeuvres de Fermat*, Vol. II, pp. 307–310.

1657, Aug. 15   Lettre à Kenelm Digby. *Oeuvres de Fermat*, Vol. II, pp. 342–346.

## II. Books Primarily on Fermat

1910 Lind, B., *Über das letzte Fermatsche Theorem*, Abh. Geschichte Math. Wiss., no. 26, 1910, pp. 23–65.

1919 Bachmann, P., *Das Fermatproblem in seiner bisherigen Entwicklung*, W. de Gruyter, Berlin, 1919; reprinted by Springer-Verlag, Berlin, 1976.

1927 Khinchin, A.I., *Velikai Teorema Ferma* (The Great Theorem of Fermat), Moskau-Leningrad, Staatsverlag, Vol. **76**, 1927.

1961 Bell, E.T., *The Last Problem*, Simon & Schuster, New York, 1961.

1966 Noguès, R., *Théorème de Fermat, son Histoire*, A. Blanchard, Paris, 1966.

1972 Mahoney, M.S., *The Mathematical Career of Pierre de Fermat, 1601–1665*, Princeton University Press, Princeton, NJ, 1973.

1977 Edwards, H.M., *Fermat's Last Theorem, A Genetic Introduction to Algebraic Number Theory*, Springer-Verlag, New York, 1977.

1979 Ribenboim, P., *13 Lectures on Fermat's Last Theorem*, Springer-Verlag, New York, 1979.

## III. Books with References to Fermat's Last Theorem

1859 Smith, H.J.S., *Report on the Theory of Numbers*, Part II, Art. 61, "Application to the last theorem of Fermat." Report of the British Association for 1859, pp. 228–267. Reprinted in *Collected Mathematical Works*, Vol. I, pp. 131–137, Clarendon Press, Oxford, 1894; reprinted by Chelsea, New York, 1965.

1897 Hilbert, D., *Die Theorie der algebraischen Zahlkörper*, Jahresber. Deutsch. Math.-Verein., **4** (1897), 175–546; reprinted in *Gesammelte Abhandlungen*, Vol. I, Chelsea, New York, 1965.

1910 Bachmann, P., *Niedere Zahlentheorie*, Vol. II, Teubner, Leipzig, 1910; reprinted by Chelsea, New York, 1968.

1920 Dickson, L.E., *History of the Theory of Numbers*, Vol. II. Carnegie Institution, Washington, DC, 1920; reprinted by Chelsea, New York, 1971.

1927 Landau, E., *Vorlesungen über Zahlentheorie*, Vol. III, Hirzel, Leipzig, 1927; reprinted by Chelsea, New York, 1947.

1928 Vandiver, H.S. and Wahlin, G.E., *Algebraic Numbers*, Vol. II. Bull. Nat. Research Council No. 62, 1928; reprinted by Chelsea, New York, 1967.

1937 Bell, E.T., *Men of Mathematics*, Simon & Schuster, New York, 1937.

1951 Nagell, T., *Introduction to Number Theory*, Wiley, New York, 1951; reprinted by Chelsea, New York, 1962.

1956 Ostmann, H.H., *Additive Zahlentheorie*, Vol. II, Springer-Verlag, Berlin, 1956.

1962 Shanks, D., *Solved and Unsolved Problems in Number Theory*, Vol. I, Spartan, Washington, DC, 1962; reprinted by Chelsea, New York, 1978.

1966 Borevich, Z.I. and Shafarevich, I.R., *Number Theory*, Academic Press, New York, 1966.

1969 Mordell, L.J., *Diophantine Equations*, Academic Press, New York, 1969.

1982 Koblitz, N. (editor), *Number Theory Related to Fermat's Last Theorem*, Birkhäuser, Boston, 1982.

1984 Weil, A., *Number Theory, An Approach Through History from Hammurapi to Legendre*, Birkhäuser, Boston, 1984.

## IV. Expository, Historical, and Bibliographic Papers

1807 Gauss, C.F., Letter to Sophie Germain (Braunschweig, 30 April, 1807). *Werke*, Vol. X, Part I, pp. 70–74. Königl. Ges. Wiss., Göttingen, Teubner, Leipzig, 1917.

1816 Gauss, C.F., Letter to Wilhelm Olbers (March 21, 1816). *Werke*, Vol. X, Part I, pp. 75–76. Königl. Ges. Wiss., Göttingen, Teubner, Leipzig, 1917.

1841 Terquem, O., *Théorème de Fermat sur un trinôme, démonstration de M. Lamé, projet de souscription*, Nouv. Ann. Math., **6** (1847), 132–134.

1879 Henry, C., *Recherches sur les manuscripts de Pierre de Fermat*, Bull. Bibliografia Storia Scienze Matem. Fis., **12** (1879), 477–568 and 619–740.

1879 Mansion, P., *Remarques sur les théorèmes arithmétiques de Fermat*, Nouv. Corr. Math., **5** (1879), 88–91 and 122–125.

1880 Germain, S., *Cinq lettres de S. Germain à C.F. Gauss, publiées par B. Boncompagni*, Arch. Math. Phys., **63** (1880), 27–31; **66** (1881), 3–10.

1883 Tannery, P., *Sur la date des principales découvertes de Fermat*, Bull. Sci. Math. Sér. 2, **7** (1883), 116–128; reprinted by Sphinx-Oedipe, **3** (1908), 169ff.

1887 Henry, C., *Lettre à M. le Prince de Boncompagni sur divers points d'histoire des mathématiques*, Bull. Bibliografia Storia Scienze Matem. Fis., **20** (1887), 389–403.

1898 Gram, J.P., *Om Fermat og haus sidote Saetning*, Forhandlingar Skandinaviska Natursforskare, Götheborg, 1898, p. 182.

1901 Gambioli D., *Memoria bibliografica sull'ultimo teorema di Fermat*, Period. Mat., **16** (1901), 145–192.

1902 Gambioli, D., *Appendice alla mia memoria bibliografica sull'ultimo teorema di Pietro Fermat*, Period. Mat., (2), IV, **17** (1902), 48–50.

1908 Hoffmann, F., *Der Satz vom Fermat. Sein seit dem Jahr 1658 gesuchter Beweis*, J. Singer, Strasbourg, 1908.

1908 Schönbaum, E., *Arbeiten von Kummer über den Fermatschen Satz*, Časopis Pěst. Mat., **37** (1908), 484–506.

1910 Gérardin, A., *État actuel de la démonstration du grand théorème de Fermat*, Assoc. Française Avanc. Sciences, Toulouse, I, **39** (1910), 55–56.

1912 *Wolfskehl Prize. Bekanntmachung*, Math. Ann., **72** (1912), 1–2.

1917 Dickson, L.E., *Fermat's last theorem and the origin and nature of the theory of algebraic numbers*, Ann. of Math., **18** (1917), 161–187.

1921 Mordell, L.J., *Three Lectures on Fermat's Last Theorem*, Cambridge University Press, Cambridge, 1921; reprinted by Chelsea, New York, 1962.

1925 Ore, O., *Fermat's theorem* (in Norwegian). Norske Mat. Tidsskrift, **7** (1925), 1–10.

1941 Brčić-Kostić, M., *Das Fermatproblem, $x^n + y^n = z^n$*, Zagreb, 1941.

1943 Hofmann, J.E., *Neues über Fermats zahlentheoretische Herausforderungen von 1657 (mit zwei bisherunbekannten Originalstuecken Fermats)*, Abh. Preussischen Akad. Wiss., Math.-Naturw. Kl., 1943, No. 9, 52 pp.

1946 Vandiver, H.S., *Fermat's last theorem*, Amer. Math. Monthly, **53** (1946), 555–578.

1948 Got, T., *Une enigme mathématique: Le dernier théorème de Fermat* (a chapter in *Les Grands Courants de la Pensées Mathématiques*, edited by F. Le Lionnais), Cahiers de Sud, Marseille, 1948. Reprinted by A. Blanchard, Paris, 1962. Translated into English (2 volumes). Dover, New York, 1971.

1948 Itard, J., *Sur la date à attribuer à une lettre de Pierre Fermat*, Rev. Historie Sci. Appl., **2** (1948), 95–98.

1957 *Un Mathématician de Génie: Pierre de Fermat, 1601–1665*, Lycée Pierre de Fermat, Toulouse, 1957.

1959 Schinzel, A., *Sur quelques propositions fausses de P. Fermat*, C. R. Acad. Sci. Paris, **249** (1959), 1604–1605.

1970 Smadja, R., *Le théorème de Fermat* (thèse de 3e cycle), Université de Paris VI, 1970.

1972 Mahoney, M.S., *Fermat's Mathematics: Proofs and Conjectures*, Science, **178** (1972), 30–36.

1973 Albis Gonzalez, V., *El señor Fermat y sus problemas*, Bol. Mat., Bogotà, **7** (1973), 219–232.

1973 Besenfelder, H.J. (now Bentz, H.J.), *Das Fermat-Problem*, Diplomarbeit, Univ. Karlsruhe, 1973, 61 pp.

1973 Fournier, J.C., *Sur le Dernier Théorème de Fermat* (thèse de 3e cycle). Université de Paris VI, 1973.

1974 Ferguson, R.P., *On Fermat's last theorem, I–II*, J. Undergrad. Math., **6** (1974), 1–14 and 85–98.

1974 Weil, A., *La cyclotomie jadis et naguère*, Enseign. Math., **20** (1974), 247–263. Reprinted in *Essais Historiques sur la Thérie des Nombres*. Monographie no. 22 de L'Enseignement Mathématique, 1975. Génève, 55 pp. Reprinted in *Collected Papers*, Vol. III, pp. 311–328. Springer-Verlag, New York, 1980.

1975 Edwards, H.M., *The background of Kummer's proof of Fermat's last theorem for regular primes*, Arch. History Exact Sci., **14** (1975), 219–236.

1975 Ferguson, R.P., *On Fermat's last theorem, III*, J. Undergrad. Math., **7** (1975), 35–45.

1976 Christy, D., *Le dernier thérème de Fermat et le théorème de Roth*, Sém. Alg. Th. Nombres, Caen, 1976/7, 12 pp.

1977 Edwards, H.M., *Postscript to: "The background of Kummer's proof of Fermat's last theorem for regular primes"*, Arch. History Exact Sci., **17** (1977), 381–394.

1980 de Rham, G., *Brève notice sur Dmitri Mirimanoff*, Cahiers Sém. Histoire Math. Paris, **1** (1980), 32–33.

1980 Cassinet, R., *La descente infinie de Campanus à Hilbert, 1260–1897*, Sém. Histoire Math. Toulouse, 1980, cahier no. 2, pp. B1 to B25.

1980 Ribenboim, P., *Les idées de Kummer sur le théorème de Fermat*, Séminaire de Philosophie et Mathématiques, École Normale Sup., Paris, 1979, 11 pp.

1980 Terjanian, G., *Fermat et son arithmétiques*, Sém. Histoire Math. Toulouse, 1980, cahier no. 2, pp. A1 to A35.

See also the articles in the following dictionaries and encyclopedias:

1961 Fermat; Fermat's last theorem. *Encyclopedia Americana*, Vol. 11, pp. 125–125, 1961 edition.

1977 Boyer, C.B., Fermat, Pierre de. *Encyclopaedia Britannica*, Macropaedia, Vol. 7, pp. 234–236, 1977 edition.

1977 Iyanaga, S. and Kawada, S. (editors), Fermat's Problem. *Encyclopedic Dictionary of Mathematics* (by The Mathematical Society of Japan), pp. 512–513. Translation reviewed by K.O. May, MIT Press, Cambridge, MA, 1977.

1979 Bouvier, A., George, M., and Le Lionnais, F., Fermat, Pierre Simon de (1606–1665). *Dictionnaire des Mathématiques*, p 296. Presses Universitaires de France, Paris, 1979.

## V. Critical Papers and Reviews

1839 Cauchy, A. and Liouville, J., *Rapport sur un mémoire de M. Lamé relatif an dernier théorème de Fermat*, C. R. Acad. Sci. Paris, **9** (1839), 359–363; also appeared in J. Math. Pures Appl., **5** (1840), 211–215 and *Oeuvres Complétes*, Sér. 1, Vol. 4, pp. 494–504, Gauthier-Villars, Paris, 1884.

1856 Cauchy, A., *Rapport sur le concours relatif au dernier théorème de Fermat (Commissaires MM. Bertrand, Liouville, Lamé, Chasles; Cauchy rapporteur)*, C. R. Acad. Sci. Paris, **44** (1856), p. 208.

1881 Catalan, E., *Jugement du concours annuel*, Bull. Acad. Roy. Sci. Belgique, (3), **6** (1881), 814–832.

1973 Weil, A., *Review of "The Mathematical Career of Pierre de Fermat" by M. S. Mahoney*, Bull. Amer. Math. Soc., **79** (1973), 1138–1149. Reprinted in *Oeuvres Scientifiques*, Vol. 3, pp. 266–277, Springer-Verlag, Berlin, 1979.

1977 Mazur, B., *Review of Kummer's "Collected Papers,"* Vols. I and II, Bull. Amer. Math. Soc., **83** (1977), 976–988.

# Name Index

# Subject Index